Springer Theses

Recognizing Outstanding Ph.D. Research

Aims and Scope

The series "Springer Theses" brings together a selection of the very best Ph.D. theses from around the world and across the physical sciences. Nominated and endorsed by two recognized specialists, each published volume has been selected for its scientific excellence and the high impact of its contents for the pertinent field of research. For greater accessibility to non-specialists, the published versions include an extended introduction, as well as a foreword by the student's supervisor explaining the special relevance of the work for the field. As a whole, the series will provide a valuable resource both for newcomers to the research fields described, and for other scientists seeking detailed background information on special questions. Finally, it provides an accredited documentation of the valuable contributions made by today's younger generation of scientists.

Theses may be nominated for publication in this series by heads of department at internationally leading universities or institutes and should fulfill all of the following criteria

- They must be written in good English.
- The topic should fall within the confines of Chemistry, Physics, Earth Sciences, Engineering and related interdisciplinary fields such as Materials, Nanoscience, Chemical Engineering, Complex Systems and Biophysics.
- The work reported in the thesis must represent a significant scientific advance.
- If the thesis includes previously published material, permission to reproduce this must be gained from the respective copyright holder (a maximum 30% of the thesis should be a verbatim reproduction from the author's previous publications).
- They must have been examined and passed during the 12 months prior to nomination.
- Each thesis should include a foreword by the supervisor outlining the significance of its content.
- The theses should have a clearly defined structure including an introduction accessible to new PhD students and scientists not expert in the relevant field.

Indexed by zbMATH.

Matteo Parisi

Combinatorial Aspects of Scattering Amplitudes

Amplituhedra, T-duality, and Cluster Algebras

Doctoral Thesis accepted by
University of Oxford, Oxford, UK

Springer

Author
Dr. Matteo Parisi
Center of Mathematical Sciences
and Applications
Harvard University
Cambridge, MA, USA

Institute for Advanced Study
School of Natural Sciences
Princeton, NJ, USA

Department of Physics
Princeton University
Princeton, NJ, USA

Supervisor
Prof. Lionel Mason
Mathematical Institute
University of Oxford
Oxford, UK

ISSN 2190-5053 ISSN 2190-5061 (electronic)
Springer Theses
ISBN 978-3-031-41068-0 ISBN 978-3-031-41069-7 (eBook)
https://doi.org/10.1007/978-3-031-41069-7

This Springer imprint is published by the registered company Springer Nature Switzerland AG
The registered company address is: Gewerbestrasse 11, 6330 Cham, Switzerland

Paper in this product is recyclable.

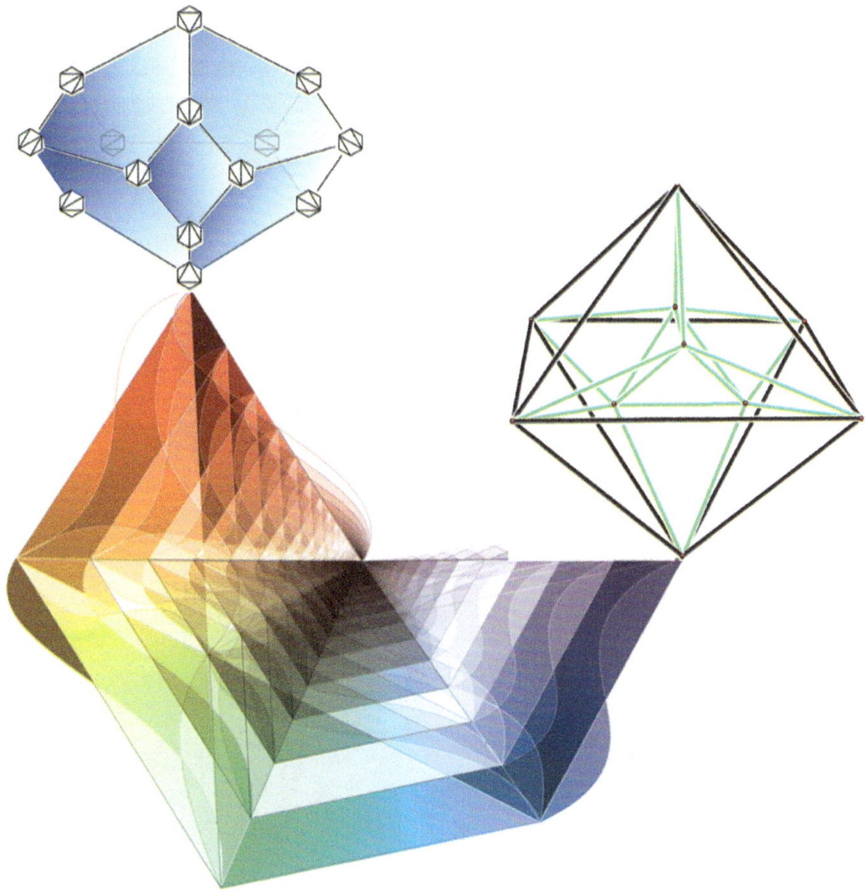

(Top left): Associahedron K_5; (bottom): artistic depiction of the amplituhedron, Gilmore; (top right): Schlegel diagram of the hypersimplex $\Delta_{2,5}$

Radical revolutions lie at the intersection of both cutting-edge pure mathematics and fundamental physics.
In this pas de deux, a novel, dynamic interaction is created,
in which each discipline learns something from the other
which neither could have learnt by dancing alone.

Pure mathematics and physics are becoming
ever more closely connected,
though their methods remain different.
Mathematicians play a game in which they
themselves invent the rules,
while physicists play a game in which the
rules are provided by Nature.
However, as time goes on, it becomes
increasingly evident
that the rules mathematicians find interesting
are the ones Nature has chosen.

Paul A. M. Dirac, 1939

Supervisor's Foreword

This thesis is a unique work at the intersection of High Energy Physics and Algebraic Combinatorics. It is based on papers published in international journals ranging from physics to mathematical physics and pure mathematics.

Scattering amplitudes generate the probabilities of the interaction processes of fundamental particles that are predicted by quantum field theories. They are central both to the understanding of the physics seen in colliders and their underlying mathematical structures. This work is at the forefront of the recent paradigm shift according to which such scattering amplitudes emerge purely from the geometry and the combinatorics of (novel) underlying mathematical objects. A key example is the 'Amplituhedron' that encodes the scattering amplitudes of maximally supersymmetric Yang–Mills theory, a model quantum field theory that is a close cousin of the theory of strong interactions of quarks and gluons.

This volume unveils many new combinatorial structures of the Amplituhedron and introduces a new important related object, the Momentum Amplituhedron, that is relevant to a much broader class of quantum field theories. Moreover, the work pioneers the connection between amplituhedra, cluster algebras and tropical geometry, central topics in both mathematical physics and combinatorics.

A striking duality is discovered between the $m = 2$ amplituhedron, a semi algebraic subset of the Grassmannian and the hypersimplex, a well-known polytope with rich combinatorics and deep connections to matroid theory. Simplices that provide a well-known triangulation of the hypersimplex into Weyl alcoves are proved to be in bijection with certain oriented uniform matroids/chambers in the $m = 2$ amplituhedron. Importantly, tiles and tilings of the two objects are proved to be in bijection, laying the groundwork to realize the tantalizing possibility of a notion of secondary polytope-like structures for Grassmann polytopes.

Remarkably, this duality can be seen as a combinatorial incarnation of T-duality from String Theory. It further predicts highly nontrivial mappings between structures in the $m = 4$ Amplituhedron and those in the Momentum Amplituhedron, mirroring the amplitude/Wilson-loop correspondence.

Finally, the work shows how Amplituhedra possesses surprisingly rich cluster structures and how they relate to their geometry and combinatorics. Cluster algebras are a remarkable class of commutative rings introduced by Fomin and Zelevinsky. Many algebraic varieties are known to have a cluster algebra structure, including the Grassmannian. They also emerged in physics in the context of scattering amplitudes, where they contributed to both conceptual and computational advances.

In summary, this thesis is a significant and substantial contribution within and across fields, and it will be a valuable resource for both physicists and mathematicians investigating geometrical structures emerging from physics for some time to come.

Oxford, UK
June 2023

Prof. Lionel Mason

Abstract

Scattering amplitudes are central in fundamental physics: they encode the probabilities of interaction among elementary particles in a Quantum Field Theory (QFT). This work is part of a new program according to which scattering amplitudes and other quantum observables in particle physics and cosmology come from underlying (novel) mathematical objects called *positive geometries*. This serves to advance both our theoretical understanding and computational frontiers in QFT. Crucially, it cross-fertilises ideas in pure mathematics, such as in combinatorics and algebraic geometry.

The *amplituhedron* $\mathcal{A}_{n,k,m}$ is the image of the positive Grassmannian $\mathrm{Gr}_{k,n}^{\geq 0}$ under a positive linear map. Regarded as a non-linear generalization of (cyclic) polytopes inside the Grassmannian, it is a semialgebraic set with beautiful combinatorics. When $m = 4$, the amplituhedron $\mathcal{A}_{n,k,4}$ is the positive geometry for scattering amplitudes in planar $\mathcal{N} = 4$ super Yang-Mills theory (SYM).

Such amplitudes have a dual formulation due to the Amplitude/Wilson Loop duality, which has roots in a certain *T-duality* from String Theory. In this work, we provide a combinatorial-geometric manifestation of T-duality by defining the T-dual of the amplituhedron $\mathcal{A}_{n,k,m}$, which we call *momentum amplituhedron* $\mathcal{M}_{n,k',m}$—also the image of the positive Grassmannian under some positive linear maps. When $m = 4$ and $k' = k + 2$, the momentum amplituhedron $\mathcal{M}_{n,k+2,4}$ is the T-dual positive geometry of $\mathcal{N} = 4$ SYM scattering amplitudes in momentum space. We conjecture that the duality between $\mathcal{M}_{n,k+2,4}$ and $\mathcal{A}_{n,k,4}$ provides a bijection between their *positroid tilings*—decompositions of the two objects into injective images of positroid cells called *tiles*. Finding tilings of a positive geometry is central, as they give different (new) expansions of the physical observable.

We then show T-duality extends beyond the $m = 4$ case and connects two other seemingly unrelated objects. The first is the $m = 2$ amplituhedron $\mathcal{A}_{n,k,2}$ —a $2k$-dimensional semialgebraic set in the Grassmannian, toy-model for the $m = 4$ case, it also enters the geometry of some loop amplitudes, tree-level form factors and correlators in planar $\mathcal{N} = 4$ SYM. The second is the *hypersimplex* $\Delta_{k+1,n}$—an $(n - 1)$-dimensional polytope which is the moment map image of the positive Grassmannian $\mathrm{Gr}_{k+1,n}^{\geq 0}$ and has nice combinatorics in connection with matroid theory and tropical geometry. One of our main results is the proof that T-duality between $\Delta_{k+1,n}$ and $\mathcal{A}_{n,k,2}$ provides a bijection between their tiles and positroid tilings. Surprisingly, simplices in the triangulation of $\Delta_{k+1,n}$ into an Eulerian number of Weyl alcoves are also T-dual to certain chambers in a decomposition of $\mathcal{A}_{n,k,2}$ related to oriented uniform matroids. Moreover, we obtain classes of positroid tilings from *BCFW recursions* and the *positive tropical Grassmannian* $\text{Trop}^{+}\text{Gr}_{k+1,n}$—which we conjecture to play the role of *secondary fan* for $\mathcal{A}_{n,k,2}$.

Concurrently with positive geometries, the understanding and computation of scattering amplitudes have also been significantly advanced by the connections with *cluster algebras*—commutative rings with a set of distinguished generators and remarkable combinatorics introduced by Fomin and Zelevinsky. In this work we pioneer the connection between amplituhedra and cluster algebras, paving the way for finding geometric origins of *cluster adjacencies* in physics. In particular, we formulate and generalize the cluster adjacency conjectures at tree-level in terms of the amplituhedron $\mathcal{A}_{n,k,m}$. We then prove them for the $m = 2$ case and find that each tile of $\mathcal{A}_{n,k,2}$ is the totally positive part of a cluster variety. Moreover, using the geometry of the *loop amplituhedron*, we make a new conjecture about cluster structures relating Leading and Landau singularities of amplitudes in planar $\mathcal{N} = 4$ SYM, providing some proves at one loop and comments on implications for the *bootstrap* program.

Publications related to this Thesis

- *The $m = 2$ amplituhedron and the hypersimplex: signs, clusters, tilings, Eulerian numbers* [1], (with Melissa Sherman-Bennett, Lauren K.Williams), Communications of the American Mathematical Society (CAMS), 3 (2023), 329–399, https://doi.org/10.1090/cams/23.[arXiv:2002.06164].
- *The positive tropical Grassmannian, the hypersimplex, and the $m = 2$ amplituhedron* [2], (with Tomasz Lukowski, Lauren K. Williams), International Mathematics Research Notices, 2023(19) 16778–16836, October 2023, https://doi.org/10.1093/imrn/rnad010 [arXiv:2002.06164].
- *Cluster patterns in Landau and Leading Singularities via the Amplituhedron* [3], (with Ömer Gürdoğan), Ann. Inst. Henri Poincaré Comb. Phys. Interact. (2023), [arXiv:2005.07154].
- *The momentum amplituhedron* [4], (with David Damgaard, Livia Ferro, Tomasz Lukowski), Journal of High Energy Physics, 2019 (2019) 042, [arXiv:1905.04216].
- *Cluster Adjacency for $m = 2$ Yangian Invariants* [5], (with Tomasz Lukowski, Marcus Spradlin, Anastasia Volovich), Journal of High Energy Physics, 2019 (2019) 158, [arXiv:1908.07618].
- *Triangulations and Canonical Forms of Amplituhedra: A Fiber-Based Approach Beyond Polytopes* [6], (with Fatemeh Mohammadi and Leonid Monin), Communications in Mathematical Physics (2021), [arXiv:2010.07254].

References

1. M. Parisi, M. Sherman-Bennett, L. Williams, The $m = 2$ amplituhedron and the hypersimplex: signs, clusters, tilings, Eulerian numbers 2023. *Comm. Amer. Math. Soc.* **3** (2023), 329–399. https://doi.org/10.1090/cams/23
2. T. Łukowski, M. Parisi, L.K. Williams, The Positive Tropical Grassmannian, the Hypersimplex, and the $m = 2$ Amplituhedron. *Int. Math. Res. Not.* **2023** (19), 16778–16836 October 2023. https://doi.org/10.1093/imrn/rnad010
3. O. Gürdoğan, M. Parisi, Cluster patterns in Landau and leading singularities via the amplituhedron. *Ann. Inst. Henri Poincar'e Comb. Phys. Interact.* **10**(2), 299–336 (2023).
4. D. Damgaard, L. Ferro, T. Lukowski, M. Parisi, The Momentum Amplituhedron. *JHEP* **08**, 042 (2019).
5. T. Łukowski, M. Parisi, M. Spradlin, A. Volovich, Cluster Adjacency for $m = 2$ Yangian Invariants. *JHEP* **10**, 158 (2019).
6. F. Mohammadi, L. Monin, and M. Parisi, Triangulations and canonical forms of amplituhedra: A fiber-based approach beyond polytopes. *Commun. Math. Phys.* **8** (2021).

Acknowledgements

First of all, I would like to thank my supervisors Lionel Mason and Tomasz Lukowski. Without them, this journey wouldn't have been possible. I would like to thank Lionel for his availability, guidance, advice, and lovely lunches and coffee breaks together. For being cheerful and welcoming. For helping me along my personal and academic path, and believing in me as a researcher and as a teacher. I would like to thank Tomek for bringing me to the core of my research field, for supporting and guiding me, especially in the early stages of my D.Phil.

I would like to address a special thank to Lauren Williams, with whom I shared the most beautiful mathematical adventure in the last couple of years. After our in-person meeting at Harvard CMSA in November 2019, we have been working at a distance and she has been *de facto* my third supervisor. I would like to thank her for the time and dedication given by her to the junior collaborators, and the opportunities she opened up in my research and life. My motivation of being at the intersection of math and physics is much stronger also thanks to her.

I would like to thank Fatemeh Mohammadi, who has invited me to Bristol many times to collaborate. I deeply thank her for taking her time to interact with me, for lunches together in Bristol, and for guiding me towards the more math-oriented path I was embarking. I also thank my collaborator and friend Leonid Monin, with whom I had good times both in Bristol and Oxford.

During my D.Phil., I had the luck to meet amazing collaborators and friends I would like to thank. Ömer Gürdoğan—for sharing the journey through cluster algebras, lovely lunches and dinners, and chats about math and life. Paolo Benincasa—for his inspiring approach to academia and science, for teaching me about cosmological polytopes, and for inviting me to spend a week at NBI just before the pandemic, during which we had the key ideas for our paper together. Melissa Sherman-Bennett—for our long and rewarding virtual collaboration and her ability to navigate new combinatorial problems with open-mindedness and flexibility. Marcus Spradlin and Anastasia Volovich—for the inspiring work together and their hospitality during my visit to Brown in my third year of D.Phil. My big thank to Livia Ferro and David

Damgaard for their collaboration. A special thank also goes to Oliver Schlotterer, Georgios Papathanasiou and Niklas Henke for discussions and their great hospitality in Uppsala and DESY, respectively. I would also like to thank my colleagues and friends at Oxford: Erik, Marko, Atul, and Anders—for nice chats and for enhancing my understanding of Landau varieties and celestial amplitudes. I would like to thank more broadly many colleagues in the Amplitude community who made the job of being a researcher much easier, more pleasant and more human. I would like to thank Nima, for being always available to chat and his enthusiasm for math and physics which keeps the community together and inspired. His weekly virtual IAS meetings—co-organised with Sebastian—kept me connected with the field in the middle of the pandemic. Finally, I would like to thank my D.Phil. examiners James Drummond and Francis Brown for their time and availability for organizing the VIVA, providing useful comments and helpful feedback on this thesis. I would like to thank the support of the Della Riccia Scholarship, the ERC grant number 724638, and the D. Sachs Scholarship at Princeton.

During my D.Phil., I had the opportunity to be a Stipendiary Lecturer at St. Peter's College. It has been an unforgettable experience, especially because of my students, and also the SCR community. A special thank goes to Balázs Szendrői, who inspired me in my teaching and outreach path. In particular, he put me in contact with a college student, Christy, whom I helped to organize the first maths camp in her hometown in Cameroon. I thank her too for the fantastic experience. I also thank all members of the math charity SAMI which made that possible. In particular, Emily for giving me the opportunity to contribute as an ambassador and her inspiring approaches to math teaching and outreach. I also thank George, who made every meal and coffee enjoyable in St. Peter's College SCR.

I would like to thank Helen Byrne, whom I met while being a lecturer at Keble College, both for being an incredible coordinator of math teaching and for her inspiring contributions as director of Equality and Diversity at MPLS. I would also like to thank Andrew Hodges, who set the model on how to both give major contributions to my field with humbleness and impact society at large with activism.

My experience and my life wouldn't be the same without my closest friends at Oxford. What I shared with them and my gratitude to them is ineffable. I start by thanking what I call my *Worcester family*. Lea—who changed my perspective of the world with her thoughts, insights, love and amazing food. I thank her also for making me join Philiminality where I could organize conferences on themes that are close to my heart and mind. Johann—whose passion, curiosity and motivation has been a source of inspiration. For our philosophical discussions about politics, world, life which opened and shaped my mind, and for showing me New Zealand. Giulia—for the infinite dedication, love and care which will always warm up my heart. Silvia—for always being my travel companion, even at a distance. Mark—for making me part of his visions about life and time, sharing unique angles to picture the inner and outer worlds. Andrea—for sharing the beauty of math, physics and music and for the hospitality and the nice time spent together in Lodrino (Switzerland) during the peak of the pandemic this year. I also thank his loving family who made me feel home. Alexey—for improvising together on the piano, sharing music and souls. Rasheed—for his witty and inspiring conversations from self-driving cars to religion. Adam—for the nice time together. Ines—for the nice dinners at Franks and the support of love at a distance. Nonie—for sharing her cheerful approach to life which brought us from exploring Plush to all the libraries of Oxford. Edward—for his sensitivity, ability to listen and the pizza nights during the pandemic. Julien—for sharing his spontaneity, positive approach to life and the fun game nights together.

I would also like to thank: Laura—for learning together about life, organizing many events together and sharing a passion for movies and tennis. Davide—for transmitting passion and commitment as a fellow activist and event organizer, and for sharing ideas and projects. Guus—for conversations on Aristotle, Plato, math, life and sharing our personal paths and passions in pursuing knowledge and beauty. Sophie—for showing me how to live with strength, sensitivity, energy and creativity, and for sharing with me the draft of her last successful book. Nesreen—for reminding me how to smile despite everything and her warm hugs. Sam—for sharing numerous events and activities within the college and in London. Max—for reminding me one can be forever young in spirit, with an active, curious and happy way of living.

I am grateful to the broader Worcester MCR communities throughout the years. It was such a valuable and rich experience to meet everyone and a rare feeling of being a part of a family to which I was happy to belong, and where I discovered who I am. A big thank goes to all the staff of Worcester College, which created such a friendly environment. A special thank goes to Kate Tunstall—for her inspiring drive towards a progressive Worcester, while provost. Her actions and ideas deeply resonated with mine and she made me proud of being in my college.

I continue by thanking the staff of the Mathematical Institute and whoever contributed to make the working place inclusive and welcoming. A special thanks to Sandy, who is always there for students. A big thank to those who organized happy hours, social and outreach activities and Mathematrix. It has been a place where I met great friends. A big thank to Luciana—for sharing our passion for outreach, teaching, fun math and her commitment to inclusion and diversity. Christine—for her curiosity, humbleness, endurance and sharing what Oxford meant for our lives. Solveig—for sharing fun game nights, delicious sweets and caring friendship. Sebastjan—for playing volleyball together, and believing in me, more than I believed in myself. Carmen—for inspiring me with her vitality and her music projects. Diego—for the fun nights we had together. Johan—for our conversations about math, languages and your hospitality in Sweden. Federico—for sharing his playful spirit and playing volleyball together.

I would like to thank Liberato for making me feel at home at his restaurant Gino's, during relaxed or intense study nights.

I would like to thank my Italian friends who always support and inspire me: Santiago, Veronica, Francesco and Federica; and my friend Nico from my Munich times.

I would also like to do some post-D.Phil. acknowledgments. I acknowledge the support of the Center of Mathematical Sciences and Applications (CMSA) at Harvard University, where I am a Postdoctoral Fellow and of the U.S. Department of Energy under the grant number DE-SC0009988 at the Institute for Advanced Study, where I am a Member. I would like to thank my Harvard friends who made Cambridge my home: Marie, Christine, Viktoria, Chiara, Christophe, Paulina, Shuangjiu, Alice, Marta, Dinesh, Sergio, Lena, Sven, and Thomas.

Thank you, Zac, for the world you opened to me, for your kindness and support. I would like to thank my Princeton friends: Lesha—for sharing an unforgettable musical and personal journey, from the tea table to the keyboard, Guillaume—for our evenings in Philadelphia and at IAS, Andrea—for our interdisciplinary friendship, Agustin—for encouraging me to play three concerts (included one on Einstein's piano), Shiyue, Jakub, ReLiang, and—for our music together—Dmitry, Nathan and Flora. Thank you to my inspiring colleagues and the kind staff at IAS.

Most importantly, I would like to thank my family. My mum Maria Luisa and dad Lorenzo—who have been supporting me in many ways, even at a distance. The freedom and happiness I gained are also due to them. My sister Angela and my brother-in-law Luca—for advice, inspiration, and the fun times spent together. My niece Giulia—for being my main source of hope for future the generation. A generation which ought to be more open-minded, aware of the issues in society and in the world, curious and always ready to learn. I thank her for being so refreshing, sensitive, understanding and, most of all, loving.

Contents

Chapter 1
Introduction

1.1 Prelude

Quantum Field Theory (QFT) is unarguably one of the most important theoretical framework in fundamental physics one has at hand to describe Nature. Among the physical observables in a QFT, *Scattering Amplitudes* are central in fundamental physics: they encode the probabilities of any interaction among elementary particles. On the experimental side, these processes are probed at colliders such as the Large Hadron Collider at CERN. On the theoretical level, Scattering Amplitudes showcase fundamental mathematical and physical properties of QFTs and they are claimed to be the only known observables of quantum gravity in asymptotically flat space-time.

Scattering Amplitudes are computed perturbatively and the standard method to perform such calculations are 'Feynman Diagrams'. On the computational side, even at leading order and for simple processes, the number of diagrams is incredibly large, challenging even the capability of modern computers. On the theoretical level, Feynman diagrams contain huge non physical redundancies and lead to much more complicated expressions than the actual final amplitude they are meant to compute.[1]

An alternative route, circumventing the standard diagrammatic, was already attempted in the sixties with the program called 'The Analytic S-matrix' [2]. Originally, the latter hasn't been successful; however, thanks to the development of the so-called 'On-Shell Methods', an *S-Matrix Program Reloaded* initiated at the beginning of the new millennium. Among these, Recursion Relations [3] and Generalized Unitarity [4] are nowadays cornerstones of modern amplitude computations. Moreover, a fundamental paradigm shift has been attempted: instead of deriving the properties of the S-Matrix from the principles of unitarity and locality, one looks for novel principles and laws, very likely associated with new mathematical structures,

[1] A very instructive example is the extremely compact form that the so-called maximally-helicity-violating (MHV) amplitudes of QCD take: Parke and Taylor discovered a formula for the scattering of any number of gluons which could fit in less than half a line [1]; while standard methods, already for the scattering of five gluons, lead to 24 pages!

M. Parisi, *Combinatorial Aspects of Scattering Amplitudes*, Springer Theses,
https://doi.org/10.1007/978-3-031-41069-7_1

that produce the S-Matrix as the answer to entirely different questions. Space-time and quantum mechanics, embodied in unitarity and locality, emerge only later as derived consequences rather than foundational principles.

One of the most studied QFT, where all these methods have been applied with great success and main subject of our research, is the *maximally supersymmetric Yang-Mills theory* ($\mathcal{N} = 4$ SYM) in four dimensions. One firstly notes that, despite being a toy-model, it is considered to be the 'supersymmetric cousin' of QCD.[2] Secondly, one of the most important features of the theory is that, in the planar limit, it enjoys the infinite-dimensional symmetry called Yangian [6], which is the hallmark of integrability, i.e. solvability of the theory. As such, it is said to be the 'Hydrogen atom' of the twenty-first century. Thirdly, it has been the protagonist of the ground-breaking works of Maldacena, discovering the AdS/CFT correspondence [7] and Witten, formulating Twistor String Theory [8].

Importantly, in more recent years, $\mathcal{N} = 4$ SYM has also been the fertile ground to pursue the *geometrization* of Scattering Amplitudes where, according to the paradigm, physical properties emerge purely from geometry. The seminal idea came from Hodges in 2009 with his attempt to give intricate algebraic identities a geometric meaning, explaining the cancellation of non-physical spurious poles [9]. This introduced the concept of 'scattering amplitudes as volumes of a geometric object', in that case a polytope, where different representations of the former are just different triangulations of the latter.

In parallel, a complete dual formulation of the S-Matrix of $\mathcal{N} = 4$ SYM had been developing [10], where the computations for n particles in the sector with k negative helicity gluons are associated with the geometric space of k-planes in n dimensions, called 'Grassmannian'. In particular, these studies led in 2012 to the remarkable connection between scattering amplitudes and the positive part of it, the *positive Grassmannian* [11]. The latter is tight to a novel diagrammatic, the *on-shell diagrams* or *plabic graphs* [12], which fully upholds the spirit of on-shell methods initiated years earlier: for planar $\mathcal{N} = 4$ SYM, to all loop orders, off-shell entities, like virtual particles familiar from Feynman diagrams, are completely replaced by internal, on-shell particles.

In 2013, these formulations and Hodges' ideas came together in Arkani-Hamed and Trnka's work [13]. They introduced a novel mathematical object, called *Amplituhedron*, which certainly represented the acme, in terms of elegance and simplicity of definition, of the geometrization of Scattering Amplitudes in planar $\mathcal{N} = 4$ SYM. This space, which will be the main subject of our research, is a Grassmannian generalization of polytopes; it indeed contains certain types of polytopes, called 'cyclic' and positive Grassmannians as special cases. Scattering amplitudes are computed from a differential form which emerges from a purely geometric definition. Neither space-time nor Hilbert space indeed make any appearance in this formulation: the associated physics of locality and unitarity arise as consequences of the geometry and combinatorics.

---

[2] See [5] for the computation of all tree-level amplitudes in massless QCD from $\mathcal{N} = 4$ SYM.

These developments inspired a much more general framework, formalized only recently in 2017, which goes under the name of *positive geometries* [14]. These are spaces which have boundaries of all dimensions and posses a differential form—called *canonical*—whose analytic properties are tightly interwoven with the geometry. The canonical form is indeed defined to have logarithmic singularities on (and only on) all the boundaries of the geometry. See Appendix B for more details. Physical observables, such as Scattering Amplitudes, can then be simply extracted from canonical forms of the corresponding positive geometries. The study of positive geometries has been increasingly acquiring relevance in physics as they appear to be the underlying mathematical structure for quantum mechanical observables in a quite large class of theories in particle physics and cosmology. They also emerged in the context of scalar scattering in the form of the *ABHY associahedron* [15] for the bi-adjoint cubic interactions, *Stokes polytopes* [16] for planar quartic interactions, and *accordiohedra* for multi scalar field amplitudes [17]. Even more surprisingly, they appeared in cosmology, where the canonical form of the so-called *cosmological polytopes* encodes the wavefunction of the universe [18], which is the relevant quantum mechanical observable, for a large class of toy models. Finally, it was recently introduced an extension of canonical forms for general polytopes, named *stringy canonical form*, which depends on a certain deformation parameter (which resembles the α' parameter in string theory) and, when applied to the ABHY associahedron returns the *Koba-Nielsen integral* known in String Theory [19]. The authors of [20] paved a way towards enhancing positive geometries which involve *non-logarithmic* singularities. Generalizing these ideas to Grassmannian geometries would be the next step to describe from less supersymmetric gauge theories [21–23] and the non-linear sigma model [15], to gravity [24, 25] and many theories in cosmology.

One parallel important trend in physics is the connection between analytic properties of scattering amplitudes and *cluster algebras* [26] and *tropical geometry* [27].

In 2013 singularities of scattering amplitudes of planar $\mathcal{N} = 4$ SYM at loop level were discovered to be describable by employing cluster algebras [28]. This enabled a powerful program of *cluster bootstrap* which pushed both the computation and the understanding of the mathematical structure of scattering amplitudes beyond the frontiers, see [29] for a recent review. In 2017 the connection with cluster algebras was deeply enhanced by the discovery of phenomena called *cluster adjacencies* [30], at loop-level, and later at tree-level [31]. Since then, cluster algebras have been taking more space on the mathematical canvas of physicists. This further prompted exploring the connection between cluster algebras and tropical geometry. In particular, the *positive tropical Grassmannian* [27] has been increasingly playing a role in different areas of scattering amplitudes: from bootstrapping loop amplitudes in $\mathcal{N} = 4$ SYM [32–35] to computing scattering amplitudes in certain scalar theories [36]. The latter led to connections with *generalized Feynman diagrams*, and, in the light of our work, with subdivisions of the *hypersimplex* [37–42]. From physics, we can also find recent works on cluster structures and plabic graphs [43–46], *wall crossing* [47], and *tensor diagrams* [48]. We note that tropical geometry found applications in String theory since [49]—where the $\alpha' \to 0$ limit of closed string theory scattering amplitudes is interpreted as a 'tropical' limit. Moreover, tropical versions of *Feynman integrals* have been considered and showed that they capture many essential properties [50] and can be used for explicit numerical evaluation [51]. More cluster and tropical structures in Feynman integrals have been explored recently in [52, 53].

In this work we will focus on positive geometries relevant for $\mathcal{N} = 4$ SYM, i.e. on amplituhedra. In the framework of positive geometries, different representations of a physical observable correspond to different ways of subdividing the associated geometry. By following this philosophy, we discover a novel geometric-combinatorial incarnation of T-duality from String Theory connecting two seemingly unrelated objects. They have been at the center of attention of both physicists and mathematicians—the *hypersimplex* and the $m = 2$ *amplituhedron*. What do these two objects have in common? The former is a polytope, whereas the latter is a subset of the Grassmannian. They are both images of some positive Grassmannians, but under very different maps, and have different dimensions. Nevertheless, we discover that the number of ways we can subdivide the hypersimplex and the amplituhedron using images of cells of the positive Grassmannian is the same! Moreover, certain classes of 'nice' ways of subdividing them is governed by *BCFW recursions* and the *positive tropical Grassmannian*—both central in computations of scattering amplitudes. We show there is a deep duality—we call *T-duality*—which allows us to both prove novel connections between the two objects and discover new properties of them.

If we generalize the story for the physical $m = 4$ amplituhedron, T-duality connects this object to a new positive geometry we introduce—the sought-after *momentum amplituhedron*—which encodes scattering amplitudes of $\mathcal{N} = 4$ SYM in momentum (spinor helicty) space. Being this space more universal, this opens the path for investigating positive geometries beyond $\mathcal{N} = 4$ SYM. For example, the authors of [54–56], building on our construction, have defined the 'momentum amplituhedron' for scattering amplitudes in *ABJM theory*.

Both positive geometries and cluster algebras have been playing an increasingly crucial role to understand and compute scattering amplitudes. However, little is known about their intersection: what can we discover of the geometry of scattering amplitudes under the guidance of cluster algebras? Can we understand cluster phenomena purely from geometry? What is the connection between amplituhedra and cluster algebras? Also leveraging our understanding of amplituhedra gained by T-duality, we will tackle these questions and pave the way for more general explorations.

1.2 Summary of the Plan

We start by providing a brief introduction to the field of *scattering amplitudes* in Chap. 1. In particular, in Sect. 1.3 we focus on the maximally-supersymmetric Yang-Mills gauge theory ($\mathcal{N} = 4$ SYM). We use this theory to explain in Sect. 1.4 the tremendous success of *on-shell methods* both in computing and in understanding the analytic structures of scattering amplitudes, and QFTs.

We then move into the worlds of geometry and combinatorics by studying *amplituhedra*. Introduced in physics, they encode all tree-level scattering amplitudes and planar loop integrands of $\mathcal{N} = 4$ SYM. In Chap. 2 we focus on the tree-level $m = 4$ amplituhedron $\mathcal{A}_{n,k,4}$ and its $m = 2$ version $\mathcal{A}_{n,k,2}$. After reviewing the definitions and known results on the *positive Grassmannian* and the amplituhedron in Sect. 2.1, we introduce the amplituhedron-analogue of the matroid stratification of the Grassmannian by considering the *sign stratification* of $\mathcal{A}_{n,k,m}$. We then focus on *positroid tiles*—full-dimensional images of positroid cells of $\text{Gr}_{k,n}^{\geq 0}$ on which the amplituhedron map is injective. These are used to 'tile' the amplituhedron and, in the physical $m = 4$ case, are associated to *Yangian invariants*—building blocks of scattering amplitudes. In Sect. 2.3 we classify all positroid tiles of the $m = 2$ amplituhedron $\mathcal{A}_{n,k,2}$ and give them a new intrinsic sign-description inside the Grassmannian. In Sect. 2.4 we use this to prove the sign-flip characterization of $\mathcal{A}_{n,k,2}$ conjectured in [57]. Finally, in Sect. 2.5 we review how to extract n-point scattering amplitudes of $\mathcal{N} = 4$ SYM in the k helicity sector from the $m = 4$ amplituhedron $\mathcal{A}_{n,k,4}$.

In Chap. 3 we then move to a seemingly unrelated object, the *hypersimplex* $\Delta_{k+1,n}$—the image of the positive Grassmannian $\text{Gr}_{k+1,n}^{\geq 0}$ under the *moment map* μ. In Sect. 3.1 we introduce and present new results about the hypersimplex and *positroid polytopes* -images of positroid cells of $\text{Gr}_{k+1,n}^{\geq 0}$ under the moment map. In particular, we show that *positroid tiles* of $\Delta_{k+1,n}$—full-dimensional positroid polytopes on which μ is injective—are in bijection with plabic trees. We then define *positroid tilings* as collections of positroid tiles whose union covers $\Delta_{k+1,n}$ and whose interiors are disjoint. In Sect. 3.2 we show that the fan structure of the positive tropical Grassmannain $\text{Trop}^+ \text{Gr}_{k+1,n}$ is the *secondary fan* of $\Delta_{k+1,n}$, i.e. it exactly regulates its regular positroid subdivisions.

In Chap. 4 we explore a striking duality—*T-Duality*—between the hypersimplex $\Delta_{k+1,n}$ (cf. Chap. 3) and the amplituhedron $\mathcal{A}_{n,k,2}$ (cf. Chap. 2). In Sect. 4.1 we define T-duality on the main combinatorial objects associated to the positive Grassmannian—decorated permutations and plabic graphs. In Sect. 4.2 we show that T-duality provides a bijection between positroid tiles of $\Delta_{k+1,n}$ and $\mathcal{A}_{n,k,2}$. Moreover, we show that inequalities cutting out positroid polytopes translate into sign conditions characterizing the T-dual Grasstopes. Both can be simply read from the statistics of an associated *plabic tiling*. Moreover, from the latter we can also read off their facets. Finally, positroid tiles are also in bijection with separable permutations—counted by (a refinement of) the *Schröder numbers*. In Sect. 4.3 we show that we can subdivide the amplituhedron $\mathcal{A}_{n,k,2}$ into w-chambers as the hypersimplex $\Delta_{k+1,n}$ can be subdivided into w-simplices—both labelled by *Eulerian numbers*, and related by T-duality. In Sect. 4.4 we define the $\mathcal{G}$-amplituhedron—an avatar of the amplituhedron in $\text{Gr}_{2,n}$. Using this we prove that w-chambers are exactly the only realizable strata in the sign stratification of $\mathcal{A}_{n,k,2}$. Finally, we show that the *total amplituhedron*—the union of $\mathcal{G}$-amplituhedra varying k—gives the amplituhedron-analogue of the hypercube—which is the union of hypersimplices varying k.

In Chap. 5 we introduce and discuss positroid tilings and *dissections*—subdivisions of the the hypersimplex or the amplituhedron into full-dimensional positroid polytopes or Grasstopes respectively. Using results from Chap. 4, in Sect. 5.1 we show that a collection of positroid polytopes is a positroid tiling of $\Delta_{k+1,n}$ if and only if the collection of T-dual Grasstopes is a positroid tiling of $\mathcal{A}_{n,k,2}(Z)$ for all Z. We also conjecture the same holds true if we replace "tilings" with "dissections". In Sect. 5.2 we show how to obtain different types of positroid tilings (and dissections) of $\Delta_{k+1,n}$ and $\mathcal{A}_{n,k,2}(Z)$ from: BCFW recursion relations, the positive tropical Grassmannian $\text{Trop}^+ \text{Gr}_{k+1,n}$, and positions of descents/sign flips. Finally, in Sect. 5.3 we discuss the relation of T-duality with cyclic symmetry and *parity duality*.

In Chap. 6 we introduce the *momentum amplituhedron* $\mathcal{M}_{n,k',m}$. The $m = 4$ momentum amplituhedron $\mathcal{M}_{n,k+2,4}$ is a new positive geometry which encodes tree-level N^kMHV_n scattering amplitudes of $\mathcal{N} = 4$ SYM in *momentum space*. In Sect. 6.1 we give the definition of $\mathcal{M}_{n,k',m}$ as the image of the non-negative Grassmannian $\text{Gr}_{k',n}^{\geq 0}$ under the momentum amplituhedron map $\Phi_{\Lambda,\tilde{\Lambda}}$. Then, as we did for $\mathcal{A}_{n,k,m}$ in Chap. 2, we define a sign stratification and discuss sign flips for $\mathcal{M}_{n,k',m}$. For the $m = 4$ momentum amplituhedron, we also discuss its facet structure in relation to the factorization channels of scattering amplitudes in $\mathcal{N} = 4$ SYM. In Sect. 6.1.4 we conjecture T-duality relates tilings (and dissections) of the momentum amplituhedron $\mathcal{M}_{n,k+2,4}$ and the amplituhedron $\mathcal{A}_{n,k,4}$. This is the combinatorial-geometric incarnation of the physical *Amplitudes/Wilson loops* duality. Finally, in Sect. 6.2 we explain how to extract N^kMHV_n scattering amplitudes in momentum space from $\mathcal{M}_{n,k+2,4}$.

In Chap. 7 we explore the intersection between amplituhedra and *cluster algebras*. In Sect. 7.1 we discuss phenomena of *cluster adjacencies* in connection with the amplituhedron, Yangian invariants, and scattering amplitudes. We formulate and generalize the cluster adjacency conjecture in terms of the amplituhedron $\mathcal{A}_{m,k,m}$. We prove this conjecture for the $m = 2$ amplituhedron $\mathcal{A}_{n,k,2}$ and provide a conjectural formula for all $m = 2$ Yangian invariants which manifest cluster adjacency. We then explore the $m = 4$ cluster adjacency conjecture—relevant for scattering amplitudes in $\mathcal{N} = 4$ SYM. Using the geometry of the amplituhedron $\mathcal{A}_{n,k,4}$, we write the poles of all Yangian Invariants in the N^2MHV sector in terms of compatible cluster variables of $\text{Gr}_{4,n}$. In Sect. 7.2 we discover cluster varieties directly inside the amplituhedron $\mathcal{A}_{n,k,2}$, using results from Sect. 2.3. In Sect. 7.3 we review how the singularity structure of scattering amplitudes—in particular, *Leading and Landau Singularities*—can be understood in terms of amplituhedra. Then in Sect. 7.4 we phrase a conjecture on novel cluster structures relating Leading and Landau singularities—*LL-cluster adjacency*. At one loop, we prove LL-cluster adjacency for all NMHV amplitudes and check it for N^2MHV amplitudes up to 9 points. We also discuss possible implications of LL-cluster adjacency for bootstrapping loop amplitudes. We end in Chap. 8 with conclusions and outlook.

1.3 Introduction to Scattering Amplitudes

In this section we will start with an introduction to scattering amplitudes (Sect. 1.3), in particular of $\mathcal{N} = 4$ SYM theory (Sect. 1.3.1), which is the main theory relevant for our work. We will briefly review *colour decomposition* techniques (Sect. 1.3.2) for gauge theories and the *super-space formalism* (Sect. 1.3.3) for super-symmetric ones. Right after, space is given to modern *on-shell methods* (Sect. 1.4): on-shell variables (Sect. 1.4.1)—such as spinor helicity and twistors—and recursion relations (Sect. 1.4.2). These allowed the computation of all tree-level amplitudes and loop-level integrands in $\mathcal{N} = 4$ SYM, and the discovery of their analytical properties and hidden symmetries.

1.3.1 $\mathcal{N} = 4$ Super Yang-Mills Theory

Let us give a brief introduction to the theory that will be the main subject of our work: $\mathcal{N} = 4$ super Yang-Mills theory, a supersymmetric four-dimensional interacting gauge theory with $SU(N)$ gauge group ($\mathcal{N} = 4$ SYM for short). In four dimensions one can construct $\mathcal{N} = 1, 2, 4$ supersymmetric theories with spins not higher than 1: we will hence be focusing on the *maximally supersymmetric* one,[3] and refer to [58, 59] for additional details.

As every traditional quantum field theorist, we start our discussion with an action. Using the standard QFT approach, which we will soon abandon from next section on, we define it for $\mathcal{N} = 4$ SYM and briefly illustrate its field content, its classical symmetries and which of them survive in the quantum regime. The action of $\mathcal{N} = 4$ SYM reads:

$$S_{\mathcal{N}=4} := \frac{1}{g^2}\int \mathrm{d}^4x \,\mathrm{Tr}\left(-\frac{1}{4}F_{\mu\nu}F^{\mu\nu} - \left(D_\mu\phi_{\mathsf{AB}}\right)\left(D^\mu\phi^{\mathsf{AB}}\right) - \frac{1}{2}[\phi_{\mathsf{AB}},\phi_{\mathsf{CD}}][\phi^{\mathsf{AB}},\phi^{\mathsf{CD}}] + \right.$$
$$\left. + i\bar{\psi}_{\mathsf{A}\dot{\alpha}}\sigma_\mu^{\dot{\alpha}\alpha}D^\mu\psi_\alpha^{\mathsf{A}} - \frac{i}{2}\bar{\psi}_{\dot{\alpha}}^{\mathsf{A}}[\phi_{\mathsf{AB}},\bar{\psi}^{\mathsf{B}\dot{\alpha}}] - \frac{i}{2}\psi_{\mathsf{A}}^{\alpha}[\phi^{\mathsf{AB}},\psi_{\mathsf{B}\alpha}]\right). \tag{1.3.1}$$

All fields are massless and transform in the adjoint representation of the gauge group $SU(N)$, with generic N. In the above expression one has: adjoint indices (the trace Tr is implicitly summing over them), Lorentz indices ($\mu, \nu = 0, 1, 2, 3$), spinor indices ($\alpha, \dot{\alpha} = 1, 2$) and the $SU(4)_R$ indices ($\mathsf{A}, \mathsf{B}, \mathsf{C}, \mathsf{D} = 1, ..., \mathcal{N} = 4$). In the spectrum of the theory one has: a gauge field A_μ, corresponding to the pure Yang-Mills *gluons*; its $\mathcal{N} = 4$ superpartners ψ^{A}, which are complex massless Weyl fermions called *gluinos*, satisfying $\bar{\psi}_{\dot{\alpha}}^{\mathsf{A}} = \left(\psi_{\mathsf{A}\alpha}\right)^*$; finally, the closure of supersymmetry algebra demands the presence of $\mathcal{N}(\mathcal{N}-1)/2 = 6$ real scalar fields $\phi^{\mathsf{AB}} = -\phi^{\mathsf{BA}}$, such

[3] If one wants to include spin 2, i.e. gravity, the maximally supersymmetric theory is $\mathcal{N} = 8$ Supergravity.

that $\phi_{AB} = \frac{1}{2}\epsilon_{ABCD}\phi^{CD}$. $\mathcal{N} = 4$ SYM has therefore 8 bosonic (2 from gluons and 6 from scalars) and 8 fermionic (2 for each gluinos) degrees of freedom, recalling that gluons and gluinos comes with two helicities ($\pm 1, \pm 1/2$ respectively). We will now list the classical symmetries of the action (1.3.1):

Conformal Symmetry. Conformal transformations are a special class of space-time transformations. They can be defined in an arbitrary space-time, but we will focus on four-dimensional Minkowski. Conformal transformations form a group named *Conformal Group*. A representation of generators $\{G_a\}$ of the group acting on fields $X \in \{A^\mu, \phi, \psi, \bar{\psi}\}$ in Minkowski space-time, i.e.

$$X(x') = X(x) + i\epsilon^a (G_a X)(x) + o(\epsilon) \tag{1.3.2}$$

where ϵ^a are the parameters of the group, is given by:

$$P_\mu = -i\partial_\mu \quad \text{(Translations)} \tag{1.3.3}$$
$$M_{\mu\nu} = -i(x_\mu \partial_\nu - x_\nu \partial_\mu) \quad \text{(Lorentz Boosts)} \tag{1.3.4}$$
$$D = ix \cdot \partial \quad \text{(Dilatations)} \tag{1.3.5}$$
$$K_\mu = -i(x^2 \partial - 2x_\mu x \cdot \partial) \quad \text{(Special Conformal Transformations)} \tag{1.3.6}$$

From this representation it easy to understand the associated algebra named *Conformal Algebra*. The subalgebra generated by $\{P_\mu, M_{\mu\nu}\}$ is exactly the Poincaré algebra. Furthermore, dilatations can be generated from Poincaré and special conformal transformations. The conformal group is isomorphic to $O(2, 4)$ which in turn is isomorphic to $SU(2, 2)$.

R-symmetry. The action in (1.3.1) is built from invariants in the indices $\mathsf{A} = 1, \ldots, \mathcal{N} = 4$, thus it has a global $SU(4)$ symmetry called *R-symmetry* and denoted as $SU(4)_R$. ψ and $\bar{\psi}$ are in the fundamental (4) and anti-fundamental ($\bar{4}$) representation of this group respectively, while ϕ is in the anti-symmetric self-dual (6) representation.

Poincaré Supersymmetry. The action in (1.3.1) has a symmetry whose infinitesimal transformations can be expressed as:

$$\delta_{\xi,\bar{\xi}} X = i\xi^{\mathsf{A}}_\alpha Q^\alpha_{\mathsf{A}}(X) + i\bar{\xi}_{\mathsf{A}\dot{\alpha}} \bar{Q}^{\mathsf{A}\dot{\alpha}}(X), \quad X \in \{A^\mu, \phi, \psi, \bar{\psi}\}$$

with the generators of *supersymmetry* satisfying[4]

$$\{Q^\alpha_{\mathsf{A}}, \bar{Q}^{\mathsf{A}\dot{\alpha}}\} = 2\delta^{\mathsf{A}}_{\mathsf{B}} \sigma^\mu_{\alpha\dot{\alpha}} P_\mu, \quad \{Q^\alpha_{\mathsf{A}}, Q^\beta_{\mathsf{B}}\} = \{\bar{Q}^{\mathsf{A}\dot{\alpha}}, \bar{Q}^{\mathsf{B}\dot{\beta}}\} = 0 \tag{1.3.7}$$

The Poincaré algebra can be extended by promoting it to a graded Lie algebra or *superalgebra* including Q^{A}_α and $\bar{Q}^{\mathsf{A}\dot{\alpha}}$. This closes into the Poincaré superalgebra.

[4] We are considering massless supersymmetric theory.

Superconformal Symmetry. The generators of supersymmetry do not commute with the full conformal group. In particular, commuting them with the generators of the special conformal transformations, one obtains the generators S^{A}_{α} and $\bar{S}_{\mathsf{A}\dot{\alpha}}$—called *conformal supercharges*—necessary for the closure of the algebra.

Thus $\{P_{\mu}, M_{muv}, D, K_{\mu}, Q^{\alpha}_{\mathsf{A}}, \bar{Q}^{\mathsf{A}\dot{\alpha}}, S^{\mathsf{A}}_{\alpha}, \bar{S}_{\mathsf{A}\dot{\alpha}}\}$ form a group which, together with R-symmetry, is called the *superconformal group* $PSU(2,2|4)$. The notation $SU(2,2|4)$ is referred to the bosonic part of the symmetry group (Poincaré and R-symmetry) $SU(2,2) \times SU(4)_R$, and the "P" refers to the fact one considers thc irreducible part of $SU(2,2|4)$ only (see [60]).

Symmetries at Quantum Level. Classical symmetries of the action might not survive in the quantum realm. This is due to the fact that a *quantum symmetry* should be a symmetry of the whole path integral. If one starts from a classical symmetry, this simply means that the path integral measure should also transform invariantly under such a symmetry. If that is not the case, one says that the symmetry is *anomalous*. In particular, conformal symmetry is usually anomalous. Indeed, when quantising a field theory, it is necessary to regularise it first in order to remove divergences and this inevitably requires the introduction of a mass scale μ. In the regularized theory, μ breaks conformal symmetry because one loses scale invariance. Of course, a physically meaningful result must not depend on the arbitrary scale. This is achieved by assuming that the parameters of the quantum theory also depend on μ in such a way that the explicit and implicit dependence cancel out. For $\mathcal{N} = 4$ SYM, the only parameter is the coupling constant g and its dependence on the scale is described by the *beta function*:

$$\beta(\mu) := \mu \frac{\partial g}{\partial \mu}. \tag{1.3.8}$$

In general, a non-vanishing beta function is related to the breakdown of scale invariance and conformal symmetry in a massless gauge theory. For $\mathcal{N} = 4$ SYM, however, the beta function is believed to vanish to all orders in perturbation theory as well as non perturbatively. This means that the superconformal symmetry $PSU(2,2,|4)$ is preserved even at the quantum level. Despite this nice property, it is not true that there are no divergences in $\mathcal{N} = 4$ SYM: infrared (IR) divergences still affect radiative corrections nonetheless. Moreover, perturbative calculations with composite operators exhibit ultraviolet (UV) divergences as well which lead to anomalous dimensions. For example, the *cusp anomalous dimension* is an important quantity in four-dimensional Yang-Mills theories (from QCD to $\mathcal{N} = 4$ SYM). It controls the leading UV divergences of Wilson loops evaluated along a closed contour in Minkowski space-time containing cusps formed by two light-like tangent vectors [61, 62].

1.3.2 Colour Decomposition and Planar Limit

Starting from the action of $\mathcal{N}=4$ SYM in (1.3.1), one can derive the *Feynman rules* of the theory and in principle construct every desired scattering amplitude. However, the elegance and feasibility of this method breaks down very soon as the complexity in computations increases very fast with the number of particles and loops. For gauge theories, Feynman rules contain two kinds of information that can be disentangled: one is the *kinematic*, depending on momenta and polarization vectors, and the other is the *colour*, carried by the structure constants of the gauge algebra.

Let $SU(N)$ be the gauge group and $\{T^a\}$ its N^2-1 generators. The *colour decomposition* can be applied to write a generic amplitude A_n, encoding also the colour of particles $\{a_i\}$, in terms of gauge-invariant *colour-ordered* (or *partial*) amplitudes A_n which depends only on the kinematics (momenta and helicity) $\{p_i, h_i\}$ as:

$$\mathsf{A}_n(\{a_i, p_i, h_i\}) = \sum_{\sigma\in S_n/\mathbb{Z}_n} \mathrm{Tr}\big(T^{a_{\sigma(1)}}...T^{a_{\sigma(n)}}\big) A_n(\{p_{\sigma(i)}, h_{\sigma(i)}\}) + \mathcal{O}(1/N). \quad (1.3.9)$$

The sum runs over non-cyclic permutations of the external legs and the extra terms containing multiple traces are suppressed when $N\to\infty$. This limit[5] is called *planar limit* and was first considered by 't Hooft in [63]. The name stems its origin in the fact that in this limit, at every loop order, non-planar diagrams are suppressed with respect to planar ones. Therefore A_n receives contributions only from planar Feynmann diagrams; moreover, they must have a specific colour-ordering, with the only poles that can occur for each order σ when a sum of momenta with indices in an interval $[\sigma(i), \sigma(j)]$ goes on shell.

Colour-ordered amplitudes are thus gauge invariant quantities encoding the full kinematics of the process which are much simpler than the complete amplitude. They satisfy even more remarkable relations, which reduce the number of independent partial amplitudes. We list the known relations for tree-level amplitudes[6]:

Cyclicity

$$A_n^{\text{tree}}(1, ..., n) = A_n^{\text{tree}}(\sigma(1, ..., n)), \quad \sigma \text{ cyclic shift} \quad (1.3.10)$$

which reduce the number by a factor of n.

Kleis-Kuijf Relations

$$A_n(1, \alpha, n, \beta) = (-1)^{|\beta|} \sum_{\sigma\in\mathrm{OP}(\alpha|\beta^\perp)} A_n(1, \sigma, n) \quad (1.3.11)$$

where α, β are two ordered set of labels such that $\alpha\cup\beta = \{2, ..., n-1\}$, $\beta^\perp$ is β with reversed order, $\mathrm{OP}(\alpha|\beta^\perp)$ is the set of ordered permutations, i.e. those which preserve the relative order of the elements of each set α, $\beta^\perp$ respectively. E.g. for

[5] With the addition of keeping the 't Hooft coupling $\lambda := g^2N$ constant.

[6] We will shortened the notation of labels by writing i instead of $\{p_i, h_i\}$.

$n = 4$ they read: $A_4(1423) = -A_4(1234) - A_4(1342)$. These relations bring the number of independent colour-order amplitudes to $(n-2)!$, see [64].

BCJ Relations

$$A_n(1, 2, \alpha, 3, \beta) = (-1)^{|\beta|} \sum_{\sigma \in \mathrm{POP}(\alpha|\beta)} \prod_{k=4}^{n} \frac{\mathcal{F}_k(3, \sigma, 1)}{s_{2,4,\ldots,k}} A_n(1, 2, 3, \sigma) \qquad (1.3.12)$$

where the sum is over partial ordered permutations, i.e. preserving the order only in β and $\mathcal{F}_k$ is a function of the Mandelstam invariants $s_\gamma := (\sum_{i \in \gamma} p_i)^2$, see [65] for complete details. E.g. for $n = 4$ they read: $A_4(1342) = s_{13}/s_{12} A_4(1234)$. These relations bring the number of independent colour-order amplitudes to $(n-3)!$.

1.3.3 The Super-Space Formalism

While supersymmetric theories enjoy a large number of symmetries, which is in general a great property in physics, their particle content is quite wide and one could think that this would make actual computations very difficult to handle. However, theories with maximal supersymmetry have the feature that all particles, of any helicity, can be packaged into a single super-multiplet, see [66], and [67] for a review. In particular, $\mathcal{N} = 4$ SYM has a single CPT-self-dual supermultiplet, involving sixteen massless states that can be grouped all in a *superfield* Φ. This is achieved by employing the auxiliary Grassmann-odd variables η^{A}, $\mathsf{A} = 1, \ldots, \mathcal{N} = 4$, which transform in the fundamental representation[7] of $SU(4)_R$:

$$\Phi(\eta) := g^+ + \eta^{\mathsf{A}} \psi^+_{\mathsf{A}} + \frac{1}{2}\eta^{\mathsf{A}}\eta^{\mathsf{B}} \phi_{\mathsf{AB}} + \frac{1}{3!}\eta^{\mathsf{A}}\eta^{\mathsf{B}}\eta^{\mathsf{C}} \epsilon_{\mathsf{ABCD}} \bar{\psi}^{-\mathsf{D}} + \frac{1}{4!}\eta^{\mathsf{A}}\eta^{\mathsf{B}}\eta^{\mathsf{C}}\eta^{\mathsf{D}} \epsilon_{\mathsf{ABCD}}\, g^-, \qquad (1.3.13)$$

where g^+ is the gluon with helicity $h = +1$, ψ^+ the gluinos ($h = +1/2$), ϕ_{AB} the scalars ($h = 0$), $\bar{\psi}^-$ the anti-gluinos ($h = -1/2$), and finally g^- the gluon with $h = -1$. Observe that if one assigns to η's helicity $+1/2$, then the superfield carries uniform helicity $+1$.

One can now define scattering amplitudes of superfields $\mathcal{A}_n(\Phi_1 \ldots \Phi_n)$, called *superamplitudes*. They are polynomials in the η's, which allow to keep track of the contribution of each state of the supermultiplet within the scattering process. The component amplitude of a specific set of states can be then extracted by integrating over suitable set of η variables. For example, one can compute the amplitude of 2 negative-helicity and $n-2$ positive-helicity gluons as

[7] An equivalent choice would have been to consider the parity-conjugate fermionic variables $\bar{\eta}^{\mathsf{A}}$, transforming in the anti-fundamental representation of $SU(4)_R$. The two choices are related to each other by (Grassmann) Fourier Transform.

$$A_n(g_1^+, ..., g_i^-, ..., g_j^-, ..., g_n^+) = \int d^4\eta_i \, d^4\eta_j \mathcal{A}_n \left(\Phi_1 ... \Phi_n\right), \tag{1.3.14}$$

which evaluates at tree-level to the famous Parke-Taylor formula.

In order for the superamplitude to be $SU(4)_R$-symmetric, $\mathcal{A}_n$ has to be a sum of monomials of degree multiple of 4 in the η's. It can be shown, see e.g. [59], that the first non-vanishing amplitude has degree 8: this is called MHV *(Maximally Helicity Violating*[8]*)*; the one with degree 12 is referred to as NMHV *(Next to Maximally Helicity Violating)* and in general amplitudes with degree $4(k+2)$ are said to belong to the N^{k}MHV *helicity sector*. By parity conjugation, one can see that the maximal degree is $k = n-4$, since $\overline{\text{MHV}} = \text{N}^{n-4}\text{MHV}$, and, more generally, $\overline{\text{N}^k\text{MHV}} = \text{N}^{n-k-4}\text{MHV}$.

1.4 Modern On-Shell Methods

In this section we will introduce the main frameworks and tools that allow both to compute scattering amplitudes in planar $\mathcal{N}=4$ SYM more efficiently and to understand their geometrization in next sections.

1.4.1 On-Shell Variables

Spinor-Helicity Variables. In this work, we will be focusing on scattering amplitudes for n massless ($m_i = 0$) particles with momenta p_i and helicities h_i, with $i = 1, ..., n$. Let us define the 2×2 matrix:

$$p_i^{\alpha\dot{\alpha}} := p_i^\mu \sigma_\mu^{\alpha\dot{\alpha}} = \begin{pmatrix} p_i^1 + p_i^3 & p_i^0 - ip_i^2 \\ p_i^0 + ip_i^2 & p_i^1 - p_i^3 \end{pmatrix}, \qquad \alpha, \dot{\alpha} = 1, 2. \tag{1.4.1}$$

where $\sigma_\mu^{\alpha\dot{\alpha}} := (\mathbb{1}_2, \sigma_1, \sigma_2, \sigma_3)^{\alpha\dot{\alpha}}$, with σ_a's being the Pauli matrices. Since the momenta are null, $p_i^{\alpha\dot{\alpha}}$ has vanishing determinant, indeed: $\det\left(p_i^{\alpha\dot{\alpha}}\right) = p_i^2 = m_i^2 = 0$. Therefore the matrix has rank less then 2 and can be written as

$$p_i^{\alpha\dot{\alpha}} = \lambda_i^\alpha \tilde{\lambda}_i^{\dot{\alpha}}, \tag{1.4.2}$$

where $\{\lambda_i, \tilde{\lambda}_i\}$ are called *spinor-helicity* variables [68–70]. If the momentum is real, one has $\tilde{\lambda}_i = \pm\lambda_i^*$; but, in general, one allows the momenta to be complex and consider them as independent complex variables. Basic Lorentz invariants can be easily built via contraction of spinor helicity variables with the ϵ-symbol:

$$\langle i \, j\rangle := \lambda_i^\alpha \lambda_{j\alpha} = \epsilon_{\alpha\beta}\lambda_i^\alpha \lambda_j^\beta, \quad [i \, j] := \tilde{\lambda}_{i\dot{\alpha}} \tilde{\lambda}_j^{\dot{\alpha}} = \epsilon_{\dot{\alpha}\dot{\beta}} \tilde{\lambda}_i^{\dot{\beta}} \tilde{\lambda}_j^{\dot{\alpha}}. \tag{1.4.3}$$

[8] The amplitude $A_n(g_1^-, g_2^-, g_3^+, ..., g_n^+)$ belongs to this sector and since $A_n(g_1^+, ..., g_n^+) = A_n(g_1^-, g_2^+, ..., g_n^+) = 0$, this is the one which maximizes the violation of helicity.

Standard Lorentz invariants, like *Mandelstam variables* s_{ij}, are easily expressed in terms of the latter anti-symmetric brackets:

$$s_{ij} = (p_i + p_j)^2 = 2p_i \cdot p_j = \langle i\, j\rangle\, [j\, i]\,. \tag{1.4.4}$$

The rescaling $\lambda_i \mapsto t_i\lambda_i$, $\tilde{\lambda}_i \mapsto t_i^{-1}\tilde{\lambda}_i$ leaves the momentum p_i invariant and thus represents the action of the *little group*. All the information about the helicities h_i of particles involved in a scattering amplitude A_n is encoded by its weights under such rescaling:

$$A_n(t_i\lambda_i, t_i^{-1}\tilde{\lambda}_i; h_i) = t_i^{-2h_i} A_n(\lambda_i, \tilde{\lambda}_i; h_i). \tag{1.4.5}$$

Spinor-helicity variables hence naturally encode both the null-ness of momenta and the information about helicity of the scattering particles.

Twistor Variables. Amplitudes are supported on momenta that, other than null-ness, satisfy momentum conservation. Clearly, it would be convenient to find variables where this constraint is trivial. In planar theories, where color ordering is available, there is a natural way to achieve this, by choosing to express the external momenta in terms of what are known [71] as *dual variables* $\{x_i\}$, writing $p_i = x_i - x_{i-1}$, with $p_1 = x_1 - x_n$. It is clear that momenta obtained in this way automatically satisfy $\sum_{i=1}^n p_i = 0$.

Now one has the variables $\{\lambda_i, \tilde{\lambda}_i\}$ which make the null condition trivial while ignoring momentum conservation, whereas the dual variables $\{x_i\}$ do the opposite. It is perfectly natural to wonder if there exists any way to combine these two constructions which makes both the null-condition and momentum conservation trivial. It turns out that such a set of variables does exist: they are known as *momentum-twistors* and were introduced by Hodges in [9]. The latter hinge on the construction called *Twistors* developed by Penrose in the '60s [72], see [73] for a review. This starts by making a connection between null rays in space-time and points in an auxiliary projective space, called twistor-space. Likewise, a complex line in twistor space is related to a point in space-time. This correspondence is realized by defining the so-called *incidence relation*, according to which points x in space-time and points $z = (\lambda, \mu) \in \mathbb{CP}^3$ in twistor-space are related via

$$\mu_{\dot{\alpha}} = x_{\dot{\alpha}\alpha}\lambda^{\alpha}, \qquad \alpha, \dot{\alpha} = 1, 2. \tag{1.4.6}$$

Given a point $x_{\dot{\alpha}\alpha}$, (1.4.6) clearly determines the line whose points are of the form $(\lambda, x \cdot \lambda) \in \mathbb{CP}^3$. Likewise, given a line in twistor-space passing through the points z_A and z_B, (1.4.6) consists in 4 relations which determine the point uniquely as

$$x_{\dot{\alpha}\alpha} = \frac{\lambda_{A,\alpha}\mu_{B,\dot{\alpha}}}{\langle\lambda_A\lambda_B\rangle} + \frac{\lambda_{B,\alpha}\mu_{A,\dot{\alpha}}}{\langle\lambda_B\lambda_A\rangle}. \tag{1.4.7}$$

There is now a crucial statement which is useful for constructing the desired variables: *two points in space-time are null separated if and only if the corresponding lines in twistor-space intersect*. Let x, y be points in space-time and $\{(\lambda, x \cdot \lambda)\}$ and $\{(\lambda, y \cdot \lambda)\}$ the corresponding lines in twistor-space and assume they intersect in z^*. Then one must have $x \cdot \lambda^* = y \cdot \lambda^*$, i.e. the 2×2 matrix $(x-y)_{\dot{\alpha}\alpha}$ has non-zero kernel ($\lambda^* \neq 0$); therefore $0 = \det((x-y)_{\dot{\alpha}\alpha}) = (x-y)^2$. It is straightforward to revert the previous reasoning.

If one denotes by z_A, z_B and z_C, z_D the points through which the lines corresponding to x and y pass respectively, one can define the following SL_4 invariant:

$$\langle z_A z_B z_C z_D \rangle := \epsilon_{abcd} z_A^a z_B^b z_C^c z_D^d. \tag{1.4.8}$$

Throughout the text, ϵ_I will denote the sign of the permutation $(1 \dots n) \to I$. For example, $\epsilon_{132} = -1$ and $\epsilon_{231} = 1$. Moreover, the summation over repeated indices is sometimes omitted. When this determinant vanishes, the four points are co-planar, i.e. the two respective lines are co-planar and in projective space this implies they intersect. The previous statement can then be compactly rephrased as:

$$(x-y)^2 = 0 \Longleftrightarrow \langle z_A z_B z_C z_D \rangle = 0. \tag{1.4.9}$$

When pairwise intersecting twistors $\{z_i\}$ are used to produce a configuration of pairwise null separated points in space-time $\{x_i \leftrightarrow (z_i, z_{i+1})\}$ which are then used to build momenta $\{p_i = x_i - x_{i-1}\}$, they are called *momentum-twistor* variables, see Fig.1.1. As desired, these encodes both momentum conservation $\sum_i^n p_i = \sum_i^n (x_i - x_{i-1}) = 0$ and null-ness $p_i^2 = (x_i - x_{i-1})^2 = 0$, since the lines associated to x_i and x_{i-1} intersect in z_i by construction.

As momentum-twistors $z_i \in \mathbb{CP}^3$ encode momenta, one can define a supersymmetric part related to super-momentum $q^{\alpha\mathsf{A}} := \sum_{i=1}^n \lambda_i^\alpha \eta_i^{\mathsf{A}}$, which is conserved by super-symmetry. As for the former case, one first defines variables $\{\theta_i\}$ in the *dual super-space* by writing $q_i^{\alpha\mathsf{A}} = \theta_i^{\alpha\mathsf{A}} - \theta_{i-1}^{\alpha\mathsf{A}}$, with $q_1 = \theta_1 - \theta_n$ [71], such that clearly $\sum_i^n q_i = 0$ is satisfied. Furthermore, one can defined the variables $\{\chi_i\}$ via the incidence relation: $\chi_i^{\mathsf{A}} = \theta_i^{\alpha\mathsf{A}} \lambda_{\alpha,i}$, which is the analogue of (1.4.6). Packing the latter with momentum twistors, one can define *Super-Momentum Twistors* $\mathcal{Z}_i^{\mathcal{A}} = (z_i^a, \chi_i^{\mathsf{A}})$ in[9] $\mathbb{CP}^{3|4}$.

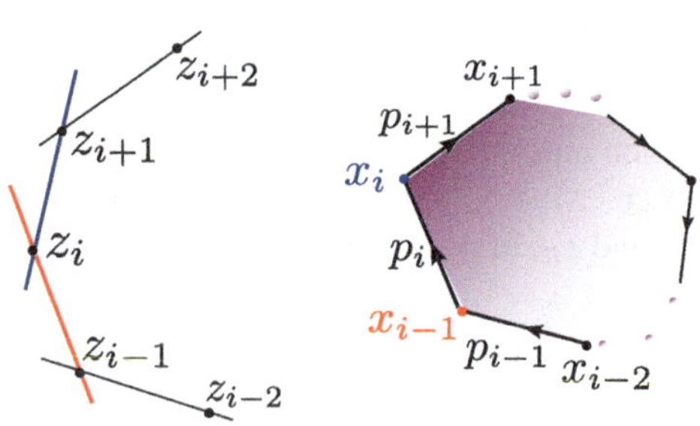

Fig. 1.1 Configuration of momentum twistors z's, dual variables x's and momenta p's

[9] The slash "|" divides the Grassmann even from the Grassmann odd part. The variables $\{\chi\}$ are protectively defined due to the little group rescaling $\eta_i \mapsto t \cdot \eta_i$, $t \in \mathbb{R}$.

1.4.2 Recursion Relations

Recursion relations are based on a remarkably simple idea of combining the *analytical properties* of functions and contour integration together with the physical concepts of *locality* and *unitarity*.

Using Feynman rules, amplitudes are constructed by connecting interaction vertices through propagators. Because of locality, such expressions are always meromorphic functions of momenta and polarization vectors, which have pole structures when propagators go on-shell. The singularity structure of tree-level amplitudes is very simple: there are only poles. Branch cuts and other singularities appear only when these expressions are integrated over. In the following we will review tree level amplitudes. For details about recursion relations for integrands of loop amplitudes see [74]. Furthermore, at the loop level another technique called *generalized unitarity* is used to probe the analytic properties of the amplitude and to extract the information one needs to construct them (see [75, 76]).

In order to exploit the analytic properties of the n-points color-ordered amplitude A_n, one performs a complex deformation into a meromorphic function $\hat{A}_n(z)$ such that $\hat{A}_n(0) = A_n$. By using the residue theorem for $\hat{A}(z)/z$, it is easy to show that

$$A_n = -\sum_{i=1}^{N} \mathrm{Res}_{z=z_i} \frac{\hat{A}(z)}{z} - \mathrm{Res}_{z=\infty} \frac{\hat{A}_n(z)}{z}, \tag{1.4.10}$$

where $\{z_1, ..., z_N\}$ is the set of isolated poles of $\hat{A}_n(z)$. The key ingredients now to turn this into a recursive formula are unitarity of scattering amplitudes and a suited choice of deformation.

Unitarity, indeed, dictates how amplitudes factorise in lower-point amplitudes when one sends one propagator on-shell:

$$\lim_{P_I^2 \to 0} \left(P_I^2 A_n\right) = \sum_s A_L(-P_I^s)\, A_R(P_I^{\bar{s}}), \tag{1.4.11}$$

where A_L, A_R are lower-point amplitudes obtained from the two diagrams (the left and the right ones) the propagator is separating, the sum is performed over all possible internal states of helicities s and $P_I := \sum_{i \in I} p_i$, $I \subset [n] := \{1, ..., n\}$.

Since one wants this to hold true for deformed amplitudes as well, one chooses deformations which preserve both the on-shell conditions $\hat{p}_i^2(z) = 0$ and the momentum conservation $\sum_i^n \hat{p}_i(z) = 0$. If one demands a linear deformation of momenta $p_i^\mu \mapsto \hat{p}_i^\mu(z) = p_i^\mu + z k_i^\mu$ to be compatible with the previous conditions, then the auxiliary vectors k_i^μ, $i = 1, ..., n$ must satisfy

$$\sum_{i=1}^{n} k_i^\mu = 0, \quad k_i \cdot k_j = 0, \quad p_i \cdot k_i = 0, \quad \forall i, j = 1, ..., n \tag{1.4.12}$$

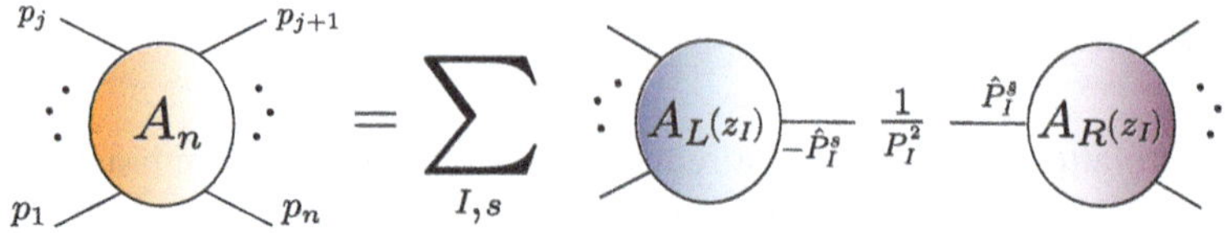

Fig. 1.2 Recursion relation for scattering amplitudes

This specific choice of deformations makes $\hat{A}_n(z)$ have only simple poles z_I, when $\hat{P}_I^2 = 0$. Furthermore, the connection between (1.4.11) and (1.4.10) stems in the fact that:

$$\text{Res} \left. \frac{\hat{A}_n(z)}{z} \right|_{z=z_I} = -\frac{1}{P_I^2} \lim_{z \to z_I} \left(\hat{P}_I^2(z) \hat{A}_n(z) \right). \tag{1.4.13}$$

With all these ingredients, one can finally state *Recursion Relations* for scattering amplitudes: if[10] $\hat{A}_n(z) \to 0$ when $z \to \infty$, then

$$A_n = \sum_I \sum_s A_L(-\hat{P}_I^s; z_I) \frac{1}{P_I^2} A_R(\hat{P}_I^{\bar{s}}; z_I), \tag{1.4.14}$$

where the sum runs over all possible $I \in [n]$ such that $\hat{P}_I^2(z) = 0$ is a pole, see Fig. 1.2. This formula is very powerful and has a wide range of applicability, in many different quantum field theories: for a particular (local) QFT one just has to find a deformation such that the deformed amplitude vanishes at infinity.[11]

BCFW shift. One of the most popular deformation is the *BCFW recursion* (after Britto, Cachazo, Feng, Witten [3]). Introducing complex shifts of the momenta using spinor-helicity variables, a $[b, a\rangle$-BCFW shift is defined by:

$$\hat{\lambda}_a = \lambda_a - z\lambda_b, \quad \hat{\tilde{\lambda}}_b = \tilde{\lambda}_b + z\tilde{\lambda}_a, \tag{1.4.15}$$

and the rest remains unshifted. It is easy to see that this shift corresponds to the following choice of auxiliary vectors: $k_b^{\alpha\dot{\alpha}} := \tilde{\lambda}_a^{\dot{\alpha}} \lambda_b^{\alpha}$, $k_a^{\alpha\dot{\alpha}} := -k_b^{\alpha\dot{\alpha}}$ and $k_i^{\alpha\dot{\alpha}} = 0$ for the rest. One observes that, in this shift, if $I \ni a, b$ then $\hat{P}_I^2(z) = P_I^2$ does not determine a pole of $\hat{A}_n(z)$. The minimum number of poles is $n - 3$ and it is obtained when one chooses a, b (cyclically) adjacent.

These recursion relations[12] were crucial to completely solve all tree-level amplitudes in $\mathcal{N} = 4$ SYM and most of the relevant scattering amplitudes in QCD too [5]. The power and elegance of these methods is that all the information needed to

[10] One must ensure that the residue at infinity of the function $\hat{A}_n(z)/z$ is zero.

[11] Even if this is not possible for every QFT, there are ways to study the behaviour of the amplitude in this case and take the residue at infinity as well into account, see [77].

[12] There is also another broadly employed shift called *MHV vertex expansion* or *CSW recursion* (after Cachazo, Svrcek, Witten [78]).

construct all scattering amplitudes at tree level are just elementary building blocks, i.e. three-point amplitudes which can be determined by fundamental properties (such as Lorentz invariance, helicities of the particles involved, etc.).

Super-BCFW shift. One can incorporate the super-space formalism within the framework of recursion relations, in order to get recursion relations for the super-amplitude $\mathcal{A}_n$ directly. If one wants a linear deformation of $\hat{\eta}_i(z) = \eta_i + z\eta_i^*$ which preserves the super-momentum conservation $\sum_{i=1}^n \hat{\lambda}_i(z)\hat{\eta}_i(z) = 0$, there is only one choice. Namely, for a $[b, a\rangle$-BCFW shift one has the following super-shift:

$$\hat{\eta}_b = \eta_b + z\eta_a, \quad \hat{\eta}_i = \eta_i, \quad \forall i \neq b \tag{1.4.16}$$

In this formalism, the sum over internal states in (1.4.14) is elegantly realized as the Grassmann integral over the variables $\eta_{\hat{P}_I}$ associated to the internal particles with momenta $\pm\hat{P}_I$:

$$\mathcal{A}_n = \sum_I \int \mathrm{d}^4\eta_{\hat{P}_I} \mathcal{A}_L(-\hat{P}_I; z_I)\frac{1}{P_I^2}\mathcal{A}_R(\hat{P}_I; z_I) \tag{1.4.17}$$

By expanding each superamplitudes as polyonmials in η's [58], one can decompose the recursion relation (1.4.17) in helicity sectors in order to obtain a recursion formula for $\mathcal{A}_n^{\mathrm{N}^k\mathrm{MHV}}$ in terms of[13] $\mathcal{A}_{n'}^{\mathrm{N}^{k'}\mathrm{MHV}}$, with $n' < n$ and $k' \leq k$.

$\mathcal{N} = 4$ **SYM Amplitudes from BCFW.** The power of recursion relations is that one can obtain all tree level amplitudes starting from the building blocks of the recursion. An advantage of having analytically continued momenta into complex variables is that the three-point amplitudes are not anymore vanishing. They are indeed the desired building blocks and, remarkably, are completely determined by symmetries. Using the super-space formalism and spinor helicity variables:

$$\mathcal{A}_3^{\mathrm{MHV}} = \frac{\delta^8(q)}{\langle 12\rangle\langle 23\rangle\langle 31\rangle}, \quad \mathcal{A}_3^{\overline{\mathrm{MHV}}} = -\frac{\delta^4([12]\eta_3 + [23]\eta_1 + [31]\eta_2)}{[12][23][31]}. \tag{1.4.18}$$

The delta functions are fermionic ones and one recalls that $q^{\alpha A} := \sum_{i=1}^n \lambda_i^\alpha \eta_i^A$ is the super-momenutum, which is conserved by super-symmetry.

For an explicit formula of all $\mathcal{A}_n^{\mathrm{N}^k\mathrm{MHV}}$, see [5]. We report here the first non-trivial cases:

$$\mathcal{A}_n^{\mathrm{MHV}} = \frac{\delta^8(q)}{\langle 12\rangle\langle 23\rangle\ldots\langle n1\rangle}, \qquad \mathcal{A}_n^{\mathrm{NMHV}} = \mathcal{A}_n^{\mathrm{MHV}} \sum_{3<i+1<j} R_{1,i,i+1,j,j+1}. \tag{1.4.19}$$

R_I are called *R-invariants* and in super momentum twistors $\mathcal{Z} = \{z, \chi\}$ are given as:

---

[13] One also needs $\overline{\mathrm{MHV}}_3$ amplitudes.

$$R_I(\mathcal{Z}) := \frac{\delta^{0|4}\left(\sum_{i\in I}\epsilon_{i,I\setminus\{i\}}\langle z_{I\setminus\{i\}}\rangle\chi_i\right)}{\prod_{i\in I}\langle z_{I\setminus\{i\}}\rangle}, \quad I \in \binom{[n]}{5}. \tag{1.4.20}$$

They play a pivotal role in the geometrization of the amplitude as well. For example:

$$R_{12345}(\mathcal{Z}) = \frac{\delta^{0|4}(\langle 1234\rangle\chi_5 + \text{cyclic})}{\langle 1234\rangle\langle 2345\rangle\langle 1345\rangle\langle 1245\rangle\langle 1235\rangle}. \tag{1.4.21}$$

Hidden Symmetries of Amplitudes. Recall from Sect. 1.3.1 that $\mathcal{N}=4$ SYM enjoys the superconformal symmetry $PSU(2,2|4)$, which has a nice realization in terms of on-shell variables $\{\lambda, \tilde{\lambda}, \eta\}$, see [58]. Remarkably, in the dual variables $\{x, \chi\}$ planar $\mathcal{N}=4$ SYM exhibit a novel, hidden $PSU(2,2|4)$ symmetry called *dual superconformal symmetry*, see [71] for details. This symmetry is linearly realized using super-momentum twistors and R-invariants (1.4.20) defined in these variables are the dual superconformal invariant building blocks of amplitudes. This is a new symmetry, and combining both one generates an infinite-dimensional symmetry called *Yangian* [79–81]. Yangian symmetry is considered to be the hallmark of *integrability*, i.e. solvability of the theory and allows to employ methods and techniques proper to (or inspired by) integrable theories, e.g. spin chains [82–84]. Yangian symmetry has been discovered at the level of scattering amplitudes thanks to the novel powerful on-shell methods [6] and a satisfactory realization at the level of the action (1.3.1) has been under recent investigation, see [85, 86].

References

1. S.J. Parke, T.R. Taylor, An amplitude for *n* gluon scattering. Phys. Rev. Lett. **56**, 2459 (1986)
2. R.J. Eden, P.V. Landshoff, D.I. Olive, J.C. Polkinghorne, *The Analytic S-Matrix* (Cambridge University Press, Cambridge, 1966)
3. R. Britto, F. Cachazo, B. Feng, E. Witten, Direct proof of tree-level recursion relation in Yang-Mills theory. Phys. Rev. Lett. **94**, 181602 (2005)
4. Z. Bern, L.J. Dixon, D.C. Dunbar, D.A. Kosower, One-loop n-point gauge theory amplitudes, unitarity and collinear limits. Nucl. Phys. B **425**, 217–260 (1994)
5. L.J. Dixon, J.M. Henn, J. Plefka, T. Schuster, All tree-level amplitudes in massless QCD. JHEP **01**, 035 (2011)
6. J.M. Drummond, J.M. Henn, J. Plefka, Yangian symmetry of scattering amplitudes in N = 4 super Yang-Mills theory. JHEP **0905**, 046 (2009)
7. J.M. Maldacena, The Large N limit of superconformal field theories and supergravity, Int. J. Theor. Phys. **38**, 1113–1133 (1999). [Adv. Theor. Math. Phys. **2**, 231 (1998)]
8. E. Witten, Perturbative gauge theory as a string theory in twistor space. Commun. Math. Phys. **252**, 189–258 (2004)
9. A. Hodges, Eliminating spurious poles from gauge-theoretic amplitudes. JHEP **05**, 135 (2013)
10. N. Arkani-Hamed, F. Cachazo, C. Cheung, J. Kaplan, A duality for the S matrix. JHEP **03**, 020 (2010)
11. N. Arkani-Hamed, J. Bourjaily, F. Cachazo, A. Goncharov, A. Postnikov, J. Trnka, *Grassmannian Geometry of Scattering Amplitudes* (Cambridge University Press, 2016)
12. A. Postnikov, Total positivity, Grassmannians, and networks, *arXiv preprint* math/0609764 (2006)

13. N. Arkani-Hamed, J. Trnka, The amplituhedron. J. High Energy Phys. (10), 33 (2014)
14. N. Arkani-Hamed, Y. Bai, T. Lam, Positive geometries and canonical forms. JHEP **11**, 039 (2017)
15. N. Arkani-Hamed, Y. Bai, S. He, G. Yan, Scattering forms and the positive geometry of kinematics, color and the worldsheet (2017)
16. P. Banerjee, A. Laddha, P. Raman, Stokes polytopes: the positive geometry for ϕ^4 interactions. JHEP **08**, 067 (2019)
17. M. Jagadale, A. Laddha, Towards positive geometry of multi scalar field amplitudes: accordiohedron and effective field theory, 4 2021
18. N. Arkani-Hamed, P. Benincasa, A. Postnikov, Cosmological polytopes and the wavefunction of the universe (2017)
19. N. Arkani-Hamed, S. He, T. Lam, Stringy canonical forms, 12 2019
20. P. Benincasa, M. Parisi, Positive geometries and differential forms with non-logarithmic singularities. Part I. J. High Energy Phys. (8), 023 (2020)
21. P. Benincasa, On-shell diagrammatics and the perturbative structure of planar gauge theories (2015)
22. P. Benincasa, D. Gordo, On-shell diagrams and the geometry of planar $\mathcal{N} < 4$ SYM theories. JHEP **11**, 192 (2017)
23. S. He, C. Zhang, Notes on scattering amplitudes as differential forms. JHEP **10**, 054 (2018)
24. E. Herrmann, J. Trnka, Gravity on-shell diagrams. JHEP **11**, 136 (2016)
25. P. Heslop, A.E. Lipstein, On-shell diagrams for $\mathcal{N} = 8$ supergravity amplitudes. JHEP **06**, 069 (2016)
26. S. Fomin, A. Zelevinsky, Cluster algebras. I. Foundations. J. Am. Math. Soc. **15**(2), 497–529 (electronic) (2002)
27. D. Speyer, L. Williams, The tropical totally positive Grassmannian. J. Algebraic Combin. **22**(2), 189–210 (2005)
28. J. Golden, A.B. Goncharov, M. Spradlin, C. Vergu, A. Volovich, Motivic amplitudes and cluster coordinates. JHEP **01**, 091 (2014)
29. S. Caron-Huot, L.J. Dixon, J.M. Drummond, F. Dulat, J. Foster, O. Gürdoğan, M. von Hippel, A.J. McLeod, G. Papathanasiou, The Steinmann cluster bootstrap for $N = 4$ super Yang-Mills amplitudes. PoS **CORFU2019**, 003 (2020)
30. J. Drummond, J. Foster, Ö. Gürdoğan, Cluster adjacency properties of scattering amplitudes in $N = 4$ supersymmetric Yang-Mills theory. Phys. Rev. Lett. **120**(16), 161601 (2018)
31. J. Drummond, J. Foster, Ö. Gürdoğan, Cluster adjacency beyond MHV. JHEP **03**, 086 (2019)
32. J. Drummond, J. Foster, Ö. Gürdoğan, C. Kalousios, Algebraic singularities of scattering amplitudes from tropical geometry, 12 2019
33. N. Arkani-Hamed, S. He, T. Lam, H. Thomas, Binary geometries, generalized particles and strings, and cluster algebras (2019)
34. N. Henke, G. Papathanasiou, How tropical are seven- and eight-particle amplitudes?, 12 2019
35. N. Henke, G. Papathanasiou, Singularities of eight- and nine-particle amplitudes from cluster algebras and tropical geometry. JHEP **21**, 007 (2020)
36. F. Cachazo, N. Early, A. Guevara, S. Mizera, Scattering equations: from projective spaces to tropical Grassmannians. JHEP **06**, 039 (2019)
37. N. Early, Planarity in generalized scattering amplitudes: PK polytope, generalized root systems and worldsheet associahedra, 6 2021
38. F. Cachazo, N. Early, Planar kinematics: cyclic fixed points, mirror superpotential, k-dimensional Catalan numbers, and root polytopes, 10 2020
39. N. Early, Weighted blade arrangements and the positive tropical Grassmannian, 5 2020
40. F. Cachazo, N. Early, Minimal kinematics: an all k and n peek into $\mathrm{Trop}^+G(k, n)$. SIGMA **17**, 078 (2021)
41. N. Early, Planar kinematic invariants, matroid subdivisions and generalized Feynman diagrams, 12 2019
42. N. Early, From weakly separated collections to matroid subdivisions, 10 2019
43. S. He, Z. Li, A note on letters of Yangian invariants. JHEP **02**, 155 (2021)

44. J. Mago, A. Schreiber, M. Spradlin, A. Volovich, Symbol alphabets from plabic graphs. JHEP **10**, 128 (2020)
45. J. Mago, A. Schreiber, M. Spradlin, A. Yelleshpur Srikant, A. Volovich, Symbol alphabets from plabic graphs II: rational letters. JHEP **04**, 056 (2021)
46. J. Mago, A. Schreiber, M. Spradlin, A. Yelleshpur Srikant, A. Volovich, Symbol alphabets from plabic graphs III: n = 9, 6 2021
47. A. Herderschee, Algebraic branch points at all loop orders from positive kinematics and wall crossing. JHEP **07**, 049 (2021)
48. L. Ren, M. Spradlin, A. Volovich, Symbol alphabets from tensor diagrams, 6 2021
49. P. Tourkine, Tropical amplitudes. Ann. Henri Poincaré **18**(6), 2199–2249 (2017)
50. E. Panzer, Hepp's bound for Feynman graphs and matroids, 8 2019
51. M. Borinsky, Tropical Monte Carlo quadrature for Feynman integrals, 8 2020
52. S. He, Z. Li, Q. Yang, Truncated cluster algebras and Feynman integrals with algebraic letters, 6 2021
53. D. Chicherin, J.M. Henn, G. Papathanasiou, Cluster algebras for Feynman integrals. Phys. Rev. Lett. **126**(9), 091603 (2021)
54. Y.-T. Huang, R. Kojima, C. Wen, S.-Q. Zhang, The orthogonal momentum amplituhedron and ABJM amplitudes. JHEP **01**, 141 (2022)
55. S. He, C.-K. Kuo, Y.-Q. Zhang, The momentum amplituhedron of SYM and ABJM from twistor-string maps. JHEP **02**, 148 (2022)
56. S. He, Y.-t. Huang, C.-K. Kuo, The ABJM amplituhedron, 6 2023
57. N. Arkani-Hamed, H. Thomas, J. Trnka, Unwinding in binary. Preprint, http://arxiv.org/abs/1704.05069
58. J.M. Henn, J.C. Plefka, Scattering amplitudes in Gauge theories. Lect. Notes Phys. **883**, 1–195 (2014)
59. H. Elvang, Y.-t. Huang, *Scattering Amplitudes in Gauge Theory and Gravity* (Cambridge University Press, 2015)
60. N. Beisert, The dilatation operator of $\mathcal{N} = 4$ super Yang-Mills theory and integrability. Phys. Rept. **405**, 1–202 (2004)
61. A.M. Polyakov, Gauge fields as rings of glue. Nucl. Phys. B **164**, 171–188 (1980)
62. G.P. Korchemsky, A.V. Radyushkin, Renormalization of the Wilson loops beyond the leading order. Nucl. Phys. B **283**, 342–364 (1987)
63. G. 't Hooft, A planar diagram theory for strong interactions. Nucl. Phys. **B72**, 461 (1974). [,337 (1973)]
64. R. Kleiss, H. Kuijf, Multigluon cross sections and 5-jet production at hadron colliders. Nuclear Phys. B **312**(3), 616–644 (1989)
65. Z. Bern, J.J.M. Carrasco, H. Johansson, New relations for gauge-theory amplitudes. Phys. Rev. D **78**, 085011 (2008)
66. V. Nair, A current algebra for some gauge theory amplitudes. Phys. Lett. B **214**(2), 215–218 (1988)
67. N. Arkani-Hamed, F. Cachazo, J. Kaplan, What is the simplest quantum field theory? JHEP **09**, 016 (2010)
68. P.D. Causmaecker, R. Gastmans, W. Troost, T.T. Wu, Multiple bremsstrahlung in gauge theories at high energies (i). general formalism for quantum electrodynamics. Nuclear Phys. B **206**(1), 53–60 (1982)
69. F. Berends, R. Kleiss, P.D. Causmaecker, R. Gastmans, W. Troost, T.T. Wu, Multiple bremsstrahlung in gauge theories at high energies (ii). single bremsstrahlung. Nuclear Phys. B **206**(1), 61–89 (1982)
70. R. Kleiss, W. Stirling, Spinor techniques for calculating $pp \to w\pm/z0$ + jets. Nuclear Phys. B **262**(2), 235–262 (1985)
71. J.M. Drummond, J. Henn, G.P. Korchemsky, E. Sokatchev, Dual superconformal symmetry of scattering amplitudes in $\mathcal{N} = 4$ super-Yang-Mills theory. Nucl. Phys. B **828**, 317–374 (2010)
72. R. Penrose, Twistor algebra. J. Math. Phys. **8**(2), 345–366 (1967)

73. S.A. Huggett, K.P. Tod, *An Introduction to Twistor Theory*, London Mathematical Society Student Texts, 2nd edn. (Cambridge University Press, 1994)
74. N. Arkani-Hamed, J.L. Bourjaily, F. Cachazo, S. Caron-Huot, J. Trnka, The all-loop integrand for scattering amplitudes in planar N = 4 SYM. JHEP **1101**, 041 (2011)
75. Z. Bern, Y.-T. Huang, Basics of generalized unitarity. J. Phys. A **44**, 454003 (2011)
76. J.J.M. Carrasco, H. Johansson, Generic multiloop methods and application to N = 4 super-Yang-Mills. J. Phys. A **44**, 454004 (2011)
77. Q. Jin, B. Feng, Recursion relation for boundary contribution. J. High Energy Phys. **2015**, 18 (2015)
78. F. Cachazo, P. Svrcek, E. Witten, MHV vertices and tree amplitudes in gauge theory. JHEP **09**, 006 (2004)
79. V.G. Drinfel'd, Hopf algebras and the quantum Yang-Baxter equation. Sov. Math. Dokl. **32**, 254–258 (1985)
80. V.G. Drinfel'd, Quantum groups. J. Math. Sci. **41**, 898 (1988)
81. L. Dolan, C. R. Nappi, E. Witten, Yangian symmetry in $d = 4$ superconformal Yang-Mills theory (2004)
82. J.A. Minahan, K. Zarembo, The Bethe-ansatz for $\mathcal{N} = 4$ super Yang-Mills. JHEP **0303**, 013 (2003)
83. B. Basso, A. Sever, P. Vieira, Space-time S-matrix and flux-tube S-matrix at finite coupling (2013)
84. N. Beisert, C. Ahn, L.F. Alday, Z. Bajnok, J.M. Drummond et al., Review of AdS/CFT integrability: an overview. Lett. Math. Phys. **99**, 3–32 (2012)
85. N. Beisert, A. Garus, M. Rosso, Yangian symmetry and integrability of planar N = 4 supersymmetric Yang-Mills theory. Phys. Rev. Lett. **118**(14), 141603 (2017)
86. N. Beisert, A. Garus, Yangian algebra and correlation functions in planar gauge theories. SciPost Phys. **5**(2), 018 (2018)

Chapter 2
The Amplituhedron

In 2009, Grassmannian formulations were introduced to describe scattering amplitudes in $\mathcal{N} = 4$ super Yang-Mills [1, 2]. Remarkably, this led to the discovery that in fact the *positive* Grassmannian $\mathrm{Gr}_{k,n}^{\geq 0}$ encodes most of the physical properties of amplitudes [3]. Building on these developments and on Hodges' ideas [4], Arkani-Hamed and Trnka arrived at the definition of the *amplituhedron* $\mathcal{A}_{n,k,m}(Z)$ [5] in 2013.

The $m = 4$ amplituhedron $\mathcal{A}_{n,k,m=4}(Z)$ encodes the geometry of tree-level scattering amplitudes in $\mathcal{N} = 4$ SYM. The $m = 2$ amplituhedron $\mathcal{A}_{n,k,m=2}(Z)$, often considered as a toy-model for the $m = 4$ case, is also relevant for physics. Firstly, it governs the geometry of planar $\mathcal{N} = 4$ SYM amplitudes at the subleading order in perturbation theory ('one-loop') of some sectors the theory, specifically the 'MHV' and 'NMHV' sector [6]. We remark its geometry also enters any loops, see Remark 7.3.9. Secondly, the authors of [7] recently found that the $m = 2$ amplituhedron encodes scattering amplitudes for a Gaussian model on a superline. Moreover, these are showed to be related to correlation functions of determinant operators in $\mathcal{N} = 4$ SYM. Finally, the $m = 2$ amplituhedron was found to describe tree-level MHV *form factors* of half-BPS operators in $\mathcal{N} = 4$ SYM [8].

In this chapter we introduce and study the amplituhedron, in particular the $m = 2$ amplituhedron $\mathcal{A}_{n,k,2}$. We focus on *positroid tiles*, which are fundamental building blocks of tilings, on the geometric side, and of representation of scattering amplitudes, on the physics side. Moreover, we adopt *twistor coordinates* to understand the geometry of positroid tiles and of $\mathcal{A}_{n,k,m=2}$. Beautiful combinatorial structures emerge by looking at twistor coordinates with a definite sign pattern. Understanding them is the key to prove major old and recent conjectures about the amplituhedron (e.g. Theorems 2.4.1, 5.1.9). Moreover, twistor coordinates are crucial to explore the emergence of 'cluster phenomena' [9, 10] from the geometry of the amplituhedron (see Chap. 7). Twistor coordinates and their sign-flips recently played a major role in providing a geometric origin of the expansion of the one-loop MHV scattering amplitudes in $\mathcal{N} = 4$ SYM in terms of 'chiral pentagons' and in probing new geometries, as in the search of the *dual amplituhedron* [6, 11].

M. Parisi, *Combinatorial Aspects of Scattering Amplitudes*, Springer Theses,
https://doi.org/10.1007/978-3-031-41069-7_2

Scattering amplitudes in planar $\mathcal{N} = 4$ SYM enjoy a remarkable duality called 'Amplitude/Wilson loop duality' [12], which was shown to arise from a more fundamental duality in String Theory called 'T-duality' [13]. The geometric counterpart of this fact is a duality between collections of $4k$-dimensional 'BCFW' cells of $\mathrm{Gr}_{k,n}^{\geq 0}$ and corresponding collections of $(2n-4)$-dimensional cells of $\mathrm{Gr}_{k+2,n}^{\geq 0}$. The former conjecturally give tilings of the $m = 4$ amplituhedron $\mathcal{A}_{n,k,4}$, whereas the former give tilings of the *momentum amplituhedron* $\mathcal{M}_{n,k+2,4}$—a new object we introduce in Chap. 6. We refer to this duality as *T-duality* and we conjecture to generalize for any (even) m. In particular, for $m = 2$, the hypersimplex $\Delta_{k+1,n}$ (a polytope) and the $m = 2$ amplituhedron $\mathcal{A}_{n,k,2}(Z)$ (not a polytope!) are T-dual. As for all dualities in mathematics and physics, the aim is to learn something new of one side from the other, trading complexity of the former with simplicity of the latter. Interestingly, T-duality has recently appeared also in the context of *critical varieties* [14], relating electric networks and Ising models.

Summary of the Chapter. This chapter is based on the following work by the author: [15, Sects. 3–5]; Sect. 2.5 contains material from [16, Sect. 4].

In this chapter we introduce and prove new properties of the *amplituhedron* $\mathcal{A}_{n,k,m}$—the image of the totally non-negative Grassmannian $\mathrm{Gr}_{k,n}^{\geq 0}$ under the amplituhedron map $\tilde{Z}$. In Sect. 2.1 we introduce the positive Grassmannian (Sect. 2.1.1), the amplituhedron (Sect. 2.1.2), and briefly report on previous works on the latter (Sect. 2.1.3). In Sect. 2.2 we introduce the amplituhedron-analogue of the matroid stratification of the Grassmannian, which we call *sign stratification* of $\mathcal{A}_{n,k,m}$. In Sect. 2.2.1 we define suitable coordinates on the amplituhedron—the *twistor coordinates*. In Sect. 2.2.2 we consider regions in the Grassmannian where all these coordinates have constant signs—the *sign strata*. Finally, in Sect. 2.2.3 we review the notions of sign variation and sign flips. In Sect. 2.3 we consider *positroid tiles* of $\mathcal{A}_{n,k,m}$—full-dimensional images of positroid cells of $\mathrm{Gr}_{k,n}^{\geq 0}$ on which the amplituhedron map is injective. In particular, we classify all the positroid tiles of the $m = 2$ amplituhedron $\mathcal{A}_{n,k,2}$ in terms of unpunctured *plabic tilings*. Then we give them an intrinsic sign-characterization in the Grassmannian as the regions where certain twistor coordinates have a fixed sign—which can be read directly from the associated plabic tilings. In Sect. 2.4, we use this result to prove the sign-flip characterization of $\mathcal{A}_{n,k,2}$, i.e. the Arkani-Hamed–Thomas–Trnka's conjecure [17]. Finally, in Sect. 2.5 we show how to extract tree-level $\mathrm{N}^k\mathrm{MHV}_n$ scattering amplitudes of $\mathcal{N} = 4$ SYM in momentum twistors from the $m = 4$ amplituhedron $\mathcal{A}_{n,k,4}$.

2.1 The Positive Grassmannian and the Amplituhedron

2.1.1 The Grassmannian and Positive Grassmannian

The *(real) Grassmannian* $\mathrm{Gr}_{k,n}$ is the space of all k-dimensional subspaces of $\mathbb{R}^n$, for $0 \le k \le n$. An element of $\mathrm{Gr}_{k,n}$ can be viewed as a $k \times n$ matrix of rank k, modulo left multiplication by invertible $k \times k$ matrices. That is, two $k \times n$ matrices of rank k represent the same point in $\mathrm{Gr}_{k,n}$ if and only if they can be obtained from each other by invertible row operations. For C a full-rank $k \times n$ matrix, we will often abuse notation and write $C \in \mathrm{Gr}_{k,n}$, identifying C with its rowspan.

Let $[n]$ denote $\{1, \dots, n\}$, and $\binom{[n]}{k}$ the set of all k-element subsets of $[n]$. We embed $\mathrm{Gr}_{k,n}$ into projective space $\mathbb{P}(\wedge^k \mathbb{R}^n)$ in the usual way. That is, choose $V \in \mathrm{Gr}_{k,n}$ and any representative matrix C with rows $C_1, \dots, C_k$. We map V to the equivalence class of $C_1 \wedge \cdots \wedge C_k$ in $\mathbb{P}(\wedge^k \mathbb{R}^n)$. This equivalence class depends only on V, not on the choice of C.

The embedding $V \mapsto C_1 \wedge \cdots \wedge C_k$ gives a natural choice of coordinates for the Grassmannian. Let $\{e_1, \dots, e_n\}$ be the standard basis of $\mathbb{R}^n$, and for $I = \{i_1 < i_2 < \cdots < i_k\} \subset \binom{[n]}{k}$, let $E_I := e_{i_1} \wedge \cdots \wedge e_{i_k}$. Writing $C_1 \wedge \cdots \wedge C_k$ in terms of the E_I, we obtain

$$C_1 \wedge \cdots \wedge C_k = \sum_{I \in \binom{[n]}{k}} p_I(V) E_I \in \wedge^k(\mathbb{R}^n), \tag{2.1.1}$$

where $p_I(V)$ is the maximal minor of C located in column set I. The $p_I(V)$ are the *Plücker coordinates* of V, and are independent of C (up to simultaneous rescaling by a constant).

We will also use the notation $\langle C_1, \dots, C_k \rangle$ for $C_1 \wedge \cdots \wedge C_k$.

Definition 2.1.2 *([18, Sect. 3])* We say that $C \in \mathrm{Gr}_{k,n}$ is *totally nonnegative* if $p_I(C) \ge 0$ for all $I \in \binom{[n]}{k}$, and *totally positive* if $p_I(C) > 0$ for all $I \in \binom{[n]}{k}$. The set of all totally nonnegative $C \in \mathrm{Gr}_{k,n}$ is the *totally nonnegative Grassmannian* $\mathrm{Gr}_{k,n}^{\ge 0}$, and the set of all totally positive C is the *totally positive Grassmannian* $\mathrm{Gr}_{k,n}^{>0}$. For $\mathcal{M} \subseteq \binom{[n]}{k}$, the *positroid cell* $S_{\mathcal{M}}$ is the set of $C \in \mathrm{Gr}_{k,n}^{\ge 0}$ such that $p_I(C) > 0$ for all $I \in \mathcal{M}$, and $p_J(C) = 0$ for all $J \in \binom{[n]}{k} \setminus \mathcal{M}$. We call $\mathcal{M}$ a *positroid* if $S_{\mathcal{M}}$ is nonempty. We let $Q_{k,n}$ denote the poset on the cells of $\mathrm{Gr}_{k,n}^{\ge 0}$ defined by $S_{\mathcal{M}} \le S_{\mathcal{M}'}$ if and only if $S_{\mathcal{M}} \subseteq \overline{S_{\mathcal{M}'}}$.

Remark 2.1.3 The positive and nonnegative part of a flag variety G/P was first introduced by Lusztig [19] (who gave a Lie-theoretic definition of $(G/P)_{>0}$ and defined $(G/P)_{\ge 0} := \overline{(G/P)_{>0}}$), and proved to have a cell decomposition by Rietsch [20]. Postnikov [18] subsequently defined the nonnegative part of the Grassmannian as in Definition 2.1.2, and independently gave the above decomposition into cells. From the beginning it was believed by experts that Postnikov's definition of $\mathrm{Gr}_{k,n}^{\ge 0}$ should agree with Lusztig's (in the case G/P is the Grassmannian); this was first proved by Rietsch [21], and reproved in [22, Corollary 1.2], where the authors additionally proved that the two cell decompositions coincide. Two subsequent proofs that the two definitions of $\mathrm{Gr}_{k,n}^{\geq 0}$ coincide were given in [23, 24].

There are many ways to index the positroid cells of $\mathrm{Gr}_{k,n}^{\geq 0}$ [18], including *decorated permutations* π, *affine permutations* f, and *plabic graphs* G. We will refer to the corresponding positroid cells using the notation S_π, S_f, S_G. For background, see Appendix A.

2.1.2 *The Amplituhedron*

Building on [25], Arkani-Hamed and Trnka [5] introduced a new mathematical object called the *(tree) amplituhedron*, which is the image of the totally nonnegative Grassmannian under a particular map. In what follows, we let $\mathrm{Mat}_{n,p}^{>0}$ denote the set of $n \times p$ matrices whose maximal minors are positive. Throughout this section we will let n, k, m be positive integers such that $k + m \leq n$.

Definition 2.1.4 *(The amplituhedron)* Let $Z \in \mathrm{Mat}_{n,k+m}^{>0}$. The *amplituhedron map* $\tilde{Z} : \mathrm{Gr}_{k,n}^{\geq 0} \to \mathrm{Gr}_{k,k+m}$ is defined by $\tilde{Z}(C) := CZ$, where C is a $k \times n$ matrix representing an element of $\mathrm{Gr}_{k,n}^{\geq 0}$, and CZ is a $k \times (k+m)$ matrix representing an element of $\mathrm{Gr}_{k,k+m}$. The *amplituhedron* $\mathcal{A}_{n,k,m} \subseteq \mathrm{Gr}_{k,k+m}$ is the image $\tilde{Z}(\mathrm{Gr}_{k,n}^{\geq 0})$.

The fact that Z has positive maximal minors ensures that $\tilde{Z}$ is well defined [5], i.e. CZ has maximal rank. See [26, Theorem 4.2] for a necessary and sufficient condition (in terms of sign-variation) for a matrix Z to give rise to a well-defined map $\tilde{Z}$. The amplituhedron $\mathcal{A}_{n,k,m}(Z)$ has full dimension km inside $\mathrm{Gr}_{k,k+m}$.

In special cases the amplituhedron recovers familiar objects. If Z is a square matrix, i.e. $k + m = n$, then $\mathcal{A}_{n,k,m}(Z)$ is isomorphic to the totally nonnegative Grassmannian $\mathrm{Gr}_{k,k+m}^{\geq}$. If $k = 1$, $\mathcal{A}_{n,1,m}(Z)$ is a *cyclic polytope* $C(n, m)$ in projective space $\mathbb{P}^m$ [27]. The cyclic polytope $C(n, d)$ is the convex hull of any n distinct points on the moment curve $\{(t, t^2, \ldots, t^d) : t \in \mathbb{R}\}$ in $\mathbb{R}^d$. In particular, when $k = 1$ and $m = 2$ $\mathcal{A}_{n,1,2}(Z)$ is an n-gon in projective space $\mathbb{P}^2$. If $m = 1$, then $\mathcal{A}_{n,k,1}(Z)$ can be identified with the complex of bounded faces of a cyclic hyperplane arrangement [28].

Example 2.1.5 (The Amplituhedron $\mathcal{A}_{4,1,2}$) The amplituhedron $\mathcal{A}_{4,1,2}(Z) \subset \mathrm{Gr}_{1,3} \cong \mathbb{P}_2$ is a (projective) 4-gon (see Fig. 2.1). It is the projectivization of the convex hull of the vertices $Z_1, Z_2, Z_3, Z_4 \in \mathbb{R}^3$, which are the rows of the matrix $Z \in \mathrm{Mat}_{4,3}^{+}$.$\diamond$

We will consider the restriction of the $\tilde{Z}$-map to positroid cells in $\mathrm{Gr}_{k,n}^{\geq 0}$.

Definition 2.1.6 *(Positroid tiles)* Fix k, n, m with $k + m \leq n$ and choose $Z \in \mathrm{Mat}_{n,k+m}^{>0}$. Given a positroid cell S_π of $\mathrm{Gr}_{k,n}^{\geq 0}$, we let $Z_\pi^\circ = \tilde{Z}(S_\pi)$ and $Z_\pi = \overline{\tilde{Z}(S_\pi)} =$

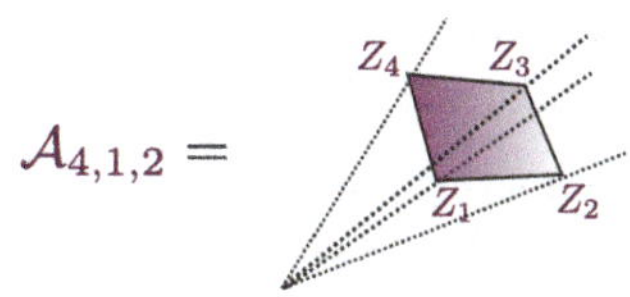

Fig. 2.1 The amplituhedron $\mathcal{A}_{4,1,2}(Z)$–a (projective) 4-gon in $\mathbb{P}^2$

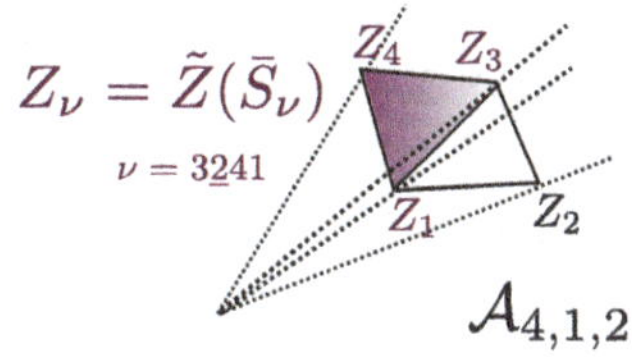

Fig. 2.2 A positroid tile of $\mathcal{A}_{4,1,2}(Z)$–a (projective) triangle in $\mathbb{P}^2$

$\tilde{Z}(\overline{S_\pi})$, and we refer to Z_π° and Z_π as *open Grasstopes* and *Grasstopes*, respectively. We call Z_π and Z_π° a *positroid tile* and an *open positroid tile* for $\mathcal{A}_{n,k,m}(Z)$ if $\dim(S_\pi) = km$ and $\tilde{Z}$ is injective on S_π.

Example 2.1.7 (Positroid tiles of $\mathcal{A}_{4,1,2}(Z)$) A positroid tile $Z_\nu \subset \mathcal{A}_{4,1,2}(Z)$ is a (projective) triangle in $\mathbb{P}^2$. Let us consider the positroid cell $S_\nu \subset \mathrm{Gr}_{1,4}^{\geq 0}$, with $\nu = 3\underline{2}41$. Its elements can be represented by row vectors $C = (c_1, c_2, c_3, c_4)$, with $p_2(C) = c_2 = 0$ and $c_1, c_3, c_4 > 0$. Then $Z_\nu = \tilde{Z}(\bar{S}_\nu)$ is the projectivization of the convex hull of the vertices Z_1, Z_3, Z_4, see Fig. 2.2. ◇

Definition 2.1.8 *(Faces of Grasstopes)* Let Z_π be a Grasstope of $\mathcal{A}_{n,k,m}(Z)$. We say that $Z_{\pi'}$ is a *face* of Z_π if it is maximal by inclusion among the Grasstopes satisfying the following properties:

1. the cell $S_{\pi'}$ is contained in $\overline{S_\pi}$
2. $Z_{\pi'}$ is contained in the boundary ∂Z_π

If $Z_{\pi'}$ has codimension 1 in Z_π, we say $Z_{\pi'}$ is a *facet* of Z_π.

Remark 2.1.9 By [23, Proposition 15.2], $\tilde{Z}(\overline{S_\pi}) = \overline{\tilde{Z}(S_\pi)}$.

If $k = 1$ and $m = 2$, the amplituhedron $\mathcal{A}_{n,1,2}(Z)$ is a convex n-gon in $\mathbb{P}^2$. The positroid tiles are exactly the triangles on vertices of the polygon.

Images of positroid cells under the map $\tilde{Z}$ have been studied since the introduction of the amplituhedron. In particular, Arkani-Hamed and Trnka [5] conjectured that the images of certain *BCFW* collections of $4k$-dimensional cells in $\mathrm{Gr}_{k,n}^{\geq 0}$ give a positroid tiling of the amplituhedron $\mathcal{A}_{n,k,4}(Z)$.

Remark 2.1.10 *Positroid tiles* were referred to as *generalized triangles* in [29]. The same terminology was then adopted in subsequent works, e.g. [16].

The terminology of Grassmann polytopes to describe images of positroid cells in the amplituhedron was used in [23]. For brevity, we prefer the term *Grasstopes*.

Remark 2.1.11 While the definition of the amplituhedron $\mathcal{A}_{n,k,m}(Z)$ depends on a choice of $Z \in \mathrm{Mat}^{>0}_{n,k+m}$, it is believed that many of its combinatorial properties do not depend on this choice. For example, whether or not $\overline{\tilde{Z}(S_\pi)}$ is a positroid tile should be independent of the choice of Z; we will see that this is true in Theorem 2.3.25 in the case that $m = 2$. It is also believed that whether or not a collection of cells in $\mathrm{Gr}^{\geq 0}_{k,n}$ gives a positroid tiling of $\mathcal{A}_{n,k,m}(Z)$ should be independent of Z.

Remark 2.1.12 We note that matrices whose maximal minors are positive (or nonnegative) have a *twisted cyclic symmetry*. If $Z \in \mathrm{Mat}^{>0}_{n,p}$ with $n \geq p$ has rows $Z_1, Z_2, \ldots, Z_n$, and if we let $\hat{Z}_i$ denote $(-1)^{p-1} Z_i$, then the matrix with rows $Z_2, \ldots, Z_n, \hat{Z}_1$ also lies in $\mathrm{Mat}^{>0}_{n,p}$. Similarly for the matrix with rows $Z_3, \ldots, Z_n, \hat{Z}_1, \hat{Z}_2$, etc.[1]

2.1.3 Previous Work on the Amplituhedron

The amplituhedron has very interesting and complicated geometric and combinatorial structure. Despite the extensive research on this topic, the geometry of $\mathcal{A}_{n,k,m}$ and its 'subdivisions' is only known for special cases, e.g. the ones mentioned in Sect. 2.1.2—when it is isomorphic to the positive Grassmannian, or cyclic polytopes [30], or the complex of bounded regions of a cyclic hyperplane arrangement [31, 32]. Conjectural descriptions in terms of *sign-flips* were given in [33]. Examples of conjectural tilings (e.g. of *BCFW type*) can be found in the seminal papers [5, 34] and some results about tilings in [35, 36]. It is also known that in special cases the amplituhedron is homeomorphic to a ball [37, 38]. Only very recently, the $m = 4$ amplituhedron was showed to be tiled by 'BCFW tiles' [39], which parallels the Britto-Cachazo-Feng-Witten recurrence [40] for computing scattering amplitudes.

Understanding how to 'subdivide' the amplituhedron is very important. Since $\mathcal{A}_{n,k,m}$ is the image of $\mathrm{Gr}^{\geq 0}_{k,n}$ which has a decomposition into positroid cells, the images of positroid cells are good candidates for decomposing $\mathcal{A}_{n,k,m}$. Hence, the main question in this context is to find collections of positroid cells whose full-dimensional images give a *positroid dissection* of the amplituhedron, i.e. they are pairwise disjoint, and together they cover a dense subset of the amplituhedron. In case the map $\tilde{Z}$ is injective on such cells, the dissection is called *positroid tiling* and its elements *positroid tiles*. Importantly, any such collection of positroid tiles comprising a tiling gives rise to an expression for the canonical form of the amplituhedron.

In this work we provide new general results for positroid tiles, tilings (dissections) and the geometry of the $m = 2$ amplituhedron $\mathcal{A}_{n,k,2}(Z)$. The original paper [5] gave a conjectural positroid tiling $\{Z_\pi\}$ of $\mathcal{A}_{n,k,2}(Z)$. [35] proved that the above collection consists of positroid tiles, that is, $\tilde{Z}$ is injective on the corresponding positroid cells. A BCFW-style recursion for producing tilings of $\mathcal{A}_{n,k,2}(Z)$ was also conjectured in [35];

[1] We will use the 'hat' notation ˆ also in the context of T-duality with a different meaning. It will be always clear from context which one we mean.

the fact that this recursion indeed produces tilings was proved in [41]. Meanwhile, [17] gave a conjectural alternative description of $\mathcal{A}_{n,k,2}(Z)$ in terms of sign flips of twistor coordinates; they gave a proof sketch of one direction of the conjecture, and an independent proof of the same direction was given in [28]. In a different direction, [42] gave a conjectural description of the boundaries of the $m = 2$ amplituhedron.

2.2 The Sign Stratification of the Amplituhedron

In this section we introduce *twistor coordinates* for the amplituhedron $\mathcal{A}_{n,k,m}(Z)$, and we use them to define the *sign stratification* of the amplituhedron. We also introduce terminology for sign variation and sign flips. We will subsequently use twistor coordinates to prove a sign flip description of $\mathcal{A}_{n,k,2}$ in Theorem 2.4.1, to characterize positroid tiles, and to describe Grasstopes.

The definitions and results in this section hold for any positive m. The subsequent sections of the chapter are mostly concerned with $m = 2$. However, many of our techniques can be applied to other m, in particular $m = 4$; we plan to investigate this in a separate work [43].

Twistor coordinates were first considered in [5], and subsequently used in [17] to give a conjectural "sign flip" description of the amplituhedron. In the case $m = 1$, [28, Corollary 3.19] studied the sign stratification and proved a sign flip description of $\mathcal{A}_{n,k,1}(Z)$.

2.2.1 Twistor Coordinates for $\mathcal{A}_{n,k,m}$

Definition 2.2.1 *(Twistor coordinates)* Fix positive $k < n$ and m such that $k + m \leq n$. Choose $Z \in \text{Mat}_{n,k+m}^{>0}$ and denote its rows by $Z_1, \ldots, Z_n \in \mathbb{R}^{k+m}$. Given a matrix Y with rows $y_1, \ldots, y_k$ representing an element of $Gr_{k,k+m}$, and $i_1, \ldots, i_m$ a sequence of elements of $[n]$, we let

$$\langle YZ_{i_1}Z_{i_2}\ldots Z_{i_m}\rangle = \langle y_1, \ldots, y_k, Z_{i_1}, \ldots, Z_{i_m}\rangle$$

denote the determinant of the $(k+m) \times (k+m)$ matrix whose rows are $y_1, \ldots, y_k$, $Z_{i_1}, \ldots, Z_{i_m}$. We call $\langle YZ_{i_1}Z_{i_2}\ldots Z_{i_m}\rangle$ a *twistor coordinate*. We abbreviate $\langle YZ_{i_1}Z_{i_2}\ldots Z_{i_m}\rangle$ by writing $\langle Yi_1i_2\ldots i_m\rangle$, when Z is understood.

Note that the twistor coordinates are a subset of the Plücker coordinates of the $(k+m) \times (k+n)$ matrix whose columns are $y_1, \ldots, y_k, Z_1, \ldots, Z_n$. There is also an interpretation of the twistor coordinates as Plücker coordinates in $\text{Gr}_{m,n}$, as we explain in Proposition 2.2.3. In the context of scattering amplitudes of n particles in SYM theory, $\mathrm{Gr}_{m,n}$ is the space of *momentum twistors*[2] for $m = 4$, which is why we call the coordinates from Definition 2.2.1 *twistor coordinates*. Remarkable connections between scattering amplitudes and the cluster algebra associated to the Grassmannian $Gr_{4,n}$ were discovered in these coordinates [44].

The fact that the twistor coordinates uniquely determine points of the amplituhedron can be deduced from some results of [28].

Definition 2.2.2 *(The $\mathcal{B}$-amplituhedron)* [28, Definition 3.8]. Given $W \in \mathrm{Gr}^{>0}_{k+m,n}$, we define the *$\mathcal{B}$ -amplituhedron*

$$\mathcal{B}_{n,k,m}(W) := \{V^{\perp} \cap W \mid V \in \mathrm{Gr}^{\geq 0}_{k,n}\} \subseteq \mathrm{Gr}_m(W),$$

where $\mathrm{Gr}_m(W) \subseteq \mathrm{Gr}_{m,n}$ denotes the subset of $\mathrm{Gr}_{m,n}$ of elements $X \in \mathrm{Gr}_{m,n}$ with $X \subseteq W$.

Proposition 2.2.3 *[28, Lemma 3.10, Proposition 3.12] Fix k, n, m and Z as in Definition 2.2.1, and let $W \in \mathrm{Gr}^{>0}_{k+m,n}$ be the column span of Z. Then the map*

$$\begin{aligned} f_Z : \mathrm{Gr}_m(W) &\to \mathrm{Gr}_{k,k+m} \\ X &\mapsto Z(X^{\perp}) = \{Z(x) \mid x \in X^{\perp}\} = rowspan(X^{\perp}Z) =: Y \end{aligned}$$

is an isomorphism. $X^{\perp} \in \mathrm{Gr}_{n-m,n}$ denotes the orthogonal complement of X in $\mathbb{R}^n$.

For $X \in \mathrm{Gr}_m(W)$, $Y := f_Z(X)$, and $I = \{i_1 < \cdots < i_m\} \subseteq [n]$, we have

$$p_I(X) = \langle Y Z_{i_1} \ldots Z_{i_m} \rangle \tag{2.2.4}$$

(we view Plücker and twistor coordinates as coordinates on points in projective space).

Finally, $f_Z : \mathcal{B}_{n,k,m}(W) \to \mathcal{A}_{n,k,m}(Z)$ is a homeomorphism sending $V^{\perp} \cap W \mapsto \tilde{Z}(V)$.

From (2.2.4) we see that $Y \in \mathrm{Gr}_{k,k+m}$ is uniquely determined by its twistor coordinates.

Remark 2.2.5 As an alternative to Proposition 2.2.3 we can consider the injective map ψ_Z

$$\begin{aligned} \psi_Z : \mathrm{Gr}_{k,k+m} &\to \mathrm{Gr}_{m,n} \\ Y &\mapsto Y^{\perp} Z^T =: z \end{aligned}$$

where $Y^{\perp}$ is any matrix representing the orthogonal complement of Y. Then it's not hard to see that for $I = \{i_1 < \cdots < i_m\} \subseteq [n]$, $p_I(z) = \langle Y Z_{i_1} \ldots Z_{i_m} \rangle$ (viewing both Plücker and twistor coordinates as coordinates on points in projective space).

---

[2] Due to *dual conformal symmetry* of scattering amplitudes in SYM theory, momentum twistors (cf. Sect. 1.4) are defined up to a PGL_4 transformation on $\mathbb{P}^3$. Therefore, they can be embedded in $Gr_{4,n}/(\mathbb{C}^*)^{n-1}$ and scattering amplitudes are functions of Plücker coordinates in $\mathrm{Gr}_{4,n}$. See [44].

The following expansion formula (2.2.7) will be useful in our proofs on positroid tiles.

Lemma 2.2.6 *Use the notation of Definition 2.2.1. If we write $Y \in \mathrm{Gr}_{k,k+m}$ as $Y = CZ$ with $C \in \mathrm{Gr}_{k,k+n}$, we can write the twistor coordinates in the form*

$$\langle CZ, Z_{i_1}, \dots, Z_{i_m}\rangle = \sum_{\{j_1<\cdots<j_k\}\in\binom{[n]}{k}} p_J(C)\langle Z_{j_1}, \dots, Z_{j_k}, Z_{i_1}, \dots, Z_{i_m}\rangle. \quad (2.2.7)$$

Proof Identifying the $k \times (k+m)$ matrix CZ with the corresponding element $\langle CZ\rangle$ of $\wedge^k(\mathbb{C}^{k+m})$, we have

$$\langle CZ\rangle = \sum_{\{j_1<\cdots<j_k\}\in\binom{[n]}{k}} p_J(C)\langle Z_{j_1}, \dots, Z_{j_k}\rangle.$$

This implies the result. □

We will give a description of positroid tiles in $\mathcal{A}_{n,k,2}$ using signs of twistor coordinates. One ingredient in our proofs is the following easy sufficient condition for a twistor coordinate to have constant sign on a Grasstope, which follows directly from (2.2.7).

Lemma 2.2.8 *Fix positive $k < n$ and m such that $k + m \le n$. Let $S_{\mathcal{M}}$ be a cell of $\mathrm{Gr}_{k,n}^{\ge 0}$. Fix $Z \in \mathrm{Mat}_{n,k+m}^{>0}$ and as usual let $Z_1, \dots, Z_n$ denote the row vectors of Z. Choose an m-element subset $1 \le i_1 < i_2 < \cdots < i_m \le n$.*

- *If $\langle Z_{j_1}, \dots, Z_{j_k}, Z_{i_1}, \dots, Z_{i_m}\rangle \ge 0$ for each $J = \{j_1 < \cdots < j_k\} \in \mathcal{M}$, then $\langle CZ, Z_{i_1}, \dots, Z_{i_m}\rangle \ge 0$ for each $C \in S_{\mathcal{M}}$.*
- *If in addition $\langle Z_{j_1}, \dots, Z_{j_k}, Z_{i_1}, \dots, Z_{i_m}\rangle > 0$ for some $J = \{j_1 < \cdots < j_k\} \in \mathcal{M}$ then $\langle CZ, Z_{i_1}, \dots, Z_{i_m}\rangle > 0$ for each $C \in S_{\mathcal{M}}$.*

2.2.2 *The Sign Stratification of $\mathcal{A}_{n,k,m}$*

Since $Y \in \mathrm{Gr}_{k,k+m}$ is uniquely determined by its twistor coordinates, it makes sense to stratify $\mathcal{A}_{n,k,m}(Z) \subset \mathrm{Gr}_{k,k+m}$ by the signs of the twistor coordinates. This was done in [28] in the case that $m = 1$. Moreover, this sign stratification is closely related to the *oriented matroid stratification* on the Grassmannian, which partitions elements of the real Grassmannian into strata based on the signs of the Plücker coordinates. By Proposition 2.2.3, the twistor coordinates of $Y \in \mathcal{A}_{n,k,m}(Z)$ are Plücker coordinates on the corresponding element of the *B-amplituhedron* [28] or *amplituhedron in momentum twistor space* [17], so this sign stratification reduces to the oriented matroid stratification in momentum twistor space.

Definition 2.2.9 *(Amplituhedron chambers)* Fix positive $k < n$ and m such that $k + m \leq n$. Let $\sigma = (\sigma_{i_1,\ldots,i_m}) \in \{0, +, -\}^{\binom{n}{m}}$ be a nonzero sign vector, considered[3] *modulo multiplication by* -1. Set

$$\mathcal{A}^{\sigma}_{n,k,m}(Z) := \{Y \in \mathcal{A}_{n,k,m}(Z) \mid \operatorname{sgn}\langle YZ_{i_1} \ldots Z_{i_m}\rangle = \sigma_{i_1,\ldots,i_m}\}.$$

We call $\mathcal{A}^{\sigma}_{n,k,m}(Z)$ an *(amplituhedron) sign stratum.* Clearly

$$\mathcal{A}_{n,k,m}(Z) = \sqcup_{\sigma} \mathcal{A}^{\sigma}_{n,k,m}(Z).$$

If $\sigma \in \{+, -\}^{\binom{n}{m}}$, we call $\mathcal{A}^{\sigma}_{n,k,m}(Z)$ an open *(amplituhedron) chamber.*[4]

For $m = 1$, all strata are nonempty [28, Definition 5.2], but this is not true for $m > 1$. Moreover, whether or not $\mathcal{A}^{\sigma}_{n,k,m}(Z)$ is empty depends on Z.

Definition 2.2.10 *(Realisable sign strata)* We say that a sign vector σ (or sign stratum $\mathcal{A}^{\sigma}_{n,k,m}$) is *realizable* for $\mathcal{A}_{n,k,m}$ if $\mathcal{A}^{\sigma}_{n,k,m}(Z)$ is nonempty for some Z.

Question 2.2.11 *What are the realisable sign strata of $\mathcal{A}_{n,k,m}$?*

Question 2.2.12 *Which of the realisable sign strata of $\mathcal{A}_{n,k,m}(Z)$ are realised for all Z?*

We will see in Sect. 4.3 that Question 2.2.11 has a beautiful combinatorial answer for the $m = 2$ amplituhedron.

Remark 2.2.13 One could introduce alternative definitions of *amplituhedron sign strata*. Given a collection $\mathcal{F} = \{P_i\}_{i \in [N]}$ of N polynomials P_i in twistor coordinates and a sign vector $\sigma \in \{0, +, -\}^N$, one could define the regions

$$\mathcal{A}^{\mathcal{F},\sigma}_{n,k,m}(Z) := \{Y \in \mathcal{A}_{n,k,m}(Z) \mid \operatorname{sgn} P_i(Y) = \sigma_i\}$$

in the amplituhedron where all the polynomials in $\mathcal{F}$ have definite signs. So if we let $\mathcal{F}$ be the collection of all twistor coordinates, then we recover Definition 2.2.9.

2.2.3 Sign Variation and Sign Flips

Signs and sign flips will be important to our description of the amplituhedron, so we introduce some useful terminology here.

Definition 2.2.14 Given $v \in \mathbb{R}^n$, let $\operatorname{var}(v)$ be the number of times v changes sign when we read the components from left to right and ignore any zeros. We also define

[3] Plücker and twistor coordinates are defined only up to multiplication by a common scalar.

[4] We borrow the word "chamber" from the theory of hyperplane arrangements.

$$\overline{\text{var}}(v) := \max\{\text{var}(w) \mid w \in \mathbb{R}^n \text{ such that } w_i = v_i \text{ for all } i \in [n] \text{ with } v_i \neq 0\}.$$

If $v \in \{0, +, -\}^n$, we define $\text{var}(v)$ and $\overline{\text{var}}(v)$ in the obvious way.

For example, if $v := (4, -1, 0, -2) \in \mathbb{R}^4$ then $\text{var}(v) = 1$ and $\overline{\text{var}}(v) = 3$.

Definition 2.2.15 If $v \in \mathbb{R}^n$, or $v \in \{+, -, 0\}^n$, we say that v has a *sign flip in position* i if $v_i, v_{i+1} \neq 0$ and they have different signs, where indices are considered modulo n. We define

$$\text{Flip}(v) = \text{Flip}(v_1, \ldots, v_n) := \{i \mid v \text{ has a sgn flip in position } i\} \subseteq [n].$$

Remark 2.2.16 We caution the reader that $|\text{Flip}(v)|$ may not equal $\text{var}(v)$. For example, the sequence $(+, 0, -, 0, +, +, -) \in \{+, -, 0\}^7$ has sign flips in positions $\{6, 7\}$, but $\text{var}(+, 0, -, 0, +, +, -)$ is 3.

2.3 Positroid Tiles in the Amplituhedron $\mathcal{A}_{n,k,2}$

Recall that a *positroid tile* of $\mathcal{A}_{n,k,m}(Z)$ is the full-dimensional image of a positroid cell on which $\tilde{Z}$ is injective. In this section, we will obtain a detailed description of the positroid tiles of $\mathcal{A}_{n,k,2}(Z)$. The main results of this section are the following:

- In Theorem 2.3.25 we classify the positroid tiles of $\mathcal{A}_{n,k,2}(Z)$, describing them as the Grasstopes $Z_{\hat{G}(\mathcal{T})}$ obtained from the $2k$-dimensional positroid cells $S_{\hat{G}(\mathcal{T})}$ associated to *unpunctured plabic tilings*, proving a conjecture of [29]. This implies that whether or not $\tilde{Z}(S_\pi)$ is a positroid tile is independent of the choice of Z.
- In Theorem 2.3.28 we characterize each (open) positroid tile $Z^\circ_{\hat{G}(\mathcal{T})}$ as the subset of $\text{Gr}_{k,k+2}$ where certain twistor coordinates have a fixed sign; this shows that each positroid tile is a union of (closures of) amplituhedron chambers.
- In Proposition 2.3.19 we solve a kind of "inverse problem" for positroid tiles: given an element $Y \in \text{Gr}_{k,k+2}$ which lies in an open positroid tile $Z^\circ_{\hat{G}(\mathcal{T})}$, we explicitly construct an element $C \in \text{Gr}_{k,n}$ whose image in $\mathcal{A}_{n,k,2}(Z)$ is Y, i.e. $CZ = Y$; the entries of C are in fact twistor coordinates.

We note that the techniques that we use in this section can be extended to give a *cell decomposition* of $\mathcal{A}_{n,k,2}(Z)$. This will be explored in a separate work.

Definition 2.3.1 *(Unpunctured plabic tilings)* Let $\mathbf{P}_n$ be a convex n-gon with vertices labeled from 1 to n in clockwise order. A *triangulated* (k, n)*-unpunctured plabic tiling*[5] is a collection of k nonoverlapping black triangles $\mathcal{T}$ inside $\mathbf{P}_n$ whose vertices are all vertices of $\mathbf{P}_n$. We say that two unpunctured triangulated plabic tilings $\mathcal{T}$ and $\mathcal{T}'$ are *equivalent* if the union of the black triangles is the same for both of them. We represent the equivalence class of an unpunctured triangulated tiling $\mathcal{T}$ by erasing

[5] Not to be confused with *plabic tilings of type* (k, n), which are dual to plabic graphs of type (k, n).

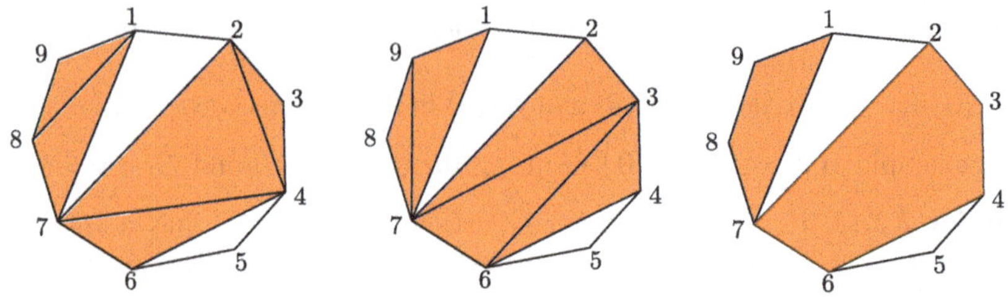

Fig. 2.3 Two equivalent triangulated (5, 9)-unpunctured plabic tilings $\mathcal{T}_1$ and $\mathcal{T}_2$, and the corresponding unpunctured plabic tiling $\overline{\mathcal{T}}_1 = \overline{\mathcal{T}}_2$

the diagonals that separate pairs of triangles that share an edge. The resulting object $\overline{\mathcal{T}}$ is a decomposition of the n-gon into white and black polygons, and is called an *unpunctured plabic tiling*. See Fig. 2.3.

The objects defined in Definition 2.3.1 were called *k nonintersecting triangles in a convex n -gon* in [29]. However, since they are special cases of the *plabic tilings* of [45], we refer to these objects as *unpunctured plabic tilings*. See Definition 4.1.14 and Remark 4.1.16.

Given a triangulated (k, n)-unpunctured plabic tiling $\mathcal{T}$, we build a corresponding bipartite graph $\hat{G}(\mathcal{T})$[6] as in Fig. 2.4, then use the recipe from Theorem A.0.7 and Remark A.0.8 to construct all points of the $2k$-dimensional cell $S_{\hat{G}(\mathcal{T})}$ of $\mathrm{Gr}_{k,n}^{\geq 0}$.

Definition 2.3.2 Given a triangulated (k, n)-unpunctured plabic tiling $\mathcal{T}$, we build a labeled bipartite graph $\hat{G}(\mathcal{T})$ by placing black boundary vertices labeled $B_1, B_2, \dots, B_n$ in clockwise order at the n vertices of the n-gon, and placing a trivalent white vertex in the middle of each black triangle, connecting it to the three vertices of the triangle. We label the k white vertices by $W_1, \dots, W_k$; we will usually label them in the order specified by Remark 2.3.5.

Remark 2.3.3 We can think of $\hat{G}(\mathcal{T})$ as a *plabic graph* (see Definition A.0.2) if we enclose it in a slightly larger disk and add n edges connecting each B_i to the boundary of the disk. We will often abuse terminology and refer to $\hat{G}(\mathcal{T})$ as a plabic graph.

Lemma 2.3.4 *If two triangulated unpunctured plabic tilings $\mathcal{T}_1$ and $\mathcal{T}_2$ are equivalent, then the plabic graphs $\hat{G}(\mathcal{T}_1)$ and $\hat{G}(\mathcal{T}_2)$ are move-equivalent (see Definition A.0.3). In other words, these two plabic graphs represent the same cell of* $\mathrm{Gr}_{k,n}^{\geq 0}$.

Proof The fact that $\mathcal{T}_1$ and $\mathcal{T}_2$ are equivalent means that we can get from $\mathcal{T}_1$ to $\mathcal{T}_2$ by performing some sequence of flips on the black triangles comprising the black polygons in $\mathcal{T}_1$. Every flip of black triangles corresponds to performing a square move on the corresponding plabic graph, so $\hat{G}(\mathcal{T}_1)$ and $\hat{G}(\mathcal{T}_2)$ are move-equivalent. □

[6] We denote this graph $\hat{G}(\mathcal{T})$ to distinguish it from the dual plabic graph $G(\mathcal{T})$ defined in Definition 4.1.14.

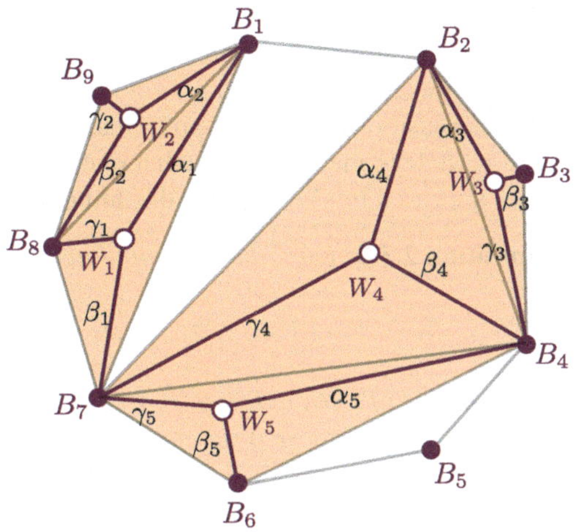

Fig. 2.4 The planar bipartite graph $\hat{G}(\mathcal{T}_1)$ together with its edge-weighting

In light of Lemma 2.3.4, we let $S_{\hat{G}(\overline{\mathcal{T}})}$ denote the cell specified by any tiling of $\overline{\mathcal{T}}$.

Remark 2.3.5 We identify each black triangle T in an unpunctured triangulated plabic tiling $\mathcal{T}$ with its three vertices $a < b < c$ listed in increasing order. We list the k black triangles

$$(T_1, \ldots, T_k) = (\{a_1 < b_1 < c_1\}, \ldots, \{a_k < b_k < c_k\})$$

in lexicographically increasing order, and label the white vertex inside of T_i by W_i.

For example, we list the five triangles of the plabic tiling $\mathcal{T}_1$ from Fig. 2.3 in the order

$$(\{1 < 7 < 8\}, \{1 < 8 < 9\}, \{2 < 3 < 4\}, \{2 < 4 < 7\}, \{4 < 6 < 7\}).$$

We label the white vertices of $\hat{G}(\mathcal{T}_1)$ in Fig. 2.4 so as to reflect this ordering on triangles.

Definition 2.3.6 *(Statistics of unpunctured plabic tilings)* Given a triangulated unpunctured plabic tiling $\mathcal{T}$ in a convex n-gon $\mathbf{P}_n$ and a pair of vertices h, j of $\mathbf{P}_n$, we say that:

- the arc $h \to j$ is *compatible* with $\mathcal{T}$ if the arc does not cross any arcs of the underlying unpunctured plabic tiling $\overline{\mathcal{T}}$, i.e. it either bounds a polygon of $\overline{\mathcal{T}}$ or it lies entirely inside a black or white polygon;
- the arc $h \to j$ is a *black arc* of $\mathcal{T}$ if it bounds a triangle of $\mathcal{T}$;
- the arc $h \to j$ is *facet-defining* if it bounds a black polygon of $\overline{\mathcal{T}}$ on its left.

In particular, each black arc of $\mathcal{T}$ is compatible with $\mathcal{T}$.

When $h \to j$ is compatible with $\mathcal{T}$, we let $\text{area}(h \to j) = \text{area}_{\mathcal{T}}(h \to j)$ denote the number of black triangles to the left of $h \to j$ in any triangulation of $\overline{\mathcal{T}}$.

For example, the arcs $1 \to 8$, $1 \to 7$ and $2 \to 6$ are compatible with the tiling $\mathcal{T}$ from Fig. 2.4, and we have $\text{area}(1 \to 8) = 4$, $\text{area}(1 \to 7) = 3$, $\text{area}(2 \to 6) = 2$. However, the arcs $2 \to 8$ and $3 \to 8$ are not compatible with $\mathcal{T}$.

We can easily write down representative matrices for points in $S_{\hat{G}(\mathcal{T})}$ using the theory of Kasteleyn matrices. Note that matrices with the same pattern of zero/nonzero entries appeared in [29] (though the authors did not prove Proposition 2.3.7 there).

Proposition 2.3.7 *Let $\mathcal{T}$ be a triangulated (k, n)-unpunctured plabic tiling. We let*

$$(\{a_1 < b_1 < c_1\}, \dots, \{a_k < b_k < c_k\})$$

denote the list of k triangles of $\mathcal{T}$, written in lexicographically increasing order, as in Remark 2.3.5. Choose a set of edge-weights for the graph $\hat{G}(\mathcal{T})$, which we write as

$$(\boldsymbol{\alpha}, \boldsymbol{\beta}, \boldsymbol{\gamma}) = ((\alpha_1, \beta_1, \gamma_1), (\alpha_2, \beta_2, \gamma_2), \dots, (\alpha_k, \beta_k, \gamma_k)) \in (\mathbb{R}_{>0})^{3k},$$

with $\alpha_i, \beta_i, \gamma_i$ denoting the weights on the edges from W_i to B_{a_i}, B_{b_i} and B_{c_i}, respectively.

Let $M_{\mathcal{T}}(\boldsymbol{\alpha}, \boldsymbol{\beta}, \boldsymbol{\gamma}) = (M_{i,j})$ be the $k \times n$ with precisely 3 nonzero entries in each row:

$$M_{i,a_i} = \alpha_i, \quad M_{i,b_i} = (-1)^{\text{area}(a_i \to b_i)} \beta_i, \quad M_{i,c_i} = (-1)^{\text{area}(a_i \to b_i) + \text{area}(b_i \to c_i)} \gamma_i. \tag{2.3.8}$$

Then the cell $S_{\hat{G}(\mathcal{T})}$ is the image of the map $(\mathbb{R}_{>0})^{3k} \to \text{Gr}_{k,n}^{\geq 0}$ sending $(\boldsymbol{\alpha}, \boldsymbol{\beta}, \boldsymbol{\gamma}) \mapsto M_{\mathcal{T}}(\boldsymbol{\alpha}, \boldsymbol{\beta}, \boldsymbol{\gamma})$.

Note that $M_{\mathcal{T}}(\boldsymbol{\alpha}, \boldsymbol{\beta}, \boldsymbol{\gamma})$ has rows and columns indexed by the white and black vertices of $\hat{G}(\mathcal{T})$. The ij-entry is nonzero if and only if there is an edge e in $\hat{G}(\mathcal{T})$ between W_i and B_j, and in that case is (up to a sign) equal to the weight of e.

Remark 2.3.9 Clearly the image of the map $(\boldsymbol{\alpha}, \boldsymbol{\beta}, \boldsymbol{\gamma}) \mapsto M_{\mathcal{T}}(\boldsymbol{\alpha}, \boldsymbol{\beta}, \boldsymbol{\gamma})$ is unchanged if we rescale each row of the matrix so that the leftmost nonzero entry is 1, i.e. set each $\alpha_i = 1$. This map is then a homeomorphism from $(\mathbb{R}_{>0})^{2k}$ to the positroid cell $S_{\hat{G}(\mathcal{T})}$.

Proof (Proof of Proposition 2.3.7) This follows from Theorem A.0.7 and Remark A.0.8. For completeness, we sketch why the choice of signs of entries is correct. For a triangle $T_i = \{a < b < c\}$ of $\mathcal{T}$, define

$$\epsilon_{i,a} := (-1)^{\#\{j<i : a_i = a_j\}}, \quad \epsilon_{i,b} := \epsilon_{i,a} \cdot (-1)^{\text{area}(a_i \to b_i)}, \quad \epsilon_{i,c} := \epsilon_{i,a} \cdot (-1)^{\text{area}(a_i \to c_i)+1}.$$

Let $(d_1, \dots, d_k)$ be a tuple of distinct vertices of triangles of $\mathcal{T}$ such that $d_i \in T_i$. The sign of the permutation σ such that $d_{\sigma(1)} < \dots < d_{\sigma(k)}$ is the product $\epsilon_{1,d_1} \cdots \epsilon_{k,d_k}$. Then:

$$p_I(M) E_I = \sum_{(d_1,\dots,d_k)} M_{1,d_1} \cdots M_{k,d_k} \langle e_{d_1}, \dots, e_{d_k} \rangle = \sum_{(d_1,\dots,d_k)} (\epsilon_{1,d_1} M_{1,d_1}) \cdots (\epsilon_{k,d_k} M_{k,d_k}) E_I$$

where the sum is over the collections defined above satisfying $\{d_1, \ldots, d_k\} = I$. A sufficient condition for $p_I(M) \geq 0$ is that $\operatorname{sgn} M_{i,d_i} = \epsilon_{i,d_i}$. Up to rescaling the row i of M by $\epsilon_{a_i,i}$, this is true, as $\text{area}(a_i \to c_i) = \text{area}(a_i \to b_i) + \text{area}(b_i \to c_i) + 1$. □

Example 2.3.10 For example, the matrix $M_{\mathcal{T}_1}$ corresponding to the unpunctured triangulated plabic tiling $\mathcal{T}_1$ from Fig. 2.4 is

$$\begin{pmatrix} \alpha_1 & 0 & 0 & 0 & 0 & 0 & -\beta_1 & -\gamma_1 & 0 \\ \alpha_2 & 0 & 0 & 0 & 0 & 0 & 0 & \beta_2 & \gamma_2 \\ 0 & \alpha_3 & \beta_3 & \gamma_3 & 0 & 0 & 0 & 0 & 0 \\ 0 & \alpha_4 & 0 & -\beta_4 & 0 & 0 & \gamma_4 & 0 & 0 \\ 0 & 0 & 0 & \alpha_5 & 0 & \beta_5 & \gamma_5 & 0 & 0 \end{pmatrix}. \tag{2.3.11}$$

Proposition 2.3.7 says that if we let the parameters $((\alpha_1, \beta_1, \gamma_1), \ldots, (\alpha_5, \beta_5, \gamma_5))$ range over all elements of $(\mathbb{R}_{>0})^{15}$, the matrices (2.3.11) will sweep out all points of the cell $S_{\hat{G}(\mathcal{T}_1)}$. ◇

Remark 2.3.12 The matrices constructed in Proposition 2.3.7 may have non-positive maximal minors rather than non-negative maximal minors. To obtain a matrix which has non-negative maximal minors, multiply row j by $(-1)^{\#\{i<j:a_i=a_j\}}$.

Lemma 2.3.13 *Let $\mathcal{T} = \{T_1, \ldots, T_k\}$ be a triangulated (k, n)-unpunctured plabic tiling. Then $p_I \neq 0$ on the positroid cell $S_{\hat{G}(\mathcal{T})}$ if and only if there is a bijection $\phi : I = \{i_1, \ldots, i_k\} \to \{T_1, \ldots, T_k\}$ with i a vertex of $\phi(i)$ for all i.*

Proof It suffices to show that the Plücker coordinate p_I is nonzero on $S_{\hat{G}(\mathcal{T})}$ if and only if there is a bijection $\phi : I = \{i_1, \ldots, i_k\} \to \{T_1, \ldots, T_k\}$ with i a vertex of $\phi(I)$ for all i.

By Theorem A.0.7, $p_I \neq 0$ on $S_{\hat{G}(\mathcal{T})}$ if and only if there is a matching M of $\hat{G}(\mathcal{T})$ such that $\partial M = I$. Note that any matching M of $\hat{G}(\mathcal{T})$ consists of k edges, obtained by pairing each white vertex W_j with one of its three incident black vertices $\{B_{a_j}, B_{b_j}, B_{c_j}\}$. The k black vertices $\{B_{i_1}, \ldots, B_{i_k}\}$ obtained in this way must be distinct (since M is a matching), so we get a bijection between $I := \{i_1, \ldots, i_k\}$ and the triangles $T_1, \ldots, T_k$. Moreover $\partial M = I$. □

Now, we turn to the open Grasstopes $Z^{\circ}_{\hat{G}(\mathcal{T})}$ and their properties.

Theorem 2.3.14 (Definite signs of twistor coordinates) *Let $\mathcal{T}$ be a (k, n)-unpunctured plabic tiling, and let $Y := CZ \in \mathrm{Gr}_{k,k+2}$, where C is a matrix representing a point of the cell $S_{\hat{G}(\mathcal{T})}$. Choose $h < j$ such that the chord $h \to j$ is compatible with $\mathcal{T}$. Then*

$$\operatorname{sgn}\langle Y Z_h Z_j \rangle = (-1)^{\text{area}(h\to j)}, \text{ or equivalently, } (-1)^{\text{area}(h\to j)}\langle Y Z_h Z_j \rangle > 0. \tag{2.3.15}$$

In other words, we have that

$$Z^{\circ}_{\hat{G}(\mathcal{T})} \subseteq \{Y \in \mathrm{Gr}_{k,k+2} \mid 2.4.15 \text{ holds for all arcs } h \to j \text{ compatible with } \mathcal{T}.\} \qquad (2.3.16)$$

Proof We start by choosing a triangulated plabic tiling $\mathcal{T}_1$ such that $\overline{\mathcal{T}_1} = \mathcal{T}$, and such that the chord $h \to j$ is one of the diagonals of $\mathcal{T}_1$. By Lemma 2.3.4, the choice of $\mathcal{T}_1$ does not affect the corresponding positroid cell. By Lemma 2.2.8, it suffices to verify (2.3.15) for each $Y := \langle Z_{i_1}, \ldots, Z_{i_k} \rangle$ indexed by $\{i_1 < \cdots < i_k\} = I$ such that $p_I \neq 0$ on the cell $S_{\hat{G}(\mathcal{T})}$. And by Lemma 2.3.13, $p_I \neq 0$ on $S_{\hat{G}(\mathcal{T})}$ if and only if there is a bijection $\phi : I = \{i_1, \ldots, i_k\} \to \{T_1, \ldots, T_k\}$ with i a vertex of $\phi(i)$ for all i.

Towards this end, choose $I = \{i_1 < \cdots < i_k\}$ such that $p_I \neq 0$ on the cell $S_{\hat{G}(\mathcal{T})}$. We need to calculate $\mathrm{sgn}\langle Z_{i_1}, \ldots, Z_{i_k}, Z_h, Z_j \rangle$.

If $h \in I$ or $j \in I$, $\langle Z_{i_1}, \ldots, Z_{i_k}, Z_h, Z_j \rangle = 0$. So without loss of generality, we can assume that h and j are not elements of I. Recall that maximal minors of Z are positive: this means that for any ordered sequence $\ell_1 < \cdots < \ell_{k+2}$, we have $\mathrm{sgn}\langle Z_{\ell_1}, \ldots, Z_{\ell_{k+2}} \rangle = 1$. To determine $\mathrm{sgn}\langle Z_{i_1}, \ldots, Z_{i_k}, Z_h, Z_j \rangle$, we need to know how many swaps are required to put the sequence $(i_1, \ldots, i_k, h, j)$ in order. Any i_ℓ which is greater than both h and j needs to get swapped past both of them, which has no effect on the sign of the determinant. Any i_ℓ which is less than both h and j does not need to get swapped past either. Each i_ℓ such that $h < i_\ell < j$ needs to get swapped past h (but not j). Therefore the parity of the number of swaps required to put the sequence $(i_1, \ldots, i_k, h, j)$ in order is the same as the parity of $\#\{i_\ell \in I : h < i_\ell < j\}$. It follows that $\mathrm{sgn}\langle Z_{i_1}, \ldots, Z_{i_k}, Z_h, Z_j \rangle = (-1)^{\#\{i_\ell \in I : h < i_\ell < j\}}$. Finally, the existence of the bijection ϕ means that $\#\{i_\ell \in I : h < i_\ell < j\}$ is the number of triangles of $\mathcal{T}_1$ which are to the left of $h \to j$.

To complete the proof, we must show that there is some $I \in \binom{[n]}{k}$ containing neither h nor j such that p_I is nonzero. Equivalently, we must find a matching of $\hat{G}(\mathcal{T}_1)$ which does not have h or j in its boundary. We do so by induction on the number of black triangles of $\mathcal{T}_1$. Clearly there is such a matching if $\hat{G}(\mathcal{T}_1)$ has a single black triangle. If $h \to j$ is contained in a white polygon of $\mathcal{T}_1$, we cut along $h \to j$ to obtain two smaller triangulated unpunctured plabic tilings $\mathcal{T}_2$ and $\mathcal{T}_3$. By induction, we can find matchings of $\hat{G}(\mathcal{T}_2)$ and $\hat{G}(\mathcal{T}_3)$ avoiding h and j; their union gives the desired matching of $\hat{G}(\mathcal{T}_1)$. Otherwise, $h \to j$ is the boundary of a black triangle T_r of $\mathcal{T}_1$. Let c be the third vertex of this triangle. Cut $\mathcal{T}_1$ along $h \to j$, $j \to c$ and $c \to h$ to obtain smaller triangulated unpunctured plabic tilings. The $\hat{G}$ plabic graphs of these tilings have matchings avoiding h, j, c by induction, since each smaller tiling contains exactly two of these vertices. The union of these matchings, together with the edge from B_c to W_r, gives the desired matching of $\mathcal{T}_1$. □

The following result solves a kind of "inverse problem:" given $Y \in Z^{\circ}_{\hat{G}(\mathcal{T})}$, we can construct a particular matrix representative $C^{\mathrm{tw}}_{\mathcal{T}}(Y)$ of $\mathrm{Gr}_{k,n}$ whose image in $\mathcal{A}_{n,k,2}(Z)$ is Y.

Definition 2.3.17 *(Twistor coordinate matrix)* Let $Y \in \mathrm{Gr}_{k,k+2}$ and let $\mathcal{T}$ be a triangulated (k, n)-unpunctured plabic tiling with triangles $T_1, \ldots, T_k$ labeled as in

Remark 2.3.5. The *twistor coordinate matrix* of Y is the $k \times n$ matrix $C^{\text{tw}}_{\mathcal{T}}(Y) = (C_{i,j})$ with precisely 3 nonzero entries in each row:

$$C_{i,a_i} = \langle Y b_i c_i \rangle, \quad C_{i,b_i} = -\langle Y a_i c_i \rangle, \quad C_{i,c_i} = \langle Y a_i b_i \rangle. \tag{2.3.18}$$

(Recall that e.g. $\langle Y b_i c_i \rangle$ is short-hand for $\langle Y Z_{b_i} Z_{c_i} \rangle$.)

Theorem 2.3.19 (Inverse problem) *Let $\mathcal{T}$ be a triangulated (k,n)-unpunctured plabic tiling with triangles $T_1, \ldots, T_k$ labeled as in Remark 2.3.5. Let $Y \in Z^{\circ}_{\hat{G}(\mathcal{T})}$, i.e. $Y := \tilde{Z}(V)$ for some $V \in S_{\hat{G}(\mathcal{T})}$. Then V is the row span of the twistor coordinate matrix $C' := C^{\text{tw}}_{\mathcal{T}}(Y)$.*

In other words, if we let $Y' = C'Z$, then there is a global scalar λ (a polynomial in $\langle Yab \rangle$'s) such that

$$\langle Y' ij \rangle = \lambda \langle Yij \rangle \text{ for all } i, j.$$

Example 2.3.20 Let $\mathcal{T}_1$ be the plabic tiling from Fig. 2.4. Proposition 2.3.19 says that if $V \in S_{\hat{G}(\mathcal{T}_1)}$ and $Y := \tilde{Z}(V)$ is the image of V in $\mathcal{A}_{n,k,2}(Z)$, then V is the row span of the following matrix:

$$\begin{pmatrix} \langle YZ_7Z_8 \rangle & 0 & 0 & 0 & 0 & 0 & -\langle YZ_1Z_8 \rangle & \langle YZ_1Z_7 \rangle & 0 \\ \langle YZ_8Z_9 \rangle & 0 & 0 & 0 & 0 & 0 & 0 & -\langle YZ_1Z_9 \rangle & \langle YZ_1Z_8 \rangle \\ 0 & \langle YZ_3Z_4 \rangle & -\langle YZ_2Z_4 \rangle & \langle YZ_2Z_3 \rangle & 0 & 0 & 0 & 0 & 0 \\ 0 & \langle YZ_4Z_7 \rangle & 0 & -\langle YZ_2Z_7 \rangle & 0 & 0 & \langle YZ_2Z_4 \rangle & 0 & 0 \\ 0 & 0 & 0 & \langle YZ_6Z_7 \rangle & 0 & -\langle YZ_4Z_7 \rangle & \langle YZ_4Z_6 \rangle & 0 & 0 \end{pmatrix}$$

◇

Proof Choose a weight vector (α, β, γ) so that the matrix $C := M_{\mathcal{T}}(\alpha, \beta, \gamma)$ from Proposition 2.3.7 represents V.

Consider a triangle $\{a < b < c\}$ of $\mathcal{T}$. Let W be the white vertex of $\hat{G}(\mathcal{T})$ in the middle of this triangle and let the edges from W to B_a, B_b, and B_c, respectively, be denoted e_a, e_b, and e_c. Say the weights of these edges are α, β, and γ, respectively.

Choose $J \in \binom{[n]}{k-1}$ which does not contain a, b, or c. Then

$$\frac{1}{\alpha} p_{J \cup \{a\}}(C) = \frac{1}{\beta} p_{J \cup \{b\}}(C) = \frac{1}{\gamma} p_{J \cup \{c\}}(C). \tag{2.3.21}$$

Indeed, each Plücker coordinate is a sum of weights of matchings. Any matching M_a contributing to $p_{J \cup \{a\}}(C)$ must include an edge covering the white vertex W. Since $b, c \notin J \cup \{a\}$, this edge must be e_a. Now, $M_b := M_a \setminus \{e_a\} \cup \{e_b\}$ is a valid matching because M_a does not include any edges covering B_b. Moreover, the boundary of M_b is $J \cup \{b\}$. This is easily seen to be a bijection between matchings with boundary $J \cup \{a\}$ and matchings with boundary $J \cup \{b\}$. It is also easy to see that $\text{wt}(M_a)/\alpha = \text{wt}(M_b)/\beta$, so the first equality above holds. The second equality is similar.

Now, we consider the twistor coordinate

$$\langle Ybc\rangle = \sum_{I\in\binom{[n]}{k}} p_I(C)\langle Z_{i_1}, Z_{i_2}, \dots, Z_{i_k}, Z_b, Z_c\rangle$$

which is nonzero by Theorem 2.3.14.

Notice that the terms in this sum indexed by I containing b or c are zero. Further, for $I \cap \{b, c\} = \emptyset$, $p_I(C)$ is zero if I does not contain a. So we can rewrite $\langle Ybc\rangle$ as

$$\langle Ybc\rangle = \alpha \cdot \sum_{\substack{J\in\binom{[n]}{k-1}:\\ \{a,b,c\}\cap J=\emptyset}} \frac{1}{\alpha} p_{J\cup a}(C)\langle Z_{j_1}, Z_{j_2}, \dots, Z_a, \dots Z_{j_{k-1}}, Z_b, Z_c\rangle \quad (2.3.22)$$

where $Z_{j_1}, \dots, Z_a, \dots, Z_{j_{k-1}}$ are ordered so the indices are increasing.

Similarly, we can write

$$\langle Yac\rangle = \beta \cdot \sum_{\substack{J\in\binom{[n]}{k-1}:\\ \{a,b,c\}\cap J=\emptyset}} \frac{1}{\beta} p_{J\cup b}(C)\langle Z_{j_1}, Z_{j_2}, \dots, Z_b, \dots Z_{j_{k-1}}, Z_a, Z_c\rangle \quad (2.3.23)$$

$$\langle Yab\rangle = \gamma \cdot \sum_{\substack{J\in\binom{[n]}{k-1}:\\ \{a,b,c\}\cap J=\emptyset}} \frac{1}{\gamma} p_{J\cup c}(C)\langle Z_{j_1}, Z_{j_2}, \dots, Z_c, \dots Z_{j_{k-1}}, Z_a, Z_b\rangle. \quad (2.3.24)$$

Consider a nonzero term in (2.3.22), which is indexed by J such that $p_{J\cup a}(C)$ is nonzero. The corresponding term in (2.3.23) is also nonzero. Because of the first equality in (2.3.21), these two terms differ only by the sign $(-1)^s$, where

$$\langle Z_{j_1}, Z_{j_2}, \dots, Z_a, \dots Z_{j_{k-1}}, Z_b, Z_c\rangle = (-1)^s\langle Z_{j_1}, Z_{j_2}, \dots, Z_b, \dots Z_{j_{k-1}}, Z_a, Z_c\rangle.$$

In other words, $s = |J \cap [a+1, b-1]| + 1 = |(J \cup a) \cap [a+1, b-1]| + 1$. Because $J \cup a$ is the boundary of some matching, the size of $(J \cup a) \cap [a+1, b-1]$ is exactly $\text{area}(a \to b)$, and in particular does not depend on J.

Similarly, consider the term of (2.3.24) indexed by J. The sign difference between this term and the corresponding one in (2.3.23) is $(-1)^s$, where

$$\langle Z_{j_1}, Z_{j_2}, \dots, Z_b, \dots Z_{j_{k-1}}, Z_a, Z_c\rangle = (-1)^s\langle Z_{j_1}, Z_{j_2}, \dots, Z_c, \dots Z_{j_{k-1}}, Z_a, Z_b\rangle.$$

It is not hard to see that $s = \text{area}(b \to c) + 1$.

Altogether, we have

$$\begin{aligned}
\langle Ybc\rangle &= \alpha \cdot Q\\
\langle Yac\rangle &= (-1)^{\text{area}(a\to b)+1}\beta \cdot Q\\
\langle Yab\rangle &= (-1)^{\text{area}(a\to b)+\text{area}(b\to c)}\gamma \cdot Q
\end{aligned}$$

where Q is a nonzero scalar. Notice that up to the factor of Q, these three twistor coordinates recover the entries of C corresponding to the edges e_a, e_b, and e_c. This means that the matrix C' with non-zero entries

$$C'_{j,a_j} = \langle Yb_jc_j\rangle, \quad C'_{j,b_j} = -\langle Ya_jc_j\rangle, \quad C'_{j,c_j} = \langle Ya_jb_j\rangle$$

is related to $M_{\hat{G}(T)}(\alpha, \beta, \gamma)$ by rescaling rows, and so also represents the subspace V. □

Using Proposition 2.3.19, we can show that $\tilde{Z}$ is injective on $S_{\hat{G}(T)}$, and moreover prove that $\tilde{Z}$ is not injective on any other $2k$-dimensional positroid cells. This will prove the conjectural characterization of positroid tiles from [29] (who used the terminology of *k non-intersecting triangles in a convex n -gon*). We note that the injectivity of $\tilde{Z}$ on $S_{\hat{G}(T)}$ was also proved rather indirectly in [46, Proposition 6.4] using results of [41].

Theorem 2.3.25 (Characterization of positroid tiles) *Fix $k < n$ and $Z \in \mathrm{Mat}^{>0}_{n,k+2}$. Then $\tilde{Z}$ is injective on the $2k$-dimensional cell $S_{\mathcal{M}}$ if and only if $S_{\mathcal{M}} = S_{\hat{G}(T)}$ for some (k, n)-unpunctured plabic tiling $\mathcal{T}$. That is, the positroid tiles for $\mathcal{A}_{n,k,2}$ are exactly the Grasstopes $Z_{\hat{G}(T)}$, where $\mathcal{T}$ is a (k, n)-unpunctured plabic tiling.*

Corollary 2.3.26 *Whether or not $\overline{\tilde{Z}(S_\pi)}$ is a positroid tile is independent of Z.*

Proof (Proof of Theorem 2.3.25) This proof uses some facts from Sect. 4.2. We first show that all cells $S_{\hat{G}(T)}$ are positroid tiles. The cell $S_{\hat{G}(T)} \subset \mathrm{Gr}^{\geq 0}_{k,n}$ is $2k$-dimensional because it is T-dual to an $(n-1)$-dimensional cell in $\mathrm{Gr}^{\geq 0}_{k+1,n}$ (see Remark 4.1.16) and T-duality preserves codimension (see Proposition 4.1.6). Say $V, V' \in S_{\hat{G}(T)}$ are represented by matrices C, C', and suppose $Y := CZ$, $Y' := C'Z$ represent the same subspace. Then by Proposition 2.3.19 V and V' are represented by the twistor coordinate matrices N and N' of Y and Y', respectively. But the twistor coordinates of Y and Y' are the same up to a global scalar, so $V = V'$.

Now, suppose a $2k$-dimensional cell $S_{\mathcal{M}}$ is not equal to $S_{\hat{G}(T)}$ for any $\mathcal{T}$. We will show $\tilde{Z}$ is not injective on $S_{\mathcal{M}}$.

First, suppose M has a *coloop* c; that is, p_I is identically 0 on $S_{\mathcal{M}}$ for all $I \in \binom{[n]}{k}$ that do not contain c. Then the twistor coordinate $\langle Yij\rangle$ is identically zero on $Z^\circ_{\mathcal{M}}$ for all j. Indeed, in the sum

$$\langle Ycj\rangle = \sum_{I \in \binom{[n]}{k}} p_I(C)\langle Z_{i_1} \ldots Z_{i_k} Z_c Z_j\rangle,$$

$p_I(C)$ is zero for $c \notin I$ and $\langle Z_{i_1} \ldots Z_{i_k} Z_c Z_j\rangle$ is zero for $c \in I$. In particular, $Z^\circ_{\mathcal{M}}$ is contained in the hypersurface $\{Y \in \mathrm{Gr}_{k,k+2} : \langle Yc(c+1)\rangle = 0\}$, and so has dimension at most $2k - 1$. So $\tilde{Z}$ is not injective on $S_{\mathcal{M}}$.

Now, if $\mathcal{M}$ does not have a coloop, then $S_{\mathcal{M}}$ is T-dual to an $(n-1)$-dimensional cell S_π of $\mathrm{Gr}^{\geq 0}_{k+1,n}$ by Proposition 4.1.6. Because $S_{\mathcal{M}}$ is not of the form $S_{\hat{G}(T)}$, a plabic graph G with trip permutation π is not a tree and so has at least one internal face. Since G has n faces total, G is not connected.

Let G be a plabic graph with trip permutation π, and say $[i, j-1]$, $[j, l]$ are the boundary vertex sets of two connected components of G. There is a single boundary face f which is adjacent to $i-1$, i, $j-1$ and j. In the plabic graph $\hat{G}$ for $S_{\mathcal{M}}$ (constructed in Proposition 4.1.12), notice that i and j are adjacent to the same black vertex, $\hat{b}(f)$. After adding bivalent white vertices to $\hat{G}$ so that every boundary vertex is adjacent to a white vertex, it is clear that all matchings of $\hat{G}$ have either i or j in the boundary. This means that if I contains neither i nor j, then p_I is identically zero on $S_{\mathcal{M}}$. Just as in the coloop case, $\langle Yij\rangle$ is identically zero on $Z^{\circ}_{\mathcal{M}}$, because all terms of

$$\sum_{I\in\binom{[n]}{k}} p_I(C)\langle Z_{i_1}\dots Z_{i_k}Z_iZ_j\rangle$$

vanish for $C \in S_{\mathcal{M}}$. So $Z^{\circ}_{\mathcal{M}}$ is contained in a hypersurface and hence $\dim Z^{\circ}_{\mathcal{M}} \leq 2k-1$. □

Remark 2.3.27 As conjectured in [29], the number of positroid tiles for $\mathcal{A}_{n,k,2}$ is sequence [47, A175124], a refinement of the *large Schröder numbers* (see Sect. 4.2.3).

Refining (2.3.16), we will now give an explicit description of each (open) positroid tile as a subset of $\mathrm{Gr}_{k,k+2}$ where certain twistor coordinates have a definite sign. In fact, since there are generally multiple triangulated plabic tilings in the equivalence class of one unpunctured plabic tiling $\mathcal{T}$, Theorem 2.3.28 gives multiple descriptions of each positroid tile—one for each triangulated unpunctured plabic tiling in the equivalence class of $\mathcal{T}$.

Theorem 2.3.28 (Sign characterization of positroid tiles) *Fix $k<n$, $m=2$, and $Z\in \mathrm{Mat}^{>0}_{n,k+2}$. Let $\mathcal{T}$ be a triangulated (k,n)-unpunctured plabic tiling. Then we have*

$$Z^{\circ}_{\hat{G}(\mathcal{T})} = \{Y \in \mathrm{Gr}_{k,k+2} \mid \mathrm{sgn}\langle Yij\rangle = (-1)^{\mathrm{area}(i\to j)} \textit{ for all black arcs } i\to j \textit{ of } \mathcal{T} \textit{ with } i<j\}$$

Moreover, if $Y\in Z^{\circ}_{\hat{G}(\mathcal{T})}$, then $C' := C^{\mathrm{tw}}_{\mathcal{T}}(Y)$ (cf. Definition 2.3.17) lies in the positroid cell $S_{\hat{G}(\mathcal{T})}$, and Y and $C'Z$ represent the same element of $\mathrm{Gr}_{k,k+2}$.

In the proof of Theorem 2.3.28, we use the notation $N_i(A) := \#\{a\in A : a<i\}$ and $N_{i,j}(A) := \#\{a \in A : i<a<j\}$. We will need the following lemmas.

Lemma 2.3.29 *Let $S\in\binom{[n]}{k+3}$, and define $\omega^S\in\mathbb{R}^n$ as*

$$\omega^S_i = \begin{cases} (-1)^{N_i(S)}\langle Z_{S\setminus\{i\}}\rangle & \textit{if } i\in S\\ 0 & \textit{else}\,. \end{cases}$$

Then ω^S is in the left kernel of Z.

Proof We have that

$$(\omega^S)^T \cdot Z = \sum_{i=1}^{n} Z_i \omega_i^S = \sum_{i \in S} (-1)^{N_i(S)} Z_i \langle Z_{S \setminus \{i\}} \rangle = \sum_{i \in S} \epsilon_{\{i\}, S \setminus \{i\}} Z_i \langle Z_{S \setminus \{i\}} \rangle.$$

From the rightmost expression, one can see that the jth coordinate of $\omega^S \cdot Z$ is the determinant of the submatrix of Z using rows S and columns $1, \ldots, j, j, \ldots, k+2$, written using Laplace expansion along column j. Therefore it is zero. □

Proposition 2.3.30 *Let $Y \in \mathrm{Gr}_{k,k+2}$ and let $\mathcal{T}$ be a (k, n)-unpunctured plabic tiling. Let $C_{\mathcal{T}}^{\mathrm{tw}} = C_{\mathcal{T}}^{\mathrm{tw}}(Y)$ be the twistor coordinate matrix of Y and let $Y' := C_{\mathcal{T}}^{\mathrm{tw}} Z$. Then*

$$\mathit{rowspan}(Y') \subseteq \mathit{rowspan}(Y).$$

Proof We start by writing $Y = CZ$, where C is a full-rank $k \times n$ matrix (we can always do this because the linear map $Z : \mathbb{R}^n \to \mathbb{R}^{k+2}$ is surjective).

Let $C_1, \ldots, C_k$ be the rows of the matrix C. We will replace each row C_i with a linear combination of the rows of C *and* a linear combination of elements of $\ker(Z)$ to obtain a new matrix C'; by construction, the rowspan of $C'Z$ is contained in the rowspan of CZ. We will show that this new matrix C' is equal to $C_{\mathcal{T}}^{\mathrm{tw}}$.

Specifically, let $T_i = \{a < b < c\}$ be a triangle in $\mathcal{T}$. The ith row C_i' of C' is

$$C_i' := \sum_{j=1}^{k} \lambda_j C_j + \sum_{S \in \binom{[n]}{k+3}} \rho_S \omega^S, \text{ where} \tag{2.3.31}$$

$$\lambda_j = \sum_{J \in \binom{[n]}{k}} (-1)^{j+1+N_a(J)+N_{b,c}(J)} p_{J \setminus \{a\}}(C_{\hat{j}}) \langle Z_{J \cup \{b,c\}} \rangle \text{ and } \rho_S = (-1)^{N_a(S)+N_{b,c}(S)} p_{S \setminus \{a,b,c\}}(C),$$

where $C_{\hat{j}}$ denotes the matrix obtained from C by removing row j, and we make the convention that $p_{A \setminus B}(C) = 0$ if B is not contained in A, and $\langle Z_{A \cup B} \rangle = 0$ if A intersects B.

Step 1. We first show that $\langle Ybc \rangle = C_{ia}'$. By (2.3.31), we have

$$C_{ia}' := \sum_{j=1}^{k} \lambda_j C_{ja} + \sum_{S \in \binom{[n]}{k+3}} \rho_S \omega_a^S. \tag{2.3.32}$$

Let us expand $\langle Ybc \rangle$ as:

$$\langle Ybc \rangle = \sum_{J \in \binom{[n]}{k}} (-1)^{N_{b,c}(J)} p_J(C) \langle Z_{J \cup \{b,c\}} \rangle. \tag{2.3.33}$$

Call the terms in this sum with $a \in J$ "type A" and the other terms "type B."

When $a \in J$, we can compute $p_J(C)$ by Laplace expansion around column a:

$$p_J(C) = \sum_{j=1}^{k} (-1)^{j+1+N_a(J)} p_{J \setminus \{a\}}(C_{\hat{j}}) \, C_{ja}.$$

Inserting this into the type A terms and summing over J, we obtain the first term in the right hand side of (2.3.32).

For the type B terms, we can change the summation index in (2.3.33) from J to $S = J \cup \{a, b, c\}$, obtaining:

$$\sum_{S \in \binom{[n]}{k+3}} (-1)^{N_{b,c}(S \setminus \{a,b,c\})} p_{S \setminus \{a,b,c\}}(C) \langle Z_{S \setminus \{a\}} \rangle$$

Since $a < b < c$, we have $N_{b,c}(S \setminus \{a, b, c\}) = N_{b,c}(S)$. This gives the second term in the right hand side of (2.3.32). Hence, summing the terms of type A and type B we get exactly C'_{ia}.

Step 2. We will show that $\langle Yac \rangle = -C'_{ib}$.

Let us consider the first term ('type A') in the right hand side of (2.3.31). We observe that:

$$\sum_{j=1}^{k} (-1)^{j+1+N_a(J)} p_{J \setminus \{a\}}(C_{\hat{j}}) C_{jb} = p_J(C^{a \to b}) = (-1)^{N_{a,b}(J)} p_{J \setminus \{a\} \cup \{b\}}(C),$$

where $C^{a \to b}$ is the matrix C with column a substituted with column b. Noting that $N_{a,b}(J) + N_{b,c}(J) = N_{a,c}(J)$ as terms with $b \in J$ do not contribute, type A reads:

$$\sum_{J \in \binom{[n]}{k}} (-1)^{N_{a,c}(J)} p_{J \setminus \{a\} \cup \{b\}}(C) \langle Z_{J \cup \{b,c\}} \rangle$$

Finally, we change summation index into $J' = J \setminus \{a\} \cup \{b\}$ and use $N_{a,c}(J') = N_{a,c}(J' \setminus \{b\} \cup \{a\}) + 1$, as $b \in J'$ and $a < b < c$, to obtain:

$$- \sum_{J' \in \binom{[n]}{k} : b \in J'} (-1)^{N_{a,c}(J')} p_{J'}(C) Z_{J' \cup \{a,c\}}. \tag{2.3.34}$$

Let us consider the first term ('type B') in the right hand side of (2.3.31). Using $N_b(S) - N_a(S) = N_{a,b}(S) - 1$ and $N_{a,b}(S) + N_{b,c}(S) = N_{a,c}(S) - 1$, as $a, b \in S$, type B reads:

$$\sum_{S \in \binom{[n]}{k+3}} (-1)^{N_{a,c}(S)} p_{S \setminus \{a,b,c\}}(C) \langle Z_{S \setminus \{b\}} \rangle$$

Finally, we perform the change of summation index into $J = S \setminus \{a, b, c\}$ and note that $N_{a,c}(S) = N_{a,c}(J \cup \{a, b, c\}) = N_{a,c}(J) + 1$, as $b \notin J$ and $a < b < c$. We obtain:

$$- \sum_{J \in \binom{[n]}{k} : b \notin J} (-1)^{N_{a,c}(J)} p_J(C) \langle Z_{J \cup \{a,c\}} \rangle. \qquad (2.3.35)$$

Hence adding together (2.3.34) and (2.3.35) we immediately get $-\langle Yac \rangle = C'_{ib}$.

Step 3. Showing that $\langle Yab \rangle = C'_{ic}$ is similar to the previous case.

Step 4. We will show that $C'_{i\ell} = 0$ for $\ell \notin \{a, b, c\}$.

Let us consider the first term ('type A') in the right hand side of (2.3.31). We observe that:

$$\sum_{j=1}^{k} (-1)^{j+1+N_a(J)} p_{J \setminus \{a\}}(C_{\hat{j}}) C_{j\ell} = p_J(C^{a \to \ell}) = (-1)^{\tilde{N}_{a,\ell}(J)} p_{J \setminus \{a\} \cup \{\ell\}}(C),$$

where $\tilde{N}_{a,\ell}(J)$ is defined as $N_{a,\ell}(J)$ if $a < \ell$ and $N_{\ell,a}(J)$ if $\ell < a$. Then type A reads:

$$\sum_{J \in \binom{[n]}{k}} (-1)^{N_{b,c}(J) + \tilde{N}_{a,\ell}(J)} p_{J \setminus \{a\} \cup \{\ell\}}(C) Z_{J \cup \{b,c\}}.$$

By changing the summation index into $J' = J \setminus \{a\} \cup \{\ell\}$ and noting that $N_{b,c}(J' \setminus \{\ell\} \cup \{a\}) = N_{b,c}(J' \setminus \{\ell\})$ and $\tilde{N}_{a,\ell}(J' \setminus \{\ell\} \cup \{a\}) = \tilde{N}_{a,\ell}(J')$, we obtain:

$$\sum_{J' \in \binom{[n]}{k}} (-1)^{N_{b,c}(J' \setminus \{\ell\}) + \tilde{N}_{a,\ell}(J')} p_{J'}(C) \langle Z_{J' \setminus \{\ell\} \cup \{a,b,c\}} \rangle. \qquad (2.3.36)$$

The Type B term can be rewritten as:

$$- \sum_{S \in \binom{[n]}{k+3}} (-1)^{N_{b,c}(S) + \tilde{N}_{a,\ell}(S)} p_{S \setminus \{a,b,c\}}(C) \langle Z_{S \setminus \{\ell\}} \rangle$$

using $(-1)^{N_a(S)+N_\ell(S)} = (-1)^{\tilde{N}_{a,\ell}(S)+1}$. Indeed, $N_a(S) - N_\ell(S) = N_{\ell,a}(S) + 1$ if $\ell < a$ and $N_\ell(S) - N_a(S) = N_{a,\ell}(S) + 1$ if $\ell > a$, since $\ell, a \in S$. Finally, we perform the change of variables $J' = S \setminus \{a, b, c\}$ and note that $N_{b,c}(J' \cup \{a, b, c\}) = N_{b,c}(J')$ and $\tilde{N}_{a,\ell}(J' \cup \{a, b, c\}) = \tilde{N}_{a,\ell}(J' \cup \{b, c\})$, as $a < b < c$, obtaining:

$$- \sum_{J' \in \binom{[n]}{k}} (-1)^{N_{b,c}(J')+\tilde{N}_{a,\ell}(J' \cup \{b,c\})} p_{J'}(C) \langle Z_{J' \setminus \{\ell\} \cup \{a,b,c\}} \rangle. \tag{2.3.37}$$

In order to complete the proof, we need to show that the sum of type A in (2.3.36) with type B in (2.3.37) is zero. Therefore it is enough to show that:

$$(-1)^{N_{b,c}(J')+\tilde{N}_{a,\ell}(J' \cup \{b,c\})} = (-1)^{N_{b,c}(J' \setminus \{\ell\})+\tilde{N}_{a,\ell}(J')}, \tag{2.3.38}$$

recalling that the only terms contributing have $\ell \in J'$ and $a, b, c \notin J'$. If $\ell < b$, then $\tilde{N}_{a,\ell}(J' \cup \{b, c\}) = \tilde{N}_{a,\ell}(J')$ and $N_{b,c}(J' \setminus \{\ell\}) = N_{b,c}(J')$. If $b < \ell < c$, then $\tilde{N}_{a,\ell}(J' \cup \{b, c\}) = \tilde{N}_{a,\ell}(J') + 1$ and $N_{b,c}(J' \setminus \{\ell\}) = N_{b,c}(J') - 1$. Finally, if $\ell > c$ then $\tilde{N}_{a,\ell}(J' \cup \{b, c\}) = \tilde{N}_{a,\ell}(J') + 2$ and $N_{b,c}(J' \setminus \{\ell\}) = N_{b,c}(J')$. Therefore (2.3.38) holds for all three cases and the proof that $C'_{i\ell} = 0$ when $\ell \notin \{a, b, c\}$ is complete. □

Proof (Proof of Theorem 2.3.28) By (2.3.16), we just need to show the inclusion

$$Z^{\circ}_{\hat{G}(\mathcal{T})} \supseteq \{Y \in \mathrm{Gr}_{k,k+2} \mid \text{(2.4.15) holds for all black chords } h \to j \text{ of } \mathcal{T}.\}$$

We will do this using the twistor coordinate matrix $C' := C^{\mathrm{tw}}_{\mathcal{T}}(Y)$ for Y in the right-hand set. First, we show that $C' \in S_{\hat{G}(\mathcal{T})}$. The nonzero entries of C' correspond to the edges of $\hat{G}(\mathcal{T})$. By Theorem A.0.7 and Remark A.0.8, whether or not C' represents an element of $S_{\hat{G}(\mathcal{T})}$ is just a question of whether or not the nonzero entries have the correct signs. By assumption, the nonzero entries of C' have the same signs as the nonzero entries of the matrix C'' from Proposition 2.3.19. Since the matrix C'' represents an element of $S_{\hat{G}(\mathcal{T})}$, so does C'.

Now, let $Y' := C'Z$. By Proposition 2.3.30, $\mathrm{rowspan} Y' \subseteq \mathrm{rowspan} Y$. Because C' is an element of $\mathrm{Gr}^{\geq 0}_{k,n}$, in fact Y' has rank k, so the two rowspans are equal. Thus, $Y = C'Z$, which shows $Y \in Z^{\circ}_{\hat{G}(\mathcal{T})}$. □

Remark 2.3.39 In the previous proof, the only place that we used the fact that the twistor coordinates $\langle Yhj \rangle$ associated to black arcs had particular signs was in showing that the matrix C' that we constructed has maximal minors all nonnegative (or all nonpositive). We will use this observation in Sect. 7.2, when we show that each positroid tile is the totally positive part of a *cluster variety*.

Corollary 2.3.40 *Let $\mathcal{T}$ be a triangulated (k, n)-unpunctured plabic tiling as in Proposition 2.3.7. The map sending the $k \times n$ matrix $M := M_{\mathcal{T}}(\alpha, \beta, \gamma)$ from* (2.3.8) *representing a point of $S_{\hat{G}(\mathcal{T})} \cong (\mathbb{R}_{>0})^{2k}$ to $Y := MZ \in Z^{\circ}_{\hat{G}(\mathcal{T})}$ is a bijection from $S_{\hat{G}(\mathcal{T})} \cong (\mathbb{R}_{>0})^{2k}$ to $Z^{\circ}_{\hat{G}(\mathcal{T})}$, and we have*

$$\frac{M_{i,b_i}}{M_{i,a_i}} = -\frac{\langle Y a_i c_i \rangle}{\langle Y b_i c_i \rangle} \quad \textit{and} \quad \frac{M_{i,c_i}}{M_{i,a_i}} = \frac{\langle Y a_i b_i \rangle}{\langle Y b_i c_i \rangle} \tag{2.3.41}$$

for all triangles $\{a_i, b_i, c_i\}$ *of* $\mathcal{T}$. *In particular, the* $2k$ *ratios of twistor coordinates* $\{\frac{\langle Ya_ic_i\rangle}{\langle Yb_ic_i\rangle}, \frac{\langle Ya_ib_i\rangle}{\langle Yb_ic_i\rangle}\}$ *are algebraically independent.*

Proof Injectivity follows from Theorem 2.3.25. Surjectivity follows from Theorem 2.3.28. Finally, (2.3.41) follows from Propositions 2.3.7 and 2.3.19. □

2.4 The Equivalence of the Two Definitions of the Amplituhedron $\mathcal{A}_{n,k,2}$

In this section we will give an alternative description of the amplituhedron $\mathcal{A}_{n,k,2}(Z)$ in terms of sign flips of twistor coordinates; this description was conjectured by Arkani-Hamed–Thomas–Trnka [17, (5.6)]. In [17, Sect. 5.4], they sketched an argument that all elements of $\mathcal{A}_{n,k,m}(Z)$ satisfy the sign flip description; a proof using a different argument was independently given in [28, Corollary 3.21]. However, the opposite inclusion remained open. We will complete the proof for $m = 2$ using the results of the previous section. Finally, we will translate the sign-flip characterization of $\mathcal{A}_{n,k,2}(Z)$ into a sign-flip characterization of the $\mathcal{B}$-amplituhedron $\mathcal{B}_{n,k,2}(W)$.

Recall the definition of $\hat{Z}_i$ from Remark 2.1.12.

Theorem 2.4.1 (Sign-flip characterization of $\mathcal{A}_{n,k,2}$)
Fix $k < n$ *and* $Z \in \mathrm{Mat}_{n,k+2}^{>0}$. *Let*

$$\mathcal{F}_{n,k,2}^{\circ}(Z) := \{Y \in Gr_{k,k+2} \mid \langle YZ_iZ_{i+1}\rangle > 0 \text{ for } 1 \le i \le n-1, \text{ and } \langle YZ_n\hat{Z}_1\rangle > 0, \\ \text{and } \operatorname{var}(\langle YZ_1Z_2\rangle, \langle YZ_1Z_3\rangle, \dots \langle YZ_1Z_n\rangle) = k.\}$$

Then $\mathcal{A}_{n,k,2}(Z) = \overline{\mathcal{F}_{n,k,2}^{\circ}(Z)}$.

Proof Let $\mathcal{A}_{n,k,m}^{\circ}(Z) := \tilde{Z}(\mathrm{Gr}_{k,n}^{>0})$. By Remarks 2.1.3 and 2.1.9, $\mathcal{A}_{n,k,2}(Z) = \overline{\mathcal{A}_{n,k,m}^{\circ}(Z)}$.

We first show that $\mathcal{A}_{n,k,2}^{\circ} \subseteq \mathcal{F}_{n,k,2}^{\circ}(Z)$. Suppose that $C \in \mathrm{Gr}_{k,n}^{>0}$ and let $Y := CZ$. Choose $1 \le i \le n-1$, and consider any $J = \{j_1 < \cdots < j_k\} \in \binom{[n]}{k}$. Since Z has maximal minors positive, the sign of $\langle Z_{j_1}, \dots, Z_{j_k}, Z_i, Z_{i+1}\rangle$ is determined by the parity of the number of swaps needed to put the sequence $\{j_1, \dots, j_k, i, i+1\}$ into increasing order. Clearly this number is even, so $\langle Z_{j_1}, \dots, Z_{j_k}, Z_i, Z_{i+1}\rangle \ge 0$ (with equality if $J \cap \{i, i+1\} \ne \emptyset$). Therefore by Lemma 2.2.8, $\langle YZ_iZ_{i+1}\rangle > 0$. The argument that $\langle YZ_n\hat{Z}_1\rangle > 0$ is similar, using the fact that the matrix with rows $Z_2, \dots, Z_n, \hat{Z}_1$ has maximal minors positive. To see that Y satisfies the sign variation condition, see the proof sketch in [17, Sect. 5.4] or [28, Corollary 3.21]. This implies that $\mathcal{A}_{n,k,2}^{\circ} \subseteq \mathcal{F}_{n,k,2}^{\circ}(Z)$ and hence $\mathcal{A}_{n,k,2} \subseteq \overline{\mathcal{F}_{n,k,2}^{\circ}(Z)}$.

For the other direction, we will show that $\mathcal{F}_{n,k,2}^{\circ}(Z) \subseteq \mathcal{A}_{n,k,2}(Z)$. Suppose $Y \in \mathcal{F}_{n,k,2}^{\circ}(Z)$. We want to show that we can write $\operatorname{rowspan} Y = \operatorname{rowspan} CZ$ for some $C \in \mathrm{Gr}_{k,n}^{\ge 0}$.

Since $\langle YZ_1Z_2\rangle > 0$, and $\operatorname{var}(\langle YZ_1Z_2\rangle, \langle YZ_1Z_3\rangle, \dots \langle YZ_1Z_n\rangle) = k$, we can find a sequence $1 = i_0 < i_1 < \dots < i_k \le n-1$ such that $\operatorname{sgn}\langle YZ_1Z_{i_\ell+1}\rangle = (-1)^\ell$ for all ℓ; choose the lexicographically minimal such sequence. Let $\mathcal{T}$ be the unpunctured triangulated plabic tiling in an n-gon consisting of the k triangles with vertices $\{1, i_\ell, i_\ell+1\}$ for $1 \le \ell \le k$. By Proposition 2.3.30, if we let $C_\mathcal{T}^{\mathrm{tw}} = C_\mathcal{T}^{\mathrm{tw}}(Y)$ be the twistor coordinate matrix of Y, and $Y' := C_\mathcal{T}^{\mathrm{tw}} Z$, then $\operatorname{rowspan}(Y') \subseteq \operatorname{rowspan}(Y)$. To complete the proof, we need to show that $C_\mathcal{T}^{\mathrm{tw}} \in \mathrm{Gr}_{k,n}^{\ge 0}$, and Y' has full rank.

Using Theorem A.0.7 (as in the proof of Theorem 2.3.28), $C_\mathcal{T}^{\mathrm{tw}}(Y)$ is a Kasteleyn matrix associated to the bipartite graph obtained from $\hat{G}(\mathcal{T})$, as in Fig. 2.4. (Some of the twistor coordinates $\langle YZ_1Z_i\rangle$ of Y may vanish, in which case we just erase some of the edges of the bipartite graph.) If none of the twistor coordinates vanish, Theorem 2.3.28 implies that all nonzero minors of $C_\mathcal{T}^{\mathrm{tw}}(Y)$ have the same sign. Erasing some of the edges of the bipartite graph preserves this property. We now claim that $C_\mathcal{T}^{\mathrm{tw}}$ has full-rank. To see this, note that if we let $I := \{i_1, \dots, i_k\}$, then $p_I(C_\mathcal{T}^{\mathrm{tw}}) \neq 0$. This is because when we restrict to columns $i_1, \dots, i_k$, the only nonzero entry in column i_ℓ (for $1 \le \ell \le k$) is the entry $\langle YZ_1Z_{i_\ell+1}\rangle$ in row ℓ, which has sign $(-1)^\ell$. Therefore $C_\mathcal{T}^{\mathrm{tw}} \in \mathrm{Gr}_{k,n}^{\ge 0}$, so $Y' = C_\mathcal{T}^{\mathrm{tw}} Z$ has full rank. □

Corollary 2.4.2 *Fix $k < n$, $m = 2$, and $Z \in \mathrm{Mat}_{n,k+2}^{>0}$. For any a with $1 \le a \le n$, we define*

$$\mathcal{F}_{n,k,2}^{\circ,a}(Z) = \{Y \in Gr_{k,k+2} \mid \langle YZ_iZ_{i+1}\rangle > 0 \text{ for } 1 \le i \le n-1, \text{ and } \langle YZ_n\hat{Z}_1\rangle > 0, \text{ and}$$
$$\operatorname{var}(\langle YZ_aZ_{a+1}\rangle, \dots \langle YZ_aZ_n\rangle, \langle YZ_a\hat{Z}_1\rangle, \dots, \langle YZ_a\hat{Z}_{a-1}\rangle) = k.\}$$

We have $\mathcal{A}_{n,k,2}(Z) = \overline{\mathcal{F}_{n,k,2}^{\circ,a}(Z)} = \overline{\mathcal{F}_{n,k,2}^{\circ}(Z)}$.

Proof The proof is nearly the same as the one for Theorem 2.4.1. To adapt it, in the second paragraph of that proof, we choose the sequence $i_0 < i_1 < \dots < i_k \le n-1$ based on examining the signs of the sequence $(\langle YZ_aZ_{a+1}\rangle, \dots \langle YZ_aZ_n\rangle, \langle YZ_a\hat{Z}_1\rangle, \dots, \langle YZ_a\hat{Z}_{a-1}\rangle)$. We then use the triangulated unpunctured plabic tiling consisting of the k triangles with vertices $\{a, i_\ell, i_\ell+1\}$ for $1 \le \ell \le k$. □

By combining 2.2.3 with Theorem 2.4.1 (or Corollary 2.4.2), we can obtain a sign-flip characterization of the $\mathcal{B}$-amplituhedron $\mathcal{B}_{n,k,2}(W)$ (see Definition 2.2.2).

Corollary 2.4.3 *Fix $k < n$ and $W \in \mathrm{Gr}_{k+2,n}^{>0}$. Let*

$$\mathcal{G}_{n,k,2}^{\circ}(W) := \{X \in \mathrm{Gr}_2(W) \mid p_{i,i+1}(X) > 0 \text{ for } 1 \le i \le n-1, \text{ and } p_{n,\hat{1}}(X) > 0,$$
$$\text{and } \operatorname{var}((p_{12}(X), p_{13}(X), \dots p_{1n}(X)) = k\}$$

where for $i < j$, $p_{j\hat{i}}(X) := (-1)^k p_{ij}(X)$. Then $\mathcal{B}_{n,k,2}(W) = \overline{\mathcal{G}_{n,k,2}^{\circ}(W)}$.

The set $\mathcal{G}_{n,k,2}^{\circ}(W)$ should agree with the set $\mathcal{G}$ from [28, Prop 3.20] when $m = 2$. For general m, the sign-characterization of the amplituhedron $\mathcal{A}_{n,k,m}$ remains still a conjecture.

Lemma 2.4.4 *Let $Z \in \mathrm{Mat}^{>0}_{n,k+m}$ and $Y = CZ$ for $C \in \mathrm{Gr}^{>0}_{k,n}$. Let $r = \lfloor \frac{m}{2} \rfloor$. Then if m is even:*

$$\langle YZ_I \rangle > 0, \quad I = \{i_1 < i_1 + 1 < i_2 < i_2 + 1 < \ldots < i_r < i_r + 1\} \in \binom{[n]}{m};$$

if m is odd:

$$(-1)^k \langle YZ_{\{1\}\cup I} \rangle > 0, \; \langle YZ_{I\cup\{n\}} \rangle > 0, \quad I = \{i_1 < i_1 + 1 < i_2 < i_2 + 1 < \ldots < i_r < i_r + 1\} \in \binom{[n]}{m-1}.$$

Conjecture 2.4.5 *(Sign-flip characterization of $\mathcal{A}_{n,k,m}$,[17])* Let $Z \in \mathrm{Mat}^{>0}_{n,k+m}$ and define

$$\mathcal{F}^{\circ}_{n,k,m}(Z) := \{Y \in \mathrm{Gr}_{k,k+m} \mid \text{the conclusions of Lemma 2.5.4 hold, and}$$
$$\mathrm{var}(\langle YZ_1 \ldots Z_{m-1} Z_m \rangle, \langle YZ_1 \ldots Z_{m-1} Z_{m+1} \rangle, \ldots, \langle YZ_1 \ldots Z_{m-1} Z_n \rangle) = k.\}$$

Then $\mathcal{A}_{n,k,m}(Z) = \overline{\mathcal{F}^{\circ}_{n,k,m}(Z)}$.

Remark 2.4.6 The fact that $\mathcal{A}_{n,k,m}(Z) \subseteq \overline{\mathcal{F}^{\circ}_{n,k,m}(Z)}$ was sketched in [17] and independently proved in [28, Corollary 3.21]. Conjecture 2.4.5 is strongly believed to hold for $m = 4$. However, in light of some work in progress [48], we expect it to fail for higher m.

2.5 Scattering Amplitudes from $\mathcal{A}_{n,k,4}$

The $m = 4$ amplituhedron $\mathcal{A}_{n,k,4}$ encodes the N^{k}MHV n-particle tree-level amplitudes in $\mathcal{N} = 4$ SYM. The starting point is the space of momentum twistors (see Sect. 1.4.1), $z_i = (\lambda_i, \tilde{\mu}_i) \in \mathbb{CP}^3$, together with the Grassmann odd parameters $\{\chi_i^{\mathcal{R}}\}$, where $\mathcal{R} \in [4]$ is the R-symmetry index. Then the amplitude is a function on n copies of the momentum twistor superspace with coordinates $\mathcal{Z}_i = \{z_i, \chi_i\}$. Let us remark that in this space both momentum and supermomentum conservation are satisfied by construction and the amplitude is a function of degree $4k$ in the χ's. We now introduce a bosonized momentum twistor space by introducing auxiliary Grassmann-odd parameters.

Let us introduce $4k$ auxiliary Grassmann-odd parameters $\phi^{\alpha}_{\mathcal{R}}$, $\alpha \in [k]$, $\mathcal{R} \in [4]$. We define *bosonized momentum twistor variables* $Z \in \mathrm{Mat}_{n,k+4}$ as

$$(Z)_{iA} = \begin{pmatrix} z_i^a \\ \phi^{\alpha} \cdot \chi_i \end{pmatrix}, \qquad A = (a|\alpha) \in [4+k], \; i \in [n]. \tag{2.5.1}$$

In order to make connection with scattering amplitudes we identify the matrices $Z \in \mathrm{Mat}^{>0}_{n,k+4}$ in Definition 2.1.4 with the above matrix. For this reason, we refer to Z as the *kinematic data*.

We refer to Appendix B for more background about this section. The next step is to find the canonical form $\mathbf{\Omega}(\mathcal{A}_{n,k,4})$ of $\mathcal{A}_{n,k,4}$, *i.e.* the differential form with logarithmic singularities on all its boundaries. One way to do this is by finding a positroid tiling of $\mathcal{A}_{n,k,4}$ (see Conjecture B.0.4). Conjectural positroid tilings of $\mathcal{A}_{n,k,4}$ are provided by the so-called BCFW tilings.[7] The canonical form on $\mathcal{A}_{n,k,4}$ is then the sum of push-forwards of canonical form of cells that give such tilings. The result is a sum of rational forms where the denominators can contain spurious singularities, corresponding to spurious boundaries in a given tiling. These singularities disappear in the complete sum and the only divergences of $\mathbf{\Omega}(\mathcal{A}_{n,k,4})$ correspond to the facets of $\mathcal{A}_{n,k,4}$.

Let us now describe how to obtain the amplitude $\mathcal{A}_{\mathrm{N}^k\mathrm{MHV}_n}$ in super momentum twistors from the canonical form $\mathbf{\Omega}(\mathcal{A}_{n,k,4})$. Let us recall that the amplituhedron $\mathcal{A}_{n,k,4}$ is $4k$-dimensional and therefore the degree of $\mathbf{\Omega}(\mathcal{A}_{n,k,4})$ is $4k$, which is top-dimensional in $\mathrm{Gr}_{k,k+4}$. Therefore we can write:

$$\mathbf{\Omega}(\mathcal{A}_{n,k,4}) = \mu_{\mathrm{Gr}_{k,k+4}}(Y)\, \Omega(\mathcal{A}_{n,k,4}), \tag{2.5.2}$$

where we refer to $\Omega(\mathcal{A}_{n,k,4})$ as the *canonical function* of $\mathcal{A}_{n,k,4}$. Then, in order to extract the amplitude from the canonical form $\mathbf{\Omega}(\mathcal{A}_{n,k,4})$ one has to localize Y on a reference subspace

$$Y^* = \begin{pmatrix} 0_{4\times k} \\ \hline 1_{k\times k} \end{pmatrix}, \tag{2.5.3}$$

and integrate out the auxiliary Grassmann-odd variables ϕ's:

$$\mathcal{A}_{\mathrm{N}^k\mathrm{MHV}_n}(\mathcal{Z}) = \int d\phi_a^1 \ldots d\phi_a^k\, \Omega_{n,k}(\mathcal{A}_{n,k,4})|_{Y=Y^*}\,, \tag{2.5.4}$$

In the following section we will show how extracting the amplitude works in practice in a few examples.

Example 2.5.5 (NMHV$_n$ Amplitudes from cyclic polytopes $\mathcal{A}_{n,1,4} \subset \mathbb{P}^4$) It is known that any cyclic polytope $\mathcal{A}_{n,1,4}(Z) \subset \mathbb{P}^4$ with n vertices Z_i, $i \in [n]$ can be triangulated by the following collection of simplices:

$$\{\Delta_{1,i,i+1,j,j+1}(Z)\}_{2\leq i<i+1<j\leq n-1}, \quad \Delta_I(Z) := \mathrm{convex}\{Z_i\}_{i\in I}, \quad I \in \binom{[n]}{5}. \tag{2.5.6}$$

Moreover, the canonical form of each simplex is:

$$\mathbf{\Omega}(\Delta_I(Z)) = \frac{\langle Z_I\rangle^4}{\prod_{i\in I}\langle Y, Z_{I\setminus\{i\}}\rangle}\mu_{\mathbb{P}^4}(Y). \tag{2.5.7}$$

[7] The package `positroid` Mathematica™ package [49] can provide the collection of BCFW cells in a tiling via the function treeContour[n,k].

Indeed, the facets of Δ_I are convex$\{Z_j\}_{j\in I\setminus\{i\}}$ and they lie on the hyperplanes $\langle Y, Z_{I\setminus\{i\}}\rangle = 0$, with $i \in I$. One can also compute $\mathbf{\Omega}(\Delta_I)(Z)$ by observing that $\Delta_I(Z) = Z_\pi$, where $S_\pi \subset \text{Gr}_{1,n}$ is a positroid cell whose elements can be represented as $C = (c_1, \ldots, c_n)$, with $c_i \neq 0$ if $i \in I$ and $c_i = 0$ otherwise. Therefore:

$$\mathbf{\Omega}(\mathcal{A}_{n,1,4}(Z)) = \sum_{2\leq i<i+1<j\leq n-1} \mathbf{\Omega}(\Delta_I(Z)). \tag{2.5.8}$$

If we identify the matrix Z with the kinematic data in Eq. 2.5.1, then:

$$\int d^4\phi\ \Omega(\Delta_I(Z))|_{Y=Y^*} = \frac{\delta^{0|4}\left(\sum_{i\in I}\epsilon_{i,I\setminus\{i\}}\langle z_{I\setminus\{i\}}\rangle\chi_i\right)}{\prod_{i\in I}\langle z_{I\setminus\{i\}}\rangle} = R_I(\mathcal{Z}), \tag{2.5.9}$$

where $R_I(\mathcal{Z})$ are the R-invariants—the building blocks of NMHV amplitudes—in super momentum twistors $\mathcal{Z}$. Applying formula Eq. 2.5.4, we then have:

$$\mathcal{A}_{\text{NMHV}_n}(\mathcal{Z}) = \int d^4\phi\Omega_{n,k}(\mathcal{A}_{n,1,4}(Z))|_{Y=Y^*} = \sum_{2\leq i<i+1<j\leq n-1} R_{1,i,i+1,j,j+1}(\mathcal{Z}), \tag{2.5.10}$$

which is known to be a representation of the NMHV$_n$ amplitude (cf. Eq. 1.4.19). ◇

Yangian Invariants from the Amplituhedron. Tree-level scattering amplitudes in $\mathcal{N} = 4$ SYM can be written as sums of rational terms called *Yangian invariants* (see for example [25, 50–56]). They can be obtained using BCFW recursion relations (see Sect. 1.4.2) or from Grassmannian formulations [1, 2]. According to the latter, a given Yangian invariant $\mathcal{Y}_a$ relevant for amplitudes $\text{A}_{\text{N}^k\text{MHV}_n}$ can be obtained from a Grassmannian integral[8] over a cycle which corresponds to a positroid cell S_a in $\text{Gr}_{k,n}^{\geq 0}$. In the framework of positive geometries—see also Fig. 2.5—such Yangian invariant $\mathcal{Y}_a$ can be extracted directly from the canonical form of the corresponding positroid tile $\tilde{Z}(\bar{S}_a)$ in the amplituhedron $\mathcal{A}_{n,k,4}$ (similarly for the amplitude using Eq. 2.5.4). Yangian invariants $\mathcal{Y}_a$ have *poles* where certain polynomials in momentum twistors $P(\langle ijls\rangle)$ vanish. They correspond to *facets* of the positroid tile $\tilde{Z}(\bar{S}_a)$, which lie on the vanishing locus of the same polynomials in twistor coordinates $P(\langle Yijls\rangle)$.

We note that positroid cells corresponding to positroid tiles are conjectured to have *intersection number* one, see [25, 57] for more details. Their corresponding Yangian invariants were called *rational* in [58] and are the building blocks for tree-level scattering amplitudes. Whereas, positroid cells with intersection number *higher* than one, are conjectured not to be mapped injectively in the amplituhedron: points in the image have a finite number (bigger than one) of pre-images in $\text{Gr}_{k,n}^{\geq 0}$. Yangian invariants associated with this type of cells do not enter representations of scattering amplitudes at tree-level, but are relevant for *Leading Singularities* (cf. Sect. 7.3.2).

---

[8] In [55], it was indeed shown that the integral enjoys an infinite dimensional symmetry, which is the Yangian of $psu(2, 2|2)$, and is simply called *the Yangian* in literature on scattering amplitudes (see Sect. 2.1.2). This symmetry is the hallmark of integrability of $\mathcal{N} = 4$ SYM.

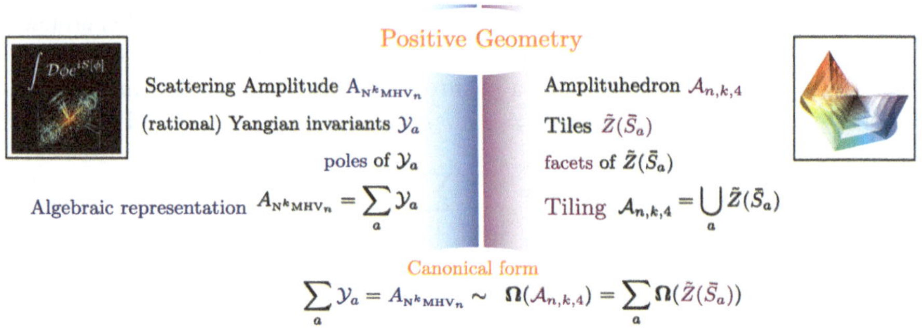

Fig. 2.5 Dictionary between scattering amplitudes and the positive geometry of the amplituhedron

These Yangian invariants can be written as a sum of terms which in general are algebraic (e.g. contain square-roots), but the sum is still rational. See Sect. 7.1.3 for a relevant example.

Integral Representation. We can introduce an integral representation of the canonical function $\Omega(\mathcal{A}_{n,k,4})$ as a contour integral (cf. Appendix B.0.14). Let us fix a point Y in $\mathcal{A}_{n,k,4}$ and let $\omega(\mathcal{A}_{n,k,4})$ and α be differential forms on $\text{Gr}_{k,n}^{\geq 0}$ such that:

$$\Omega(\text{Gr}_{k,n}^{\geq 0}) = \omega(\mathcal{A}_{n,k,4}) \wedge \mu^*_{\text{Gr}_{k,k+4}} + \alpha \tag{2.5.11}$$

and the restriction of α vanishes on the amplituhedron fiber $\tilde{Z}^{-1}(Y)$. Here $\mu^*_{\text{Gr}_{k,k+4}}$ is the pull-back of $\mu_{\text{Gr}_{k,k+4}}(Y)$ under $\tilde{Z}$. Then the *fiber volume form* of $\mathcal{A}_{n,k,4}$ of degree $k(n-k-4)$ is the restriction of $\omega(\mathcal{A}_{n,k,4})$ on the fiber $\tilde{Z}^{-1}(Y)$. By choosing an appropriate contour γ, one obtains the canonical function as a contour integral of the fiber volume form (see Appendix B.0.12):

$$\Omega(\mathcal{A}_{n,k,4})(Y) = \int_\gamma \omega(\mathcal{A}_{n,k,4})|_{\tilde{Z}^{-1}(Y)}. \tag{2.5.12}$$

Remark 2.5.13 As explained in [16], this procedure is equivalent to the formulas in [59], where $\Omega(\mathcal{A}_{n,k,4})$ is obtained from the following contour integral:

$$\int_\gamma \mathbf{\Omega}(\text{Mat}_{k,n}^{\geq 0})\, \delta^{k(k+4)}(Y - CZ)\,, \tag{2.5.14}$$

where

$$\mathbf{\Omega}(\text{Mat}_{k,n}^{\geq 0}) := \frac{d^{k \cdot n} C}{p_{[1,k]}(C)\, p_{[2,k+1]}(C) \ldots p_{[n,k-1]}(C)}. \tag{2.5.15}$$

References

1. N. Arkani-Hamed, F. Cachazo, C. Cheung, J. Kaplan, A duality for the S matrix. JHEP **03**, 020 (2010)
2. M. Bullimore, L.J. Mason, D. Skinner, Twistor-strings, grassmannians and leading singularities. JHEP **03**, 070 (2010)
3. N. Arkani-Hamed, J. Bourjaily, F. Cachazo, A. Goncharov, A. Postnikov, J. Trnka, *Grassmannian Geometry of Scattering Amplitudes*. (Cambridge University Press, 2016)
4. A. Hodges, Eliminating spurious poles from gauge-theoretic amplitudes. JHEP **05**, 135 (2013)
5. N. Arkani-Hamed, J. Trnka, The amplituhedron. J. High Energy Phys. **10**, 33 (2014)
6. Kojima, R., Langer, C. Sign flip triangulations of the amplituhedron. J. High Energ. Phys. **2020**, 121 (2020). https://doi.org/10.1007/JHEP05(2020)121
7. S. Caron-Huot, F. Coronado, B. Mühlmann, Determinants in self-dual $N = 4$ SYM and twistor space. **4** (2023)
8. B. Basso, A.G. Tumanov, Wilson Loop Duality and OPE for Super Form Factors of Half-BPS Operators. arXiv preprint http://arxiv.org/abs/2308.08432, (2023)
9. J. Drummond, J. Foster, Ö. Gürdoğan, Cluster adjacency properties of scattering amplitudes in $N = 4$ supersymmetric Yang-Mills theory. Phys. Rev. Lett. **120**(16), 161601 (2018)
10. J. Drummond, J. Foster, Ö. Gürdoğan, Cluster adjacency beyond MHV. JHEP **03**, 086 (2019)
11. E. Herrmann, C. Langer, J. Trnka, M. Zheng, Positive geometry, local triangulations, and the dual of the Amplituhedron. JHEP **01**, 035 (2021)
12. L.F. Alday, R. Roiban, Scattering amplitudes, wilson loops and the string/gauge theory correspondence. Phys. Rept. **468**, 153–211 (2008)
13. N. Berkovits, J. Maldacena, Fermionic T-duality, dual superconformal symmetry, and the amplitude/wilson loop connection. JHEP **09**, 062 (2008)
14. P. Galashin, *Critical Varieties in the Grassmannian* (2021)
15. M. Parisi, M. Sherman-Bennett, L. Williams, The $m = 2$ amplituhedron and the hypersimplex: signs, clusters, tilings, Eulerian numbers" 2023. Accepted and to be published in *Comm. Amer. Math. Soc. 3* (2023), 329–399. https://doi.org/10.1090/cams/23
16. F. Mohammadi, L. Monin, M. Parisi, Triangulations and canonical forms of amplituhedra: A fiber-based approach beyond polytopes. Commun. Math. Phys. **8** (2021)
17. N. Arkani-Hamed, H. Thomas, J. Trnka, Unwinding in binary.' Preprint http://arxiv.org/abs/1704.05069
18. A. Postnikov, Total positivity, Grassmannians, and networks. Preprint http://math.mit.edu/~apost/papers/tpgrass.pdf
19. G. Lusztig, Total positivity in reductive groups, in *Lie Theory and Geometry*, vol. 123 of *Progr. Math.* (Birkhäuser Boston, Boston, MA, 1994), pp. 531–568
20. K.C. Rietsch, *Total Positivity and Real Flag Varieties*. Ph.D. thesis (Massachusetts Institute of Technology, 1998)
21. K. Rietsch, *Private Communication* (2009)
22. K. Talaska, L. Williams, Network parametrizations for the grassmannian. Algebra Numb. Theor. **7**(9), 2275–2311 (2013)
23. T. Lam, *Totally nonnegative Grassmannian and Grassmann polytopes, in Current Developments in Mathematics 2014* (Int. Press, SÖmerville, MA, 2016), pp.51–152
24. G. Lusztig, *On the Totally Positive Grassmannian* (2019)
25. N. Arkani-Hamed, J.L. Bourjaily, F. Cachazo, A.B. Goncharov, A. Postnikov, J. Trnka, *Scattering Amplitudes and the Positive Grassmannian*. (Cambridge University Press, 2012)
26. S.N. Karp, Sign variation, the Grassmannian, and total positivity. J. Combin. Theor. Ser. A **145**, 308–339 (2017)
27. B. Sturmfels, Totally positive matrices and cyclic polytopes, in *Proceedings of the Victoria Conference on Combinatorial Matrix Analysis (Victoria, BC, 1987)*, vol. 107, pp. 275–281 (1988)
28. S.N. Karp, L.K. Williams, *The $m = 1$ amplituhedron and cyclic hyperplane arrangements* Int. Math. Res. Not, IMRN (to appear) (2017)

29. T. Łukowski, M. Parisi, M. Spradlin, A. Volovich, Cluster adjacency for $m = 2$ yangian invariants. JHEP **10**, 158 (2019)
30. B. Sturmfels, Totally positive matrices and cyclic polytopes. Linear Algebra Appl. **107**, 275–281 (1988)
31. S. Karp, L. Williams, The amplituhedron and cyclic hyperplane arrangements. Int. Math. Res. Not. **2019**(5), 1401–1462 (2019)
32. S.N. Karp, J. Machacek, Shelling the $m = 1$ amplituhedron. **4** (2021)
33. N. Arkani-Hamed, H. Thomas, J. Trnka, Unwinding in binary. JHEP **01**, 016 (2018)
34. N. Arkani-Hamed, Y. Bai, T. Lam, Positive geometries and canonical forms. J. High Energy Phys. **2017**(11), 39 (2017)
35. S.N. Karp, L.K. Williams, Y.X. Zhang, *Decompositions of Amplituhedra* (2017)
36. P. Galashin, T. Lam, Parity duality for the amplituhedron. Compositio Math. **156**(11), 2207–2262 (2020)
37. P. Galashin, S.N. Karp, T. Lam, The totally nonnegative Grassmannian is a ball. **7** (2017)
38. P.V.M. Blagojević, P. Galashin, N. Palić, G.M. Ziegler, Some more amplituhedra are contractible. ArXiv e-prints arXiv:1806.00827 (2018)
39. C. Even-Zohar, T. Lakrec, R. Tessler, The amplituhedron BCFW triangulation. http://arxiv.org/abs/2112.02703 (2021)
40. R. Britto, F. Cachazo, B. Feng, E. Witten, Direct proof of the tree-level scattering amplitude recursion relation in Yang-Mills theory. Phys. Rev. Lett. **94**(18), 181602–4 (2005)
41. H. Bao, X. He, *The* $m = 2$ *Amplituhedron* (2019)
42. T. Łukowski, *On the Boundaries of the* $m = 2$ *Amplituhedron* (2019)
43. C. Even-Zohar, M. Parisi, M. Sherman-Bennett, R.J. Tessler, T. Lakrec, L. Williams, Cluster algebras and the m = 4 amplituhedron, In preparation
44. J. Golden, A.B. Goncharov, M. Spradlin, C. Vergu, A. Volovich, Motivic amplitudes and cluster coordinates. JHEP **01**, 091 (2014)
45. S. Oh, A. Postnikov, D.E. Speyer, Weak separation and plabic graphs. Proc. Lond. Math. Soc. **110**(3), 721–754 (2015)
46. T. Łukowski, M. Parisi, L.K. Williams, The positive tropical grassmannian, the hypersimplex, and the $m = 2$ amplituhedron. Int. Math. Res. Not. **03**, rnad010 (2023)
47. N.J. Sloane et al., *The On-line Encyclopedia of Integer Sequences* (2003)
48. N. Arkani-Hamed, C. Even-Zohar, M. Parisi, M. Sherman-Bennett, R.J. Tessler, T. Lakrec, L. Williams, The $m = 6$ amplituhedron. In preparation
49. J.L. Bourjaily, Positroids, plabic graphs, and scattering amplitudes in mathematica **12** (2012)
50. L. Mason, D. Skinner, Dual superconformal invariance, momentum twistors and grassmannians. JHEP **0911**, 045 (2009)
51. N. Arkani-Hamed, F. Cachazo, C. Cheung, The grassmannian origin Of dual superconformal invariance. JHEP **1003**, 036 (2010)
52. N. Arkani-Hamed, J. Bourjaily, F. Cachazo, J. Trnka, Unification of residues and grassmannian dualities. J. High Energy Phys. **2011**, 49 (2011)
53. J.M. Drummond, L. Ferro, The yangian origin of the grassmannian integral. JHEP **12**, 010 (2010)
54. S.K. Ashok, E. Dell'Aquila, On the classification of residues of the grassmannian. J. High Energy Phys. **2011**, 97 (2011)
55. J. Drummond, L. Ferro, Yangians, grassmannians and T-duality. JHEP **1007**, 027 (2010)
56. L. Ferro, T. Łukowski, A. Orta, M. Parisi, Yangian symmetry for the tree amplituhedron. J. Phys. A **50**(29), 294005 (2017)
57. T. Lam, *Totally Nonnegative Grassmannian and Grassmann Polytopes* (2015)
58. J. Mago, A. Schreiber, M. Spradlin, A. Volovich, Yangian invariants and cluster adjacency in $\mathcal{N} = 4$ Yang-Mills. JHEP **10**, 099 (2019)
59. L. Ferro, T. Lukowski, M. Parisi, Amplituhedron meets Jeffrey-Kirwan residue. J. Phys. A **52**(4), 045201 (2018)

Chapter 3
The Hypersimplex

In 1987, the foundational work of Gelfand-Goresky-MacPherson-Serganova [1] initiated the study of the Grassmannian and torus orbits in the Grassmannian via the *moment map* and *matroid polytopes*, which arise as moment map images of (closures of) torus orbits. Classifying points of the Grassmannian based on the moment map images of the corresponding torus orbits leads naturally to the *matroid stratification* of the Grassmannian. The moment map image of the entire Grassmannian $\mathrm{Gr}_{k+1,n}$ is the $(n-1)$-dimensional hypersimplex $\Delta_{k+1,n} \subseteq \mathbb{R}^n$, the convex hull of the indicator vectors $e_I \in \mathbb{R}^n$ where $I \in \binom{[n]}{k+1}$. Over the last decades there has been a great deal of work on matroid subdivisions of the hypersimplex [2–4]; these are closely connected to the *tropical Grassmannian* [4–6] and the *Dressian* [6], which parametrizes regular matroidal subdivisions of the hypersimplex.

The matroid stratification of the real Grassmannian is notoriously complicated: Mnev's universality theorem says that the topology of the matroid strata can be as bad as that of any algebraic variety. However, as we saw in Sect. 2.1, there is a subset of the Grassmannian called the *totally nonnegative Grassmannian* or (informally) the *positive Grassmannian* [7, 8], where these difficulties disappear: the restriction of the matroid stratification to the positive Grassmannian gives a cell complex [8–10], whose cells S_π are called *positroid cells* and labelled by (among other things) *decorated permutations*. Since the work of Postnikov [8], there has been an extensive study of *positroids* [11–13] – the matroids associated to the positroid cells. The moment map images of positroid cells are precisely the *positroid polytopes* [14], and as we will discuss in this chapter, the *positive tropical Grassmannian* [15] (which equals the *positive Dressian* [16]) parametrizes the regular positroid subdivisions of the hypersimplex.

Besides the moment map, in Sect. 2.1 we discussed another interesting map on the positive Grassmannian—the *amplituhedron map*. It was introduced by Arkani-Hamed and Trnka [17] in the context of *scattering amplitudes* in $\mathcal{N}=4$ SYM (cf. Chap. 2). In particular, we recall that any $n \times (k+m)$ matrix Z with maximal minors positive induces a map $\tilde{Z}$ from $\mathrm{Gr}_{k,n}^{\geq 0}$ to the Grassmannian $\mathrm{Gr}_{k,k+m}$, whose image has

M. Parisi, *Combinatorial Aspects of Scattering Amplitudes*, Springer Theses,
https://doi.org/10.1007/978-3-031-41069-7_3

full dimension mk and is the *amplituhedron* $\mathcal{A}_{n,k,m}$. The case $m = 4$ is most relevant to physics: in this case, the *BCFW recursions* give rise to collections of $4k$-dimensional cells in $\mathrm{Gr}_{k,n}^{\geq 0}$, whose images "triangulate" the amplituhedron.

Given that the hypersimplex and the amplituhedron are images of the positive Grassmannian, which has a nice decomposition into positroid cells, one can ask the following questions: when does a collection of positroid cells give—via the moment map—a *positroid dissection* of the hypersimplex? By *dissection*, we mean that the images of these cells are disjoint and cover a dense subset of the hypersimplex.[1] And when does a collection of positroid cells give—via the amplituhedron map—a dissection of the amplituhedron? We can also ask about *positroid tilings*, which are dissections in which the moment map (respectively, the amplituhedron map) are injections on the various cells. See Definitions 3.2.1 and 5.1.1.

The combinatorics of positroid tilings for both the hypersimplex and the amplituhedron is very interesting: Speyer's f-vector theorem [4, 18] gives an upper bound on the number of matroid polytopes of each dimension in a matroidal subdivision coming from the tropical Grassmannian. In particular, it says that the number of top-dimensional matroid polytopes in such a subdivision of $\Delta_{k+1,n}$ is at most $\binom{n-2}{k}$. This number is in particular achieved by finest positroid subdivisions [16]. Meanwhile, it is conjectured [19] that the number of cells in a tiling of the amplituhedron $\mathcal{A}_{n,k,m}$ for even m is precisely $M(k, n-k-m, \frac{m}{2})$, where

$$M(a, b, c) := \prod_{i=1}^{a}\prod_{j=1}^{b}\prod_{k=1}^{c} \frac{i+j+k-1}{i+j+k-2}$$

is the number of plane partitions contained in an $a \times b \times c$ box. Note that when $m = 2$, this conjecture says that the number of cells in a tiling of $\mathcal{A}_{n,k,2}$ equals $\binom{n-2}{k}$.

In Chap. 5 we will show that the appearance of the number $\binom{n-2}{k}$ in the context of both the hypersimplex $\Delta_{k+1,n}$ and the amplituhedron $\mathcal{A}_{n,k,2}$ is not a coincidence!

Summary of the Chapter. This chapter is based on the following works by the author: [20, Sect. 3,9] and [21, Sect. 7].

In this chapter we discuss the *hypersimplex* $\Delta_{k+1,n}$—the image under the *moment map* of the totally non-negative Grassmannian $\mathrm{Gr}_{k+1,n}^{\geq 0}$. In Sect. 3.1 we introduce and present new results about the hypersimplex and *positroid polytopes* -images of positroid cells of $\mathrm{Gr}_{k+1,n}^{\geq 0}$ under the moment map. In Sect. 3.1.1 we define $\Delta_{k+1,n}$ and in Sect. 3.1.2 we introduce matroid and positroid polytopes and present a new characterization of the latter. In Sect. 3.1.3 we give a full characterization of positroid polytopes which are images of positroid cells where the moment map is injective. In particular, the full-dimensional ones are referred to as *positroid tiles*—which we show to be in bijection with plabic trees. In Sect. 3.2 we show that the *positive tropical Grassmannain* $\mathrm{Trop}^{+}\,\mathrm{Gr}_{k+1,n}$ exactly regulates the regular positroid subdivisions of $\Delta_{k+1,n}$. In particular, in Sect. 3.2.1 we introduce the tropical Grassmannian, the Dressian, and their positive analogues. In Sects. 3.2.2 and 3.2.3 we show that the fan

[1] We do not put any constraints on how their boundaries match up.

structure of $\text{Trop}^+ \text{Gr}_{k+1,n}$ coincides with the secondary fan for (regular) positroid subdivisions of $\Delta_{k+1,n}$. In particular, maximal cones of $\text{Trop}^+ \text{Gr}_{k+1,n}$ are in bijection with regular positroid tilings of $\Delta_{k+1,n}$. Finally, we provide some data on f-vectors of the secondary fans.

3.1 The Hypersimplex and Positroid Polytopes

In this section we study *positroid polytopes*, which are images of positroid cells of $\text{Gr}_{k+1,n}$ under the moment map $\mu : \text{Gr}_{k+1,n} \to \mathbb{R}^n$. We recall some of the known properties of matroid and positroid polytopes, we give a new characterization of positroid polytopes (Theorem 3.1.14), and we describe when the moment map is an injection on a positroid cell, or equivalently, when the moment map restricts to a homeomorphism from the closure of a positroid cell to the corresponding positroid polytope (see Lemma 3.1.24 and Theorem 3.1.23).

3.1.1 The Hypersimplex

Throughout, for $x \in \mathbb{R}^n$ and $I \subset [n]$, we use the notation $x_I := \sum_{i \in I} x_i$.

Definition 3.1.1 *(The Hypersimplex)* Let $e_I := \sum_{i \in I} e_i \in \mathbb{R}^n$, where $\{e_1, \ldots, e_n\}$ is the standard basis of $\mathbb{R}^n$. The $(k+1, n)$*-hypersimplex* $\Delta_{k+1,n}$ is the convex hull of the points e_I where I runs over $\binom{[n]}{k+1}$.

Remark 3.1.2 Under the projection $p : (x_1, \ldots, x_n) \mapsto (x_1, \ldots, x_{n-1})$, we see that the hypersimplex is linearly equivalent to the polytope in $\mathbb{R}^{n-1}$ defined by

$$\{(x_1, \ldots, x_{n-1}) \mid 0 \leq x_i \leq 1; k \leq x_{[n-1]} \leq k+1\}.$$

This shows that the projected hypersimplex $\Delta_{k+1,n}$ can be thought of as the slice of the unit hypercube $\square_{n-1_{n-1}}$ contained between the two hyperplanes $x_{[n-1]} = k$ and $x_{[n-1]} = k+1$.

Example 3.1.3 *(The hypersimplex $\Delta_{2,4}$)* The vertices of $\Delta_{2,4} \subset \mathbb{R}^4$ in are: $e_{12} = (1, 1, 0, 0)$, $e_{13} = (1, 0, 1, 0)$, $e_{14} = (1, 0, 0, 1)$, $e_{23} = (0, 1, 1, 0)$, $e_{24}{=}(0, 1, 0, 0)$, $e_{34} = (0, 0, 1, 1)$. By Remark 3.1.2, it can be thought of as the slice of the unit cube $\square_3 \subset \mathbb{R}^3$ contained between the two hyperplanes $x_1 + x_2 + x_3 = 1$ and $x_1 + x_2 + x_3 = 2$ (see Fig. 3.1). ◇

The torus $T = \mathbb{R}^n$ acts on $\text{Gr}_{k+1,n}$ by scaling the columns of a matrix representative A. This is an $(n-1)$-dimensional torus since the Grassmannian is a projective variety. We let TA denote the orbit of A under the action of T, and $\overline{TA}$ its closure.

The *moment map* from the Grassmannian $\text{Gr}_{k+1,n}$ to $\mathbb{R}^n$ is defined as follows.

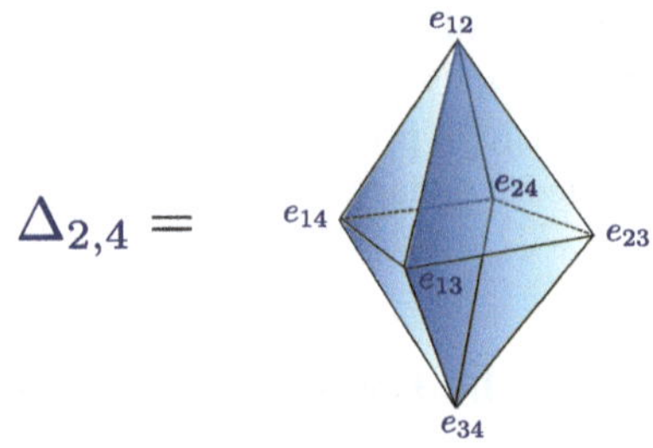

Fig. 3.1 The hypersimplex $\Delta_{2,4}$ projected in $\mathbb{R}^3$

Definition 3.1.4 *(The Moment Map)* Let A be a $(k+1) \times n$ matrix representing a point of $\mathrm{Gr}_{k+1,n}$. The *moment map* $\mu : \mathrm{Gr}_{k+1,n} \to \mathbb{R}^n$ is defined by

$$\mu(A) = \frac{\sum_{I \in \binom{[n]}{k+1}} |p_I(A)|^2 e_I}{\sum_{I \in \binom{[n]}{k+1}} |p_I(A)|^2}.$$

It is well-known that the image of the Grassmannian $\mathrm{Gr}_{k+1,n}$ under the moment map is the hypersimplex $\Delta_{k+1,n}$. If one restricts the moment map to $\mathrm{Gr}_{k+1,n}^{\geq 0}$ then the image is again the hypersimplex $\Delta_{k+1,n}$ [14, Proposition 7.10].

Remark 3.1.5 Results in this chapter that hold for the moment map, analogously hold for the algebraic moment map in [22].

3.1.2 *Matroid and Positroid Polytopes*

In general, it follows from classical work of Atiyah [23] and Guillemin-Sternberg [24] that the image $\mu(\overline{TA})$ is a convex polytope, whose vertices are the images of the torus-fixed points, i.e. the vertices are the points e_I such that $p_I(A) \neq 0$. This motivates the notion of *matroid polytope*. Recall that any full rank $(k+1) \times n$ matrix A gives rise to a matroid $\mathcal{M}(A) = ([n], \mathcal{B})$, where $\mathcal{B} = \{I \in \binom{[n]}{k+1} \mid p_I(A) \neq 0\}$.

Definition 3.1.6 *(Matroid polytopes)* Given a matroid $\mathcal{M} = ([n], \mathcal{B})$, the (basis) *matroid polytope* $\Gamma_{\mathcal{M}}$ of $\mathcal{M}$ is the convex hull of the indicator vectors of the bases of $\mathcal{M}$:

$$\Gamma_{\mathcal{M}} := \mathrm{convex}\{e_B : B \in \mathcal{B}\} \subset \mathbb{R}^n.$$

The following elegant characterization of matroid polytopes is due to Gelfand, Goresky, MacPherson, and Serganova.

Theorem 3.1.7 *([1]) Let $\mathcal{B}$ be a collection of subsets of $[n]$ and let $\Gamma_{\mathcal{B}} := \mathrm{convex}\{e_B \mid B \in \mathcal{B}\} \subset \mathbb{R}^n$. Then $\mathcal{B}$ is the collection of bases of a matroid if and only if every edge of $\Gamma_{\mathcal{B}}$ is a parallel translate of $e_i - e_j$ for some $i, j \in [n]$.*

The dimension of a matroid polytope is determined by the number of connected components of the matroid. Recall that a matroid which cannot be written as the direct sum of two nonempty matroids is called *connected.*

Proposition 3.1.8 ([25]) *Let $\mathcal{M}$ be a matroid on E. For two elements $a, b \in E$, we set $a \sim b$ whenever there are bases $\mathcal{B}_1, \mathcal{B}_2$ of $\mathcal{M}$ such that $\mathcal{B}_2 = (\mathcal{B}_1 - \{a\}) \cup \{b\}$. The relation $\sim$ is an equivalence relation, and the equivalence classes are precisely the connected components of $\mathcal{M}$.*

Proposition 3.1.9 ([26]) *For any matroid, the dimension of its matroid polytope is $\dim \Gamma_{\mathcal{M}} = n - c$, where c is the number of connected components of $\mathcal{M}$.*

Matroid polytopes also have a straightforward description in terms of inequalities.

Proposition 3.1.10 ([27]) *Let $\mathcal{M} = ([n], \mathcal{B})$ be any matroid of rank k, and let $r_{\mathcal{M}} : 2^{[n]} \to \mathbb{Z}_{\geq 0}$ be its rank function. Then the matroid polytope $\Gamma_{\mathcal{M}}$ can be described as*

$$\Gamma_{\mathcal{M}} = \{\mathbf{x} \in \mathbb{R}^n : x_{[n]} = k,\ x_A \leq r_{\mathcal{M}}(A) \text{ for all } A \subset [n]\}$$
$$\Gamma_{\mathcal{M}} = \{\mathbf{x} \in \mathbb{R}^n : x_{[n]} = k,\ x_A \geq k - r_{\mathcal{M}}([n] \setminus A) \text{ for all } A \subset [n]\}.$$

In this work, we are interested in *positroid polytopes*, that is, matroid polytopes $\Gamma_{\mathcal{M}}$ where $\mathcal{M}$ is a positroid. They arise as $\mu(\overline{TA})$ where A is a totally nonnegative matrix. Of more interest to us, they can also be obtained as moment map images of positroid cells. That is, for a positroid cell $S_{\mathcal{M}}$ associated to a positroid $\mathcal{M}$, one can consider the closure of $\mu(S_{\mathcal{M}})$. This coincides exactly with the corresponding positroid polytope:

Proposition 3.1.11 ([14, Proposition 7.10]) *Let $\mathcal{M}$ be the positroid associated to the positroid cell S_π. Then $\Gamma_{\mathcal{M}} = \mu(\overline{S_\pi}) = \overline{\mu(S_\pi)}$.*

The first statement in Theorem 3.1.12 below was proved in [12, Corollary 5.4] (and generalized to the setting of Coxeter matroids in [14, Theorem 7.13].) The second statement follows from the proof of [14, Theorem 7.13].

Theorem 3.1.12 *Every face of a positroid polytope is a positroid polytope. Moreover, every face $\Gamma_{\pi'}$ of a positroid polytope Γ_π has the property that $S_{\pi'} \subset \overline{S_\pi}$.*

There is a simple inequality characterization of positroid polytopes, which specialises the one of matroid polytopes in Proposition 3.1.10.

Proposition 3.1.13 ([12, Proposition 5.7]) *A matroid $\mathcal{M}$ of rank k on $[n]$ is a positroid if and only if its matroid polytope $\Gamma_{\mathcal{M}}$ can be described by the equality $x_{[n]} = k$ and inequalities of the form*

$$x_{[i,j]} \leq r_{\mathcal{M}}([i, j]), \quad \text{with } i, j \in [n].$$

Here $[i, j]$ is the cyclic interval given by $[i, j] = \{i, i+1, \ldots, j\}$ if $i < j$ and $[i, j] = \{i, i+1, \ldots, n, 1, \ldots, j\}$ if $i > j$.

We now give a new characterization of positroid polytopes. In what follows, we use Sab as shorthand for $S \cup \{a, b\}$, etc.

Theorem 3.1.14 *Let $\mathcal{M}$ be a matroid of rank k on the ground set $[n]$, and consider the matroid polytope $\Gamma_{\mathcal{M}}$. It is a positroid polytope (i.e. $\mathcal{M}$ is a positroid) if and only if all of its two-dimensional faces are positroid polytopes.*

Moreover, if $\mathcal{M}$ fails to be a positroid polytope, then $\Gamma_{\mathcal{M}}$ has a two-dimensional face F with vertices $e_{Sab}, e_{Sad}, e_{Sbc}, e_{Scd}$, for some $1 \leq a < b < c < d \leq n$ and S of size $k-2$ disjoint from $\{a, b, c, d\}$.

Remark 3.1.15 We note that a different characterization of positroids in terms of faces of their matroid polytopes was given in [28, Proposition 6.4], see also [28, Lemmas 6.2 and 6.3]. There are also some related ideas in the proof of [29, Lemma 30].

By Theorem 3.1.12, every two-dimensional face of $\Gamma_{\mathcal{M}}$ is a positroid polytope. To prove the other half of Theorem 3.1.14, we use the following lemma.

Lemma 3.1.16 *Let $\mathcal{M}$ be a matroid of rank k on $[n]$ which has two connected components, i.e. $\mathcal{M} = \mathcal{M}_1 \oplus \mathcal{M}_2$ such that the ground sets of $\mathcal{M}_1$ and $\mathcal{M}_2$ are S and $T = [n] \setminus S$. Suppose that $\{S, T\}$ fails to be a noncrossing partition of $[n]$, in other words, there exists $a < b < c < d$ (in cyclic order) such that $a, c \in S$ and $b, d \in T$. Then $\Gamma_{\mathcal{M}}$ has a two-dimensional face which is not a positroid polytope; in particular, that face is a square with vertices $e_{Sab}, e_{Sad}, e_{Sbc}, e_{Scd}$, for some $1 \leq a < b < c < d \leq n$ and S of size $k-2$ disjoint from $\{a, b, c, d\}$.*

Proof By Proposition 3.1.8, we have bases Aa and Ac of $\mathcal{M}_1$ and also bases Bb and Bd of $\mathcal{M}_2$. We can find a linear functional on $\Gamma_{\mathcal{M}_1}$ given by a vector in $\mathbb{R}^S$ whose dot product is maximized on the convex hull of e_{Aa} and e_{Ac} (choose the vector w such that $w_h = 1$ for $h \in A$, $w_h = \frac{1}{2}$ for $h = a$ or $h = c$, and $w_h = 0$ otherwise); therefore there is an edge in Γ_{M_1} between e_{Aa} and e_{Ac}. Similarly, there is an edge in $\Gamma_{\mathcal{M}_2}$ between e_{Bb} and e_{Bd}. Therefore $\Gamma_M = \Gamma_{\mathcal{M}_1} \times \Gamma_{\mathcal{M}_2}$ has a two-dimensional face whose vertices are $e_{ABab}, e_{ABad}, e_{ABbc}, e_{ABcd}$. This is not a positroid polytope because $\{ab, ad, bc, cd\}$ are not the bases of a rank 2 positroid. □

Proposition 3.1.17 *Let $\mathcal{M}$ be a connected matroid. If all of the two-dimensional faces of $\Gamma_{\mathcal{M}}$ are positroid polytopes, then $\Gamma_{\mathcal{M}}$ is a positroid polytope (i.e. $\mathcal{M}$ is a positroid).*

Proof Suppose for the sake of contradiction that $\Gamma_{\mathcal{M}}$ is not a positroid polytope.

Since $\Gamma_{\mathcal{M}}$ is not a positroid polytope, then by Proposition 3.1.13, it has a facet F of the form $\sum_{i \in S} x_i = r_{\mathcal{M}}(S)$, where S is not a cyclic interval. In other words, S and $T = [n] \setminus S$ fail to form a noncrossing partition. Each facet of $\Gamma_{\mathcal{M}}$ is the matroid polytope of a matroid with two connected components, so by the greedy algorithm for matroids (see e.g. [12, Proposition 2.12]), F must be the matroid polytope of $\mathcal{M}|S \oplus \mathcal{M}/S$. But now by Lemma 3.1.16, F has a two-dimensional face which is not a positroid polytope. □

We now complete the proof of Theorem 3.1.14.

Proof We start by writing $\mathcal{M}$ as a direct sum of connected matroids $\mathcal{M} = \mathcal{M}_1 \oplus \cdots \oplus \mathcal{M}_l$. Let $S_1, \dots, S_l$ be the ground sets of $\mathcal{M}_1, \dots, \mathcal{M}_l$. By [12, Lemma 7.3], either one of the $\mathcal{M}_i$'s fails to be a positroid, or $\{S_1, \dots, S_l\}$ fails to be a non-crossing partition of $[n]$. If one of the $\mathcal{M}_i$'s fails to be a positroid, then by Proposition 3.1.17, $\Gamma_{\mathcal{M}_i}$ has a two-dimensional face which fails to be a positroid. But then so does $\Gamma_{\mathcal{M}} = \Gamma_{\mathcal{M}_1} \times \cdots \times \Gamma_{\mathcal{M}_l}$. On the other hand, if $\{S_1, \dots, S_l\}$ fails to be a non-crossing partition of $[n]$, then by Lemma 3.1.16, $\Gamma_{\mathcal{M}}$ has a two-dimensional face which fails to be a positroid. This completes the proof. □

3.1.3 Positroid Tiles of the Hypersimplex

As for the amplituhedron (see Sect. 2.3), we will be particularly interested in the cells on which the moment map is injective. For this purpose, we fix some notation for *positroid polytopes* and we introduce *positroid tiles* for the hypersimplex:

Definition 3.1.18 *(Positroid Polytopes and Positroid Tiles)* Given a positroid cell S_π of $\mathrm{Gr}_{k+1,n}^{\geq 0}$, we let $\Gamma_\pi^\circ = \mu(S_\pi)$ and $\Gamma_\pi = \overline{\mu(S_\pi)}$, and we refer to Γ_π° and Γ_π as *open positroid polytopes* and *positroid polytopes*, respectively. We call Γ_π a *positroid tile* for $\Delta_{k+1,n}$ if $\dim(S_\pi) = n-1$, and μ is injective on S_π.

Example 3.1.19 *(Positroid tiles of $\Delta_{2,4}$)* Let us consider the positroid cell $S_\pi \subset \mathrm{Gr}_{2,4}^{\geq 0}$, with $\pi = 2413$. Its elements can be represented by matrices $C \in \mathrm{Mat}_{2,4}$, with $p_{34}(C) = 0$ and $p_I(C) > 0$, $I \in \binom{[4]}{2} \setminus \{3,4\}$. Then the positroid polytope $\Gamma_\pi = \mu(\bar{S}_\pi) \subset \Delta_{2,4}$ is the convex hull of the vertices $e_{12}, e_{13}, e_{14}, e_{23}, e_{24}$, see Fig. 3.2. Γ_π is a positroid tile of $\Delta_{2,4}$. ◇

Example 3.1.20 *(Non positroid polytope in $\Delta_{2,4}$)* Let us consider the polytope P which is the convex hull of the vectors $\{e_I\}$, with $I \in \binom{[4]}{2} \setminus \{2,4\}$, see Fig. 3.3. By Proposition 3.1.11, if P is a positroid polytope $\Gamma_{\mathcal{M}}$, then the basis $\mathcal{B}$ of the associated positroid $\mathcal{M} = (\mathcal{B}, [4])$ should be $\mathcal{B} = \binom{[4]}{2} \setminus \{2,4\}$. Elements C of the associated

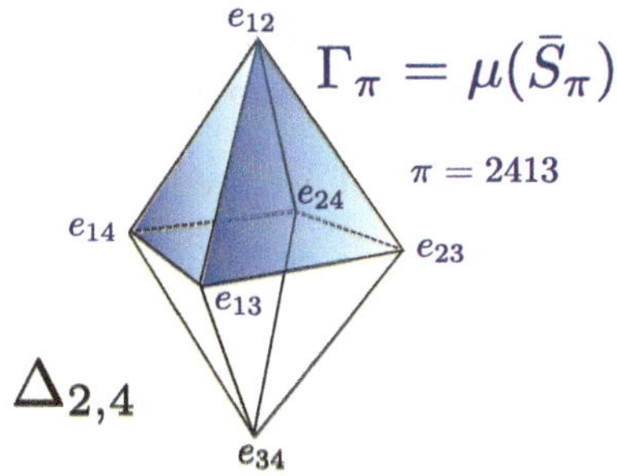

Fig. 3.2 A positroid tile of $\Delta_{2,4}$–a 3-dimensional positroid polytope

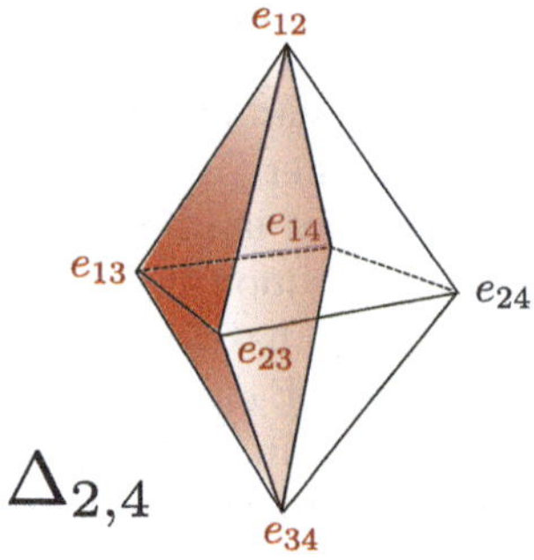

Fig. 3.3 Example of a pyramid in $\Delta_{2,4}$ which is *not* a positroid polytope

positroid cell $S_{\mathcal{M}} \subset \mathrm{Gr}_{2,4}^{\geq 0}$ would have $p_{24}(C) = 0$ and the other Plücker coordinates positive. By the Plücker relation $p_{12}p_{34} - p_{24}p_{13} + p_{14}p_{23} = 0$ this is not possible. Therefore $P = \Gamma_{\mathcal{M}}$ is not a positroid polytope. ◇

The following result comes from [12, Theorem 10.7] and its proof.

We say that a permutation π of $[n]$ is *stabilized-interval-free (SIF)* if it does not stabilize any proper interval of $[n]$; that is, $\pi(I) \neq I$ for all intervals $I \subsetneq [n]$.

Proposition 3.1.21 *Let S_π be a positroid cell of $\mathrm{Gr}_{k+1,n}^{\geq 0}$ and let $\mathcal{M}_\pi$ be the corresponding positroid. Then $\mathcal{M}_\pi$ is connected if and only if π is a SIF permutation of $[n]$. More generally, the number of connected components of $\mathcal{M}_\pi$ equals the number of connected components of any reduced plabic graph associated to π.*

Example 3.1.22 Consider the permutation $\pi = (5, 3, 4, 2, 6, 7, 1)$, which in cycle notation is $(234)(1567)$. Then there are two minimal-by-inclusion cyclic intervals such that $\pi(I) = I$, namely $[2, 4]$ and $[5, 1]$, and hence the matroid M_π has two connected components. (Note that $[1, 7]$ is also a cyclic interval with $\pi([1, 7]) = [1, 7]$ but it is not minimal-by-inclusion.) ◇

Theorem 3.1.23 (Injectivity of Moment Map) *Consider a positroid cell $S_G \subset \mathrm{Gr}_{k+1,n}^{\geq 0}$, with G any reduced plabic graph associated to the positroid cell. Then the moment map is injective on S_G if and only if G is a* forest. *When G is a forest, μ is moreover a stratification-preserving homeomorphism from $\overline{S_G}$ to the polytope $\Gamma_G \subset \mathbb{R}^n$. We have that* $\dim S_G = \dim \Gamma_G = n - c$, *where c is the number of trees comprising the forest G.*

In particular, given an $(n-1)$-dimensional cell $S_G \subset \mathrm{Gr}_{k+1,n}^{\geq 0}$, Γ_G is a positroid tile *for $\Delta_{k+1,n}$ if and only if G is a* tree.

For the proof, we need the following.

Lemma 3.1.24 *Consider a positroid cell $S_\pi \subset \mathrm{Gr}_{k+1,n}^{\geq 0}$ and let $\mathcal{M}_\pi$ be the corresponding positroid. Then the following statements are equivalent:*

1. *the moment map restricts to an injection on S_π*
2. *the moment map is a homeomorphism from $\overline{S_\pi}$ to Γ_π*

3. $\dim S_\pi = \dim \Gamma_\pi = n - c$, *where* c *is the number of connected components of the matroid* $\mathcal{M}_\pi$.

Proof Suppose that (1) holds, i.e. that the moment map is an injection when restricted to a cell S_π. Then $\dim \Gamma_\pi = \dim S_\pi$. By [14, Proposition 7.12], the positroid variety X_π is a toric variety if and only if $\dim \Gamma_\pi = \dim S_\pi$, so this implies that X_π is a toric variety, and $\overline{S_\pi}$ is its nonnegative part. It is well-known that the moment map is a homeomorphism when restricted to the nonnegative part of a toric variety [30, Sect. 4.2], so it follows that μ is a homeomorphism on $\overline{S_\pi}$. Therefore (1) implies (2). But obviously (2) implies (1).

Now suppose that (2) holds. Since Γ_π is the moment map image of $\overline{S_\pi}$, it follows that $\dim \Gamma_\pi = \dim S_\pi$, and by Proposition 3.1.9, we have that $\dim \Gamma_\pi = n - c$, where c is the number of connected components of the matroid $\mathcal{M}_\pi$. Therefore (2) implies (3).

Now suppose (3) holds. Then by [14, Proposition 7.12], X_π must be a toric variety, and so the moment map restricts to a homeomorphism from $\overline{S_\pi}$ to Γ_π. So (3) implies (2). □

Proof (Proof of Theorem 3.1.23) This follows from Lemma 3.1.24 and Proposition 3.1.21, together with the fact that we can read off the dimension of a positroid cell from any reduced plabic graph G for it as the number of regions of G minus 1. □

3.2 The Hypersimplex and the Positive Tropical Grassmannian

The goal of this section is to use the positive tropical Grassmannian to understand the regular positroid subdivisions of the hypersimplex.

The *tropical Grassmannian*—or rather, an outer approximation of it called the *Dressian* – controls the regular matroidal subdivisions of the hypersimplex [2, 4, Proposition 2.2]. There is a positive subset of the tropical Grassmannian, called the *positive tropical Grassmannian*, which was introduced by Speyer and Williams in [15]. The positive tropical Grassmannian equals the positive Dressian [31], and as we will show in Proposition 3.2.14, it controls the regular *positroid* subdivisions of the hypersimplex.

Definition 3.2.1 *(Positroid Tilings/Dissections of* $\Delta_{k+1,n}$*)* Let $\mathcal{C} = \{\Gamma_\pi\}$ be a collection of positroid polytopes, with $\{S_\pi\}$ positroid cells of $\mathrm{Gr}_{k+1,n}^{\geq 0}$. We say that $\mathcal{C}$ is a *positroid tiling* of $\Delta_{k+1,n}$ if we have that:

1. each Γ_π is a positroid tile (μ is injective on S_π, and $\dim S_\pi = n - 1$)
2. pairs of distinct open positroid polytopes Γ_π° and $\Gamma_{\pi'}^\circ$ in the collection are disjoint
3. $\bigcup_\pi \Gamma_\pi = \Delta_{k+1,n}$.

We say that $\mathcal{C}$ is a *positroid dissection* of $\Delta_{k+1,n}$ if each positroid polytope Γ_π is full-dimensional (i.e. $\dim \Gamma_\pi = n - 1$) and (2) and (3) are satisfied.

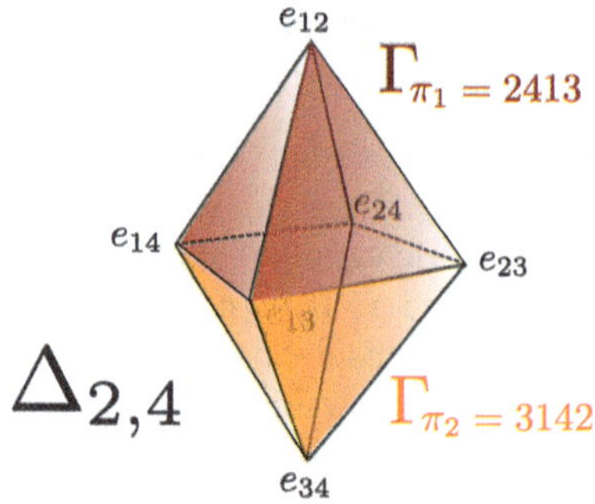

Fig. 3.4 A positroid tiling of $\Delta_{2,4}$

Example 3.2.2 *(Positroid tiling of $\Delta_{2,4}$)* Figure 3.4 shows a positroid tiling $\{\Gamma_{\pi_1}, \Gamma_{\pi_2}\}$ of $\Delta_{2,4}$, where Γ_{π_1} is the positroid tile considered in Example 3.1.19, and Γ_{ν_2} is the positroid tile obtained from the positroid cell S_{π_2}. Elements of the cell can be represented by a matrix $C \in \text{Mat}_{2,4}$, with $p_{12}(C) = 0$ and $p_I > 0$, $I \in \binom{[4]}{2}$. Therefore Γ_{π_2} is the positroid polytope with vertices $e_{13}, e_{14}, e_{23}, e_{24}, e_{34}$. ◇

Remark 3.2.3 We have learned since circulating the first draft of the paper [20] from which this section is based, that some of our results in this section regarding positroid subdivisions of the hypersimplex and the positive tropical Grassmannian, though not previously in the literature, were known or anticipated by various other experts including David Speyer, Nima Arkani-Hamed, Thomas Lam, Marcus Spradlin, Nick Early, Felipe Rincon, Jorge Olarte. There was some related work in [29], and [32] then in preparation. ◇

3.2.1 *The Tropical Grassmannian, the Dressian, and Their Positive Analogues*

Definition 3.2.4 Given $e = (e_1, \ldots, e_N) \in \mathbb{Z}^N_{\geq 0}$, we let $\mathbf{x}^e$ denote $x_1^{e_1} \ldots x_N^{e_N}$. Let $E \subset \mathbb{Z}^N_{\geq 0}$. For $f = \sum_{e \in E} f_e \mathbf{x}^e$ a nonzero polynomial, we denote by $\text{Trop}(f) \subset \mathbb{R}^N$ the set of all points $(X_1, \ldots, X_N)$ such that, if we form the collection of numbers $\sum_{i=1}^N e_i X_i$ for e ranging over E, then the minimum of this collection is not unique. We say that $\text{Trop}(f)$ is the *tropical hypersurface associated to* f.

In our examples, we always consider polynomials f with real coefficients. We also have a positive version of Definition 3.2.4.

Definition 3.2.5 Let $E = E^+ \sqcup E^- \subset \mathbb{Z}^N_{\geq 0}$, and let f be a nonzero polynomial with real coefficients which we write as $f = \sum_{e \in E^+} f_e \mathbf{x}^e - \sum_{e \in E^-} f_e \mathbf{x}^e$, where all of the coefficients f_e are nonnegative real numbers. We denote by $\text{Trop}^+(f) \subset \mathbb{R}^N$ the set of all points $(X_1, \ldots, X_N)$ such that, if we form the collection of numbers $\sum_{i=1}^N e_i X_i$ for e ranging over E, then the minimum of this collection is not unique and furthermore is achieved for some $e \in E^+$ and some $e \in E^-$. We say that $\mathrm{Trop}^+(f)$ is the *positive part of* $\mathrm{Trop}(f)$.

The Grassmannian $\mathrm{Gr}_{k+1,n}$ is a projective variety which can be embedded in projective space $\mathbb{P}^{\binom{[n]}{k+1}-1}$, and is cut out by the *Plücker ideal*, that is, the ideal of relations satisfied by the Plücker coordinates of a generic $(k+1) \times n$ matrix. These relations include the *three-term Plücker relations* defined below.

Definition 3.2.6 Let $1 < a < b < c < d \leq n$ and choose a subset $S \in \binom{[n]}{k-1}$ which is disjoint from $\{a, b, c, d\}$. Then $p_{Sac}p_{Sbd} = p_{Sab}p_{Scd} + p_{Sad}p_{Sbc}$ is a *three-term Plücker relations* for the Grassmannian $\mathrm{Gr}_{k,n}$. Here Sac denotes $S \cup \{a, c\}$, etc.

Definition 3.2.7 Given S, a, b, c, d as in Definition 3.2.6, we say that the *tropical three-term Plücker relation holds* if

- $P_{Sac} + P_{Sbd} = P_{Sab} + P_{Scd} \leq P_{Sad} + P_{Sbc}$ or
- $P_{Sac} + P_{Sbd} = P_{Sad} + P_{Sbc} \leq P_{Sab} + P_{Scd}$ or
- $P_{Sab} + P_{Scd} = P_{Sad} + P_{Sbc} \leq P_{Sac} + P_{Sbd}$.

And we say that the *positive tropical three-term Plücker relation holds* if either of the first two conditions above holds.

Definition 3.2.8 *((Positive) Tropical Grassmannian)* The *tropical Grassmannian* $\mathrm{Trop}\,\mathrm{Gr}_{k+1,n} \subset \mathbb{R}^{\binom{[n]}{k+1}}$ is the intersection of the tropical hypersurfaces $\mathrm{Trop}(f)$, where f ranges over all elements of the Plücker ideal. The *Dressian* $\mathrm{Dr}_{k+1,n} \subset \mathbb{R}^{\binom{[n]}{k+1}}$ is the intersection of the tropical hypersurfaces $\mathrm{Trop}(f)$, where f ranges over all three-term Plücker relations.

Similarly, the *positive tropical Grassmannian* $\mathrm{Trop}^+\,\mathrm{Gr}_{k+1,n} \subset \mathbb{R}^{\binom{[n]}{k+1}}$ is the intersection of the positive tropical hypersurfaces $\mathrm{Trop}^+(f)$, where f ranges over all elements of the Plücker ideal. The *positive Dressian* $\mathrm{Dr}^+_{k+1,n} \subset \mathbb{R}^{\binom{[n]}{k+1}}$ is the intersection of the positive tropical hypersurfaces $\mathrm{Trop}^+(f)$, where f ranges over all three-term Plücker relations.

Note that the Dressian $Dr_{k+1,n}$ (respectively, the positive Dressian $\mathrm{Dr}^+_{k+1,n}$) is the subset of $\mathbb{R}^{\binom{[n]}{k+1}}$ where the tropical (respectively, positive tropical) three-term Plücker relations hold.

In general, the Dressian $\mathrm{Dr}_{k+1,n}$ is much larger than the tropical Grassmannian $\mathrm{Trop}\,\mathrm{Gr}_{k+1,n}$—for example, the dimension of the Dressian $\mathrm{Dr}_{3,n}$ grows quadratically is n, while the dimension of the tropical Grassmannian $\mathrm{Trop}\,\mathrm{Gr}_{3,n}$ is linear in n [33]. However, the situation for their positive parts is different.

Theorem 3.2.9 *([31]) The positive tropical Grassmannian* $\mathrm{Trop}^+\,\mathrm{Gr}_{k+1,n}$ *equals the positive Dressian* $\mathrm{Dr}^+_{k+1,n}$.

Definition 3.2.10 *((Positive) Plücker Vectors)* We say that a point $\{P_I\}_{I \in \binom{[n]}{k+1}} \in \mathbb{R}^{\binom{[n]}{k+1}}$ is a *(finite) tropical Plücker vector* if it lies in the Dressian $\mathrm{Dr}_{k+1,n}$, i.e. for every three-term Plücker relation, it lies in the associated tropical hypersurface. And we say that $\{P_I\}_{I\in\binom{[n]}{k+1}}$ is a *positive tropical Plücker vector*, if it lies in the positive Dressian $\mathrm{Dr}^+_{k+1,n}$ (equivalently, the positive tropical Grassmannian $\mathrm{Trop}^+\,\mathrm{Gr}_{k+1,n}$), i.e. for every three-term Plücker relation, it lies in the positive part of the associated tropical hypersurface.

Example 3.2.11 For $\mathrm{Gr}_{2,4}$, there is only one Plücker relation, $p_{13}p_{24} = p_{12}p_{34} + p_{14}p_{23}$. The Dressian $\mathrm{Dr}_{2,4} \subset \mathbb{R}^{\binom{[4]}{2}}$ is defined to be the set of points $(P_{12}, P_{13}, P_{14}, P_{23}, P_{24}, P_{34}) \in \mathbb{R}^6$ such that

- $P_{13} + P_{24} = P_{12} + P_{34} \le P_{14} + P_{23}$ or
- $P_{13} + P_{24} = P_{14} + P_{23} \le P_{12} + P_{34}$ or
- $P_{12} + P_{34} = P_{14} + P_{23} \le P_{13} + P_{24}$.

And $\mathrm{Dr}^+_{2,4} = \mathrm{Trop}^+\,\mathrm{Gr}_{2,4} \subset \mathbb{R}^{\binom{[4]}{2}}$ is defined to be the set of points $(P_{12}, P_{13}, P_{14}, P_{23}, P_{24}, P_{34}) \in \mathbb{R}^6$ such that

- $P_{13} + P_{24} = P_{12} + P_{34} \le P_{14} + P_{23}$ or
- $P_{13} + P_{24} = P_{14} + P_{23} \le P_{12} + P_{34}$. ◇

3.2.2 *The Positive Tropical Grassmannian and Positroid Subdivisions*

Recall that $\Delta_{k+1,n}$ denotes the $(k+1, n)$-hypersimplex, defined as the convex hull of the points e_I where I runs over $\binom{[n]}{k+1}$. Consider a real-valued function $\{I\} \mapsto P_I$ on the vertices of $\Delta_{k+1,n}$. We define a polyhedral subdivision $\mathcal{D}_P$ of $\Delta_{k+1,n}$ as follows: consider the points $(e_I, P_I) \in \Delta_{k+1,n} \times \mathbb{R}$ and take their convex hull. Take the lower faces (those whose outwards normal vector have last component negative) and project them back down to $\Delta_{k+1,n}$; this gives us the subdivision $\mathcal{D}_P$. We will omit the subscript P when it is clear from context. A subdivision obtained in this manner is called *regular*.

Remark 3.2.12 A lower face F of the regular subdivision defined above is determined by some vector $\lambda = (\lambda_1, \ldots, \lambda_n, -1)$ whose dot product with the vertices of the F is maximized. So if F is the matroid polytope of a matroid M with bases $\mathcal{B}$, this is equivalent to saying that $\lambda_{i_1} + \cdots + \lambda_{i_{k+1}} - P_I = \lambda_{j_1} + \cdots + \lambda_{j_{k+1}} - P_J > \lambda_{h_1} + \cdots + \lambda_{h_{k+1}} - P_H$ for any two bases $I, J \in \mathcal{B}$ and $H \notin \mathcal{B}$.

Given a subpolytope Γ of $\Delta_{k+1,n}$, we say that Γ is *matroidal* if the vertices of Γ, considered as elements of $\binom{[n]}{k+1}$, are the bases of a matroid M, i.e. $\Gamma = \Gamma_M$.

The following result is originally due to Kapranov [2]; it was also proved in [4, Proposition 2.2].

Theorem 3.2.13 (Regular matroid subdivisions) *The following are equivalent.*

- *The collection* $\{P_I\}_{I\in\binom{[n]}{k+1}}$ *is a tropical Plücker vector.*

- *The one-skeleta of $\mathcal{D}_P$ and $\Delta_{k+1,n}$ are the same.*
- *Every face of $\mathcal{D}_P$ is matroidal.*

Given a subpolytope Γ of $\Delta_{k+1,n}$, we say that Γ is *positroid* if the vertices of Γ, considered as elements of $\binom{[n]}{k+1}$, are the bases of a positroid M, i.e. $\Gamma = \Gamma_M$. We now give a positroid version of Proposition 3.2.13.

Theorem 3.2.14 (Regular positroid subdivisions) *The following are equivalent.*

- *The collection $\{P_I\}_{I\in\binom{[n]}{k+1}}$ is a positive tropical Plücker vector.*
- *Every face of $\mathcal{D}_P$ is positroid.*

Proof Suppose that the collection $\{P_I\}_{I\in\binom{[n]}{k+1}}$ are positive tropical Plücker coordinates. Then in particular they are tropical Plücker coordinates, and so by Proposition 3.2.13, every face of $\mathcal{D}_P$ is matroidal.

Suppose that one of those faces Γ_M fails to be positroid. Then by Theorem 3.1.14, Γ_M (and hence $\mathcal{D}_P$) has a two-dimensional face with vertices $e_{Sab}, e_{Sad}, e_{Sbc}, e_{Scd}$, for some $1 \le a < b < c < d \le n$ and S of size $k-1$ disjoint from $\{a, b, c, d\}$. By Remark 3.2.12, this means that there is a vector $\lambda = (\lambda_1, \ldots, \lambda_n, -1)$ whose dot product is maximized at the face F. In particular, if we compare the value of the dot product at vertices of F versus e_{Sac} and e_{Sbd}, we get $\lambda_a + \lambda_b - P_{Sab} = \lambda_c + \lambda_d - P_{Scd} = \lambda_a + \lambda_d - P_{Sad} = \lambda_b + \lambda_c - P_{Sbc}$ is greater than either $\lambda_a + \lambda_c - P_{Sac}$ or $\lambda_b + \lambda_d - P_{Sbd}$. But then

$$\lambda_a + \lambda_b - P_{Sab} + \lambda_c + \lambda_d - P_{Scd} = \lambda_a + \lambda_d - P_{Sad} + \lambda_b + \lambda_c - P_{Sbc} > \lambda_a + \lambda_c - P_{Sac} + \lambda_b + \lambda_d - P_{Sbd},$$

which implies that

$$P_{Sab} + P_{Scd} = P_{Sad} + P_{Sbc} < P_{Sac} + P_{Sbd},$$

which contradicts the fact that $\{P_I\}$ is a collection of positive tropical Plücker coordinates.

Suppose that every face of $\mathcal{D}_P$ is positroid. Then every face is in particular matroidal, and so by Proposition 3.2.13, the collection $\{P_I\}_{I\in\binom{[n]}{k+1}}$ are tropical Plücker coordinates. Suppose that they fail to be positive tropical Plücker coordinates. Then there is some $S \in \binom{[n]}{k-1}$ and $a < b < c < d$ disjoint from S such that $P_{Sab} + P_{Scd} = P_{Sad} + P_{Sbc} < P_{Sac} + P_{Sbd}$. We will obtain a contradiction by showing that $\mathcal{D}_P$ has a two-dimensional (non-positroid) face with vertices $e_{Sab}, e_{Sad}, e_{Sbc}, e_{Scd}$, for some $1 \le a < b < c < d \le n$ and S of size $k-1$ disjoint from $\{a, b, c, d\}$.

To show that these vertices form a face, choose some large number N which is greater than the absolute value of any of the tropical Plücker coordinates, i.e. $N > \max\{|P_I|\}_{I\in\binom{[n]}{k+1}}$. We define a vector $\lambda \in \mathbb{R}^n$ by setting

$$\lambda_i = \begin{cases} P_{Sab} + P_{Sac} + P_{Sad} & \text{for i=a} \\ P_{Sab} + P_{Sbc} + P_{Sbd} & \text{for i=b} \\ P_{Sac} + P_{Sbc} + P_{Scd} & \text{for i=c} \\ P_{Sad} + P_{Sbd} + P_{Scd} & \text{for i=d} \\ 3N & \text{for } i \in S \\ -3N & \text{for } i \notin S \cup \{a, b, c, d\} \end{cases}$$

Then the vertices of $\Delta_{k+1,n}$ maximizing the dot product with λ will certainly be some subset of $e_{Sab}, e_{Sac}, e_{Sad}, e_{Sbc}, e_{Sbd}, e_{Scd}$. And in fact the relation $P_{Sab} + P_{Scd} = P_{Sad} + P_{Sbc} < P_{Sac} + P_{Sbd}$ implies that the vertices maximizing the dot product will be exactly $e_{Sab}, e_{Sad}, e_{Sbc}, e_{Scd}$. □

It follows from Proposition 3.2.14 that the regular subdivisions of $\Delta_{k+1,n}$ consisting of positroid polytopes are precisely those of the form $\mathcal{D}_P$, where $P = \{P_I\}$ is a positive tropical Plücker vector. This motivates the following definition.

Definition 3.2.15 *(Regular Positroid Subdivision)* We say that a positroid dissection of $\Delta_{k+1,n}$ is a *regular positroid subdivision* if it has the form $\mathcal{D}_P$, where $P = \{P_I\} \in \mathbb{R}^{\binom{[n]}{k+1}}$ is a positive tropical Plücker vector.

Remark 3.2.16 Every regular subdivision of a polytope is a polytopal subdivision, and so in particular it is a good dissection (see Definition 5.2.14).

3.2.3 Fan Structures on the Dressian and Positive Dressian

As described in [33], there are two natural fan structures on the (positive) Dressian: the *Plücker fan*, and the *secondary fan*.

We say that two elements of the Dressian, i.e. two tropical Plücker vectors $\{P_I\}_{I \in \binom{[n]}{k+1}}$ and $\{P'_I\}_{I \in \binom{[n]}{k+1}} \in \mathbb{R}^{\binom{[n]}{k+1}}$, lie in the same cone of the *Plucker fan* if for each S, a, b, c, d as in Remark 3.2.7, the same inequality holds for both $\{P_{Sac}, P_{Sbd}, P_{Sab}, P_{Scd}, P_{Sad}, P_{Sbc}\}$ and $\{P'_{Sac}, P'_{Sbd}, P'_{Sab}, P'_{Scd}, P'_{Sad}, P'_{Sbc}\}$. In particular, the maximal cones in the Plücker fan structure are the cones where the inequalities from Remark 3.2.7 are all strict.

On the other hand, using Propositions 3.2.13 and 3.2.14, we say that two elements of the Dressian, i.e. two tropical Plücker vectors $\{P_I\}_{I \in \binom{[n]}{k+1}}$ and $\{P'_I\}_{I \in \binom{[n]}{k+1}} \in \mathbb{R}^{\binom{[n]}{k+1}}$, lie in the same cone of the *secondary fan* if the matroidal subdivisions $\mathcal{D}_P$ and $\mathcal{D}_{P'}$ coincide. In particular, the maximal cones in the secondary fan structure are the cones corresponding to the unrefinable positroid subdivisions.

In [33] it was shown that for the Dressian $Dr_{3,n}$, the Plücker fan structure and the secondary fan structure coincide. And in [34, Theorem 14] it was shown that the fan structures coincide for general Dressians $Dr_{k+1,n}$. We can now just refer to the fan structure on $\mathrm{Dr}^+_{k+1,n} = \mathrm{Trop}^+ \mathrm{Gr}_{k+1,n}$ without specifying either "Plücker fan" or "secondary fan."

We have the following result.

Corollary 3.2.17 *A collection $\mathcal{C} = \{S_\pi\}$ of positroid cells of $\mathrm{Gr}^{\geq 0}_{k+1,n}$ gives a regular positroid tiling of $\Delta_{k+1,n}$ (see Definition 3.2.1) if and only if this tiling has the form $\mathcal{D}_P$, for $P = \{P_I\}_{I \in \binom{[n]}{k+1}}$ a positive tropical Plücker vector from a maximal cone of $\mathrm{Trop}^+ \mathrm{Gr}_{k+1,n}$.*

Proof Suppose that a collection $\{S_\pi\}$ of positroid cells of $\mathrm{Gr}^{\geq 0}_{k+1,n}$ is a regular positroid tiling; in other words, the images of the cells $\{S_\pi\}$ under the moment map are the top-dimensional positroid polytopes in the subdivision $\mathcal{D}_P$ of $\Delta_{k+1,n}$, and the moment map is an injection on each S_π. Therefore by Lemma 3.1.24 and Theorem 3.1.23, $\dim S_\pi = n - 1$, each positroid M_π is connected, and the reduced plabic graph associated to π is a (planar) tree.

We claim that the collection $\{S_\pi\}$ gives an unrefineable possible positroid subdivision of the hypersimplex. That is, there is no nontrival way to subdivide one of the positroid polytopes Γ_π into two full dimensional positroid polytopes. If we *can* subdivide Γ_π as above, and there is another full-dimensional positroid polytope $\Gamma_{\pi'}$ strictly contained in Γ_π, then the bases of $M_{\pi'}$ are a subset of the bases of Γ_π, and hence the cell $S_{\pi'}$ lies in the closure of S_π. But then a reduced plabic graph G' for $S_{\pi'}$ can be obtained by deleting some edges from a reduced plabic graph G for S_π; this means that G' has fewer faces than G and hence has the corresponding cell has smaller dimension, which is a contradiction, so the claim is true.

But now the fact that $\{S_\pi\}$ gives an unrefineable positroid subdivision means that it came from a maximal cone of $\mathrm{Trop}^+ \mathrm{Gr}_{k+1,n}$.

Conversely, consider a regular positroid subdivision $\mathcal{D}_P$ coming from a maximal cone of $\mathrm{Trop}^+ \mathrm{Gr}_{k+1,n}$. Then the subdivision $\mathcal{D}_P$ (which we identify with its top-dimensional pieces $\{S_\pi\}$) is an unrefineable positroid subdivision. In other words, none of the positroid polytopes Γ_π can be subdivided into two full-dimensional positroid polytopes, which in turn means that the reduced plabic graph corresponding to π must be a tree. This implies that the moment map is an injection on each S_π and hence $\{S_\pi\}$ gives a regular positroid tiling of $\Delta_{k+1,n}$. □

Corollary 3.2.18 *The number of regular positroid tilings of the hypersimplex $\Delta_{k+1,n}$ equals the number of maximal cones in the positive tropical Grassmannian $\mathrm{Trop}^+ \mathrm{Gr}_{k+1,n}$.*

The fact that the Plücker fan structure and the secondary fan structure on $\mathrm{Trop}^+ \mathrm{Gr}_{k+1,n}$ coincide also implies that the f-vector of $\mathrm{Trop}^+ \mathrm{Gr}_{k+1,n}$ reflects the number of positroid subdivisions of $\Delta_{k+1,n}$ (with maximal cones corresponding to unrefineable subdivisions and rays corresponding to coarsest subdivisions).

The f-vector of $\mathrm{Trop}^+ \mathrm{Gr}_{k+1,n}$. We report here what is known about the f-vector of the positive tropical Grassmannian. This is the vector $(f_0, f_1, \ldots, f_d)$ whose components compute the number of cones of fixed dimension.

As shown in [15], the positive tropical Grassmannian has an n-dimensional lineality space coming from the torus action. However, one may mod out by this torus action and study the resulting fan. The method used in [15] was to show that $\mathrm{Trop}^+ \mathrm{Gr}_{k+1,n}$ (a polyhedral subcomplex of $\mathbb{R}^{\binom{[n]}{k+1}}$) is combinatorially equivalent to an $k(n-k-2)$-dimensional fan $F_{k+1,n}$, obtained by using an "X-cluster" or "web" parametrization of the positive Grassmannian, and modding out by the torus action. As explained in [15, Sect. 6], $F_{k+1,n}$ is the dual fan to the Minkowski sum of the $\binom{n}{k+1}$ Newton polytopes obtained by writing down each Plücker coordinate in the X-cluster parametrization.

Using this technique, [15] computed the f-vector of $\mathrm{Trop}^+ \mathrm{Gr}_{2,n}$ (which is the f-vector of the associahedron, with maximal cones corresponding to triangulations of a polygon) $\mathrm{Trop}^+ \mathrm{Gr}_{3,6}$, and $\mathrm{Trop}^+ \mathrm{Gr}_{3,7}$. The above f-vector computations were recently extended in [35] using the notion of "stringy canonical forms" and in [36, 37] using planar arrays and matrices of Feynman diagrams. See also [38–40] for recent, physics-inspired developments in this direction. We list all known results about maximal cones in the positive tropical Grassmannian $\mathrm{Trop}^+ \mathrm{Gr}_{k+1,n}$ and their relation to tilings of hypersimplex $\Delta_{k+1,n}$ in Table 5.2.

Apart from the f-vector of $\mathrm{Trop}^+ \mathrm{Gr}_{2,n}$, the known f-vectors of positive tropical Grassmannians $\mathrm{Trop}^+ \mathrm{Gr}_{k,n}$ (with $k \leq \frac{n}{2}$) are the following:

$$\mathrm{Trop}^+ \mathrm{Gr}_{3,6} : (1, 48, 98, 66, 16, 1)$$
$$\mathrm{Trop}^+ \mathrm{Gr}_{3,7} : (1, 693, 2163, 2583, 1463, 392, 42, 1)$$
$$\mathrm{Trop}^+ \mathrm{Gr}_{3,8} : (1, 13612, 57768, 100852, 93104, 48544, 14088, 2072, 120, 1)$$
$$\mathrm{Trop}^+ \mathrm{Gr}_{4,8} : (1, 90608, 444930, 922314, 1047200, 706042, 285948, 66740, 7984, 360, 1)$$

For $\mathrm{Trop}^+ \mathrm{Gr}_{4,9}$ it is also known that the second component of the f-vector is 30659424 [37].

References

1. I.M. Gelfand, R.M. Goresky, R.D. MacPherson, V.V. Serganova, Combinatorial geometries, convex polyhedra, and Schubert cells. Adv. Math. **63**(3), 301–316 (1987)
2. M.M. Kapranov, Chow quotients of Grassmannians. I, in *I. M. Gelfand Seminar*, vol. 16 of *Adv. Soviet Math.* Amer. Math. Soc. (Providence, RI, 1993), pp. 29–110
3. L. Lafforgue, *Chirurgie des grassmanniennes, CRM Monograph Series*, vol. 19 (American Mathematical Society, Providence, RI, 2003)
4. D.E. Speyer, Variations on a theme of Kasteleyn, with application to the totally nonnegative Grassmannian. Electron. J. Combin. **23**(2), 2.24–7 (2016)
5. D. Speyer, B. Sturmfels, The tropical Grassmannian. Adv. Geom. **4**(3), 389–411 (2004)
6. S. Herrmann, M. Joswig, D.E. Speyer, Dressians, tropical Grassmannians, and their rays. Forum Math. **26**(6), 1853–1881 (2014)
7. G. Lusztig, Total positivity in reductive groups, in *Lie Theory and Geometry*, vol. 123 of *Progr. Math.* (Birkhäuser Boston, Boston, MA, 1994), pp. 531–568
8. A. Postnikov, *Total Positivity, Grassmannians, and Networks*.' Preprint http://math.mit.edu/~apost/papers/tpgrass.pdf

9. K.C. Rietsch, *Total Positivity and Real Flag Varieties*. Ph.D. thesis (Massachusetts Institute of Technology, 1998)
10. A. Postnikov, D. Speyer, L. Williams, Matching polytopes, toric geometry, and the totally non-negative Grassmannian. J. Algebraic Combin. **30**(2), 173–191 (2009)
11. S. Oh, Positroids and schubert matroids. J. Combin. Theory Ser. A **118**(8), 2426–2435 (2011)
12. F. Ardila, F. Rincón, L. Williams, Positroids and non-crossing partitions. Trans. Amer. Math. Soc. **368**(1), 337–363 (2016)
13. F. Ardila, F. Rincón, L. Williams, Positively oriented matroids are realizable. J. Eur. Math. Soc. (JEMS) **19**(3), 815–833 (2017)
14. E. Tsukerman, L. Williams, Bruhat interval polytopes. Adv. Math. **285**, 766–810 (2015)
15. D. Speyer, L. Williams, The tropical totally positive Grassmannian. J. Algebraic Combin. **22**(2), 189–210 (2005)
16. D. Speyer, L. Williams, The positive Dressian equals the positive tropical Grassmannian. To appear in Transactions of the AMS
17. N. Arkani-Hamed, J. Trnka, The amplituhedron. J. High Energy Phys. **10**, 33 (2014)
18. D.E. Speyer, A matroid invariant via the K-theory of the Grassmannian. Adv. Math. **221**(3), 882–913 (2009)
19. S.N. Karp, L.K. Williams, Y.X. Zhang, *Decompositions of Amplituhedra* (2017)
20. T. Łukowski, M. Parisi, L.K. Williams, The positive tropical Grassmannian, the hypersimplex, and the $m = 2$ amplituhedron. Int. Math. Res. Not. **03**, rnad010 (2023)
21. M. Parisi, M. Sherman-Bennett, L. Williams, The $m = 2$ amplituhedron and the hypersimplex: signs, clusters, triangulations, Eulerian numbers, 2023. Accepted and to be published in *Communications of the American Mathematical Society*
22. F. Sottile, Toric ideals, real toric varieties, and the moment map. In Topics in algebraic geometry and geometric modeling, volume 334 of Contemp. Math., pages 225–240. Amer. Math. Soc., Providence, RI, 2003
23. M.F. Atiyah, Convexity and commuting Hamiltonians. Bull. London Math. Soc. **14**(1), 1–15 (1982)
24. V. Guillemin, S. Sternberg, Convexity properties of the moment mapping. Invent. Math. **67**(3), 491–513 (1982)
25. J. Oxley, *Matroid Theory, Oxford Graduate Texts in Mathematics*, vol. 21, 2nd edn. (Oxford University Press, Oxford, 2011)
26. A.V. Borovik, I.M. Gelfand, N. White, *Coxeter Matroids, Progress in Mathematics*, vol. 216 (Birkhäuser Boston Inc, Boston, MA, 2003)
27. D.J.A. Welsh, *Matroid Theory* (Academic Press [Harcourt Brace Jovanovich Publishers], London, 1976). L. M. S. Monographs, No. 8
28. F. Rincón, C. Vinzant, J. Yu, Positively hyperbolic varieties, tropicalization, and positroids. Adv. Math. **383**(107677), 35 (2021)
29. N. Early, *From Weakly Separated Collections to Matroid Subdivisions* (2019). Preprint http://arxiv.org/abs/1910.11522
30. W. Fulton, *Introduction to Toric Varieties*, vol. 131 of *Annals of Mathematics Studies*. (Princeton University Press, Princeton, NJ, 1993). The William H. Roever Lectures in Geometry
31. D. Speyer, L.K. Williams, The positive Dressian equals the positive tropical Grassmannian. Trans. Amer. Math. Soc. Ser. B **8**, 330–353 (2021)
32. N. Arkani-Hamed, T. Lam, M. Spradlin, Positive configuration space. Comm. Math. Phys. **384**(2), 909–954 (2021)
33. S. Herrmann, A.N. Jensen, M. Joswig, B. Sturmfels, How to draw tropical planes. Electr. J. Comb. **16** (2008)
34. J.A. Olarte, M. Panizzut, B. Schröter, *On local Dressians of matroids, in Algebraic and Geometric Combinatorics on Lattice Polytopes* (World Sci. Publ, Hackensack, NJ, 2019), pp.309–329
35. N. Arkani-Hamed, T. Lam, M. Spradlin, Non-perturbative geometries for planar $\mathcal{N} = 4$ SYM amplitudes. **12** (2019)
36. F. Borges, F. Cachazo, *Generalized Planar Feynman Diagrams: Collections* (2019)

37. F. Cachazo, A. Guevara, B. Umbert, Y. Zhang, *Planar Matrices and Arrays of Feynman Diagrams* (2019)
38. N. Henke, G. Papathanasiou, How tropical are seven- and eight-particle amplitudes? **12** (2019)
39. J. Drummond, J. Foster, Ö. Gürdoğan, C. Kalousios, Algebraic singularities of scattering amplitudes from tropical geometry. **12** (2019
40. N. Early, Planar kinematic invariants, matroid subdivisions and generalized Feynman diagrams. **12** (2019)

Chapter 4
T-Duality: The Hypersimplex Versus the Amplituhedron

There are some interesting maps which one can apply to the positive Grassmannian $\mathrm{Gr}_{r,n}^{\geq 0}$ and its cells—as we saw in Chaps. 2 and 3. The first map is the *amplituhedron map*—a map $\tilde{Z}$ from $\mathrm{Gr}_{k,n}^{\geq 0}$ to the Grassmannian $\mathrm{Gr}_{k,k+m}$, whose image of full dimension mk is the amplituhedron $\mathcal{A}_{n,k,m}(Z)$. The second map is the *moment map*—a map from $\mathrm{Gr}_{k+1,n}^{\geq 0}$ to $\mathbb{R}^n$, whose image of dimension $n-1$ is the hypersimplex $\Delta_{k+1,n}$.

At first glance, the $(n-1)$-dimensional hypersimplex $\Delta_{k+1,n} \subset \mathbb{R}^n$ doesn't seem to have any relation to the $2k$-dimensional amplituhedron $\mathcal{A}_{n,k,2}(Z) \subset \mathrm{Gr}_{k,k+2}$. Nevertheless, in this chapter we draw striking parallels between $\Delta_{k+1,n}$ and $\mathcal{A}_{n,k,2}(Z)$, some of which are illustrated in Table 4.1. The key to connect the two worlds is *T-duality*. One of its combinatorial realization is a simple map on decorated permutations labelling positroid cells (Definition 4.1.1). Surprisingly, T-duality provides very deep insights also at the level of the geometry itself (e.g. Sect. 4.2). It can shed light on the sign stratification of the amplituhedron (Sect. 4.3), and open connections to *tropical geometry* (Sect. 5.2.2) and *cluster algebras* (Sect. 7.1.2). Our discussion will culminate in Chap. 5, where—using results from this chapter—we show that T-duality also relates the *secondary geometry* of the hypersimplex and the amplituhedron—i.e. their tilings (dissections).

T-duality is part of a broader picture: in Chap. 6 we see how T-duality can be thought of as a generalization of the homonymous duality in physics coming from String Theory. In particular, the *momentum amplituhedron* is conjectured to be T-dual to the amplituhedron for $m = 4$ (Conjecture 6.1.70). Our work serves therefore also as a model for what still has to be explored in that context.

Summary of the Chapter. This chapter is based on the following works by the author: [1, Sects. 8–10,12] and [2, Sect. 5].

In this chapter we explore a duality between two seemingly unrelated objects: the hypersimplex $\Delta_{k+1,n}$ (cf. Chap. 3) and the amplituhedron $\mathcal{A}_{n,k,2}(Z)$ (cf. Chap. 2). We refer to it as *T-Duality* and it is one of the main topics of this work. In Sect. 4.1 we define T-duality on the main combinatorial objects associated to the totally

M. Parisi, *Combinatorial Aspects of Scattering Amplitudes*, Springer Theses,
https://doi.org/10.1007/978-3-031-41069-7_4

Table 4.1 Table of correspondences via t-duality: the hypersimplex versus the amplituhedron

The Hypersimplex $\Delta_{k+1,n}$	Versus	The Amplituhedron $\mathcal{A}_{n,k,2}$
$\Delta_{k+1,n} = \mu(\mathrm{Gr}_{k+1,n}^{\geq 0})$ (moment map)		(Amplituhedron map) $\mathcal{A}_{n,k,2} = \tilde{Z}(\mathrm{Gr}_{k,n}^{\geq 0})$
$\dim(\Delta_{k+1,n}) = n-1$ in $\mathbb{R}^n$		$\dim(\mathcal{A}_{n,k,2}) = 2k$ in $\mathrm{Gr}_{k,k+2}$
	Positroid Tiles	
$\Gamma_{G(\mathcal{T})}$ (Positroid Polytope)	Unpunctured plabic tiling $\mathcal{T}$ in an n-gon	(Grasstope) $Z_{\hat{G}(\mathcal{T})}$
$x_{[h,j-1]} \geq \mathrm{area}_{\mathcal{T}}(h \to j)$	Compatible arc $h \to j$	$\mathrm{sgn}\langle Yhj \rangle = (-1)^{\mathrm{area}_{\mathcal{T}}(h \to j)}$
$x_{[h,j-1]} = \mathrm{area}_{\mathcal{T}}(h \to j)$	Facet defining arc $h \to j$	$\langle Yhj \rangle = 0$
	w-Simplices and w-Chambers	
w-simplex $\Delta_w \subset \Delta_{k+1,n}$ such that:	$w \in S_n$: $w_n = n$, $\#\,\mathrm{Des}_{\mathrm{L}}(w) = k$	w-chamber $\hat{\Delta}_w(Z) \subset \mathcal{A}_{n,k,2}$ such that:
$\Delta_w = \mathrm{conv}\{e_{I_1}, \dots, e_{I_n}\}$	$I_a = I_a(w) := \mathrm{cDes}_{\mathrm{L}}(w^{(a-1)})$	$\mathrm{Flip}(\langle Ya\hat{1}\rangle, \langle Ya\hat{2}\rangle, \dots, \langle Yan\rangle) = I_a \setminus \{a\}$
Hypersimplex $\Delta_{k+1,n} = \bigcup_w \Delta_w$		Amplituhedron $\mathcal{A}_{n,k,2}(Z) = \bigcup_w \hat{\Delta}_w(Z)$
Tile $\Gamma_\pi = \bigcup_{\Delta_w \cap \Gamma_\pi^\circ \neq \emptyset} \Delta_w$		Tile $Z_{\hat{\pi}} = \bigcup_{\hat{\Delta}_w(Z) \cap Z_{\hat{\pi}}^\circ \neq \emptyset} \hat{\Delta}_w(Z)$
$\Delta_w \subset \Gamma_\pi$	$\Leftrightarrow$	$\hat{\Delta}_w(Z) \subset Z_{\hat{\pi}}$
Hypercube $\square_{n-1} = \bigcup_k \Delta_{k+1,n} = \bigcup_w \Delta_w$	$w \in S_n : w_n = n$	Total Amplituhedron $\mathcal{G}_n^{(2)} = \bigcup_k \mathcal{G}_{n,k,2} = \bigcup_w \hat{\Delta}_w(\mathcal{G})$
	Positroid Tilings	
$\{\Gamma_\pi\}$ tiles $\Delta_{k+1,n}$	$\Leftrightarrow$	$\{Z_{\hat{\pi}}\}$ tiles $\mathcal{A}_{n,k,2}(Z)$, for all Z
BCFW tilings $\{\Gamma_\pi\}$	BCFW recurrences	BCFW tiling $\{Z_{\hat{\pi}}\}$
Regular tiling $\{\Gamma_\pi\}$	max cones of $\mathrm{Trop}^+ \mathrm{Gr}_{k+1,n}$	Regular tiling $\{Z_{\hat{\pi}}\}$
Catalan tiling $\{\Gamma_\pi\}$	Colouring vertices of a fixed tree	Catalan tiling $\{Z_{\hat{\pi}}\}$
Descent tiling $\{\Gamma_\pi\}$	Positions of descents/sign flips	Sign-flip tiling $\{Z_{\hat{\pi}}\}$

non-negative Grassmannian: decorated permutations (Sect. 4.1.1) and plabic graphs (Sect. 4.1.2). In Sect. 4.2 we explore how positroid polytopes in $\Delta_{k+1,n}$ and Grasstopes in $\mathcal{A}_{n,k,2}(Z)$ are related via T-Duality. In Sect. 4.2.1 we show that T-duality provides a bijection between positroid tiles of $\Delta_{k+1,n}$ and $\mathcal{A}_{n,k,2}$. Moreover, we show that inequalities cutting out a positroid polytope translates into sign conditions on twistor coordinates characterizing the T-dual Grasstope. Both can be simply read from the statistics of the associated *plabic tiling*. Moreover, we show in Sect. 4.2.2 that we can also read off their facets. Finally, in Sect. 4.2.3 we provide a bijection between positroid tiles of $\Delta_{k+1,n}$ and $\mathcal{A}_{n,k,2}(Z)$ with separable permutations on $[n-1]$ refined by numbers of descents k—counted by (a refinement of) the *Schröder numbers*. In Sect. 4.3 we show that we can subdivide the amplituhedron $\mathcal{A}_{n,k,2}(Z)$ into w-chambers as the hypersimplex $\Delta_{k+1,n}$ can be subdivided into w-simplices—both labelled by permutations w on $[n-1]$ with k descents, and therefore enumerated by Eulerian numbers $E_{n-1,k}$. Moreover, we show that w-simplices and w-chambers can be really thought as T-dual of each other. In Sect. 4.4 we define the $\mathcal{G}$-amplituhedron—a Z-independent, full-dimensional analogue of the amplituhedron inside $\mathrm{Gr}_{2,n}$. Using this, we show that all w-chambers of $\mathcal{A}_{n,k,2}(Z)$ are realizable (i.e. are non-empty for some Z). We then define the *total amplituhedron* in $\mathrm{Gr}_{2,n}$ as the union of $\mathcal{G}$-amplituhedra with different k. This is the amplituhedron-analogue of the hypercube—which is the union of different hypersimplices by varying k.

4.1 Combinatorial T-Duality

4.1.1 T-Duality on Decorated Permutations

Definition 4.1.1 (*T-duality on decorated permutations*) Let $\pi = a_1 a_2 \dots a_n$ be a loopless decorated permutation (written in one- line notation). The *T-dual* decorated permutation $\hat{\pi}$ is defined as the map $i \mapsto \pi(i-1)$, so that $\hat{\pi} = a_n a_1 a_2 \dots a_{n-1}$. Any fixed points in $\hat{\pi}$ are declared to be loops.

Remark 4.1.2 This map was previously defined in [3, Definition 4.5]) and can be regarded as an '$m = 2$' version of the map that appeared in [4] for the $m = 4$ amplituhedron. The T-duality map was also recently studied in [5, 6].

Recall that an anti-excedance of a decorated permutation is a position i such that $\pi(i) < i$, or $\pi(i) = i$ and i is a coloop. Our first result shows that T-duality is a bijection between loopless cells of $\mathrm{Gr}_{k+1,n}^{\geq 0}$ and coloopless cells of $\mathrm{Gr}_{k,n}^{\geq 0}$.

Lemma 4.1.3 *The T-duality map $\pi \mapsto \hat{\pi}$ is a bijection between the loopless permutations on $[n]$ with $k+1$ anti-excedances, and the coloopless permutations on $[n]$ with k anti-excedances. Equivalently, the T-duality map is bijection between loopless positroid cells of $\mathrm{Gr}_{k+1,n}^{\geq 0}$ and coloopless positroid cells of $\mathrm{Gr}_{k,n}^{\geq 0}$.*

Proof The second statement follows from the first by Appendix A, so it suffices to prove the first statement. Let $\pi = (a_1, \dots, a_n)$ be a loopless permutation on $[n]$ with $k+1$ anti-excedances; then $\hat{\pi} = (a_n, a_1, \dots, a_{n-1})$. Consider any i such that $1 \le i \le n-1$. Suppose i is a position of a anti-excedance, i.e. either $a_i < i$ or $a_i = \overline{i}$. Then the letter a_i appears in the $(i+1)$st position in $\hat{\pi}$, and since $a_i < i+1$, we again have an anti-excedance. On the other hand, if i is *not* a position of an anti-excedance, i.e. $a_i > i$ (recall that π is loopless), then in the $(i+1)$st position of $\hat{\pi}$ we have $a_i \ge i+1$. By Definition 4.1.1 if we have a fixed point in position $i+1$ (i.e. $a_i = i+1$) this is a loop, and so position $i+1$ of $\hat{\pi}$ will not be a anti-excedance. Therefore if $I \subset [n-1]$ is the positions of the anti-excedances located in the first $n-1$ positions of π, then $I+1$ is the positions of the anti-excedances located in positions $\{2, 3, \dots, n\}$ in $\hat{\pi}$.

Now consider position n of π. Because π is loopless, n will be the position of a anti-excedance in π. And because $\hat{\pi}$ is defined to be coloopless, 1 will never be the position of a anti-excedance in $\hat{\pi}$. Therefore the number of anti-excedances of $\hat{\pi}$ will be precisely one less than the number of anti-excedances of π.

It is easy to reverse this map so it is a bijection. □

Remark 4.1.4 Since by Lemma 4.1.3 the map $\pi \mapsto \hat{\pi}$ is a bijection, we can also talk about the inverse map from coloopless permutations on $[n]$ with k anti-excedances to loopless permutations on $[n]$ with $k+1$ anti-excedances. We denote this inverse map by $\pi \mapsto \check{\pi}$.

Remark 4.1.5 Our map $\pi \mapsto \hat{\pi}$ is in fact a special case of the map ρ_A introduced by Benedetti-Chavez-Tamayo in [5, Definition 23] (in the case where $A = \emptyset$).

In Lemma 4.1.3 we showed that T-duality gives a bijection between loopless cells S_π of $Gr_{k+1,n}^{\ge 0}$ and coloopless cells $S_{\hat{\pi}}$ of $\mathrm{Gr}_{k,n}^{\ge 0}$. We now show that this map is a poset isomorphism. Abusing notation, in this subsection we use π, ν to denote bounded affine permutations rather than decorated permutations.

Proposition 4.1.6 (T-duality as a poset isomorphism) *T-duality gives a codimension-preserving poset isomorphism between loopless cells of $Gr_{k+1,n}^{\ge 0}$ and coloopless cells of $\mathrm{Gr}_{k,n}^{\ge 0}$. That is, for π, ν loopless decorated permutations of type $(k+1, n)$, $S_\nu \subset \overline{S_\pi}$ if and only if $S_{\hat{\nu}} \subset \overline{S_{\hat{\pi}}}$, and, further the codimensions are the same.*

Proof Recall that the poset $Q(k, n)$ is dual to the poset $\mathrm{Bound}(k, n)$ of bounded affine permutations with respect to the Bruhat order [7], so we will work with $\mathrm{Bound}(k, n)$ instead. In $\mathrm{Bound}(k, n)$, $\pi \gtrdot \nu$ if $\pi = \tau \circ \nu$ for some transposition τ and $\mathrm{inv}(\pi) = \mathrm{inv}(\nu) + 1$.

Let $\delta : \mathbb{Z} \to \mathbb{Z}$ be the map $i \mapsto i-1$. Translating Definition 4.1.1 into the language of bounded affine permutations, we see that for loopless $\pi \in \mathrm{Bound}(k+1, n)$, the T-dual of π is $\hat{\pi} = \pi \circ \delta$. Fix loopless $\pi, \nu \in \mathrm{Bound}(k+1, n)$. We show that $\pi \ge \nu$ if and only if $\hat{\pi} \ge \hat{\nu}$.

First, note that π and $\pi \circ \delta$ have the same length. Further, $\widehat{\nu \circ \pi} = \nu \circ \pi \circ \delta = \nu \circ \hat{\pi}$. So $\pi \gtrdot \nu$ if and only if $\hat{\pi} \gtrdot \hat{\nu}$.

To extend this beyond cover relations, notice that if $\nu \in \text{Bound}(k+1, n)$ has $\nu(i) = i$ or $\nu(i) = i + n$, then for all $\pi \in \text{Bound}(k+1, n)$ with $\pi > \nu$, we have $\pi(i) = \nu(i)$. Indeed, in matroidal terms, this is saying that if the positroid $\mathcal{M}_\nu$ has a loop (resp. coloop), then so does $\mathcal{M}_\pi$ for all $\mathcal{M}_\pi \subset \mathcal{M}_\nu$.

Now, $\pi \geq \nu$ if and only if there exists a maximal chain $\pi \gtrdot \pi_1 \gtrdot \cdots \gtrdot \pi_r \gtrdot \nu$. Since ν is loopless, the observation in the previous paragraph shows that π_i is loopless for $i = 1, \ldots, r$. Since T-duality and its inverse preserve cover relations, we have such a chain if and only if we have the chain $\hat{\pi} \gtrdot \hat{\pi}_1 \gtrdot \cdots \gtrdot \hat{\pi}_r \gtrdot \hat{\nu}$ in $\text{Bound}(k, n)$, which is equivalent to $\hat{\pi} \geq \hat{\nu}$.

The codimension statement follows from the fact that the codimension of S_ν in $\overline{S_\pi}$ is the length of any maximal chain from ν to π in $\text{Bound}(k, n)$. □

T-duality can also be defined for arbitrary even m, as in [2, Eq. 5.13], and is also of interest for understanding the $m = 4$ amplituhedron. As is clear from [2, Eq. 5.13], the T-duality map for even m is simply composing the "$m = 2$" T-duality map $m/2$ times. Proposition 4.1.6 also gives us information about this composition.

Definition 4.1.7 Let $L^r \text{Gr}_{k',n}^{\geq 0}$ be the set of cells $S_\pi \subset \text{Gr}_{k',n}^{\geq 0}$ such that $\pi(i) \geq i + r$ for all i. Analogously, let us define $CL^{-r} \text{Gr}_{k,n}^{\geq 0}$ to be the set of cells $S_\nu \subset \text{Gr}_{k,n}^{\geq 0}$ such that $\nu(i) \leq i + n - r$ for all i. Each is ordered by inclusion on the closures of cells.

Remark 4.1.8 The composition of T-duality r times is a well-defined map from $L^r \text{Gr}_{k+r,n}^{\geq 0}$ to $CL^{-r} \text{Gr}_{k,n}^{\geq 0}$. Indeed, if $\pi(i) \geq i + r$, then applying T-duality s times gives a loopless bounded affine permutation for $s = 1, \ldots, r-1$. Moreover, it is easy to see that applying T-duality r times to such a π gives a bounded affine permutation ν with $\nu(i) \leq i + n - r$.

From Proposition 4.1.6 we immediately have the following:

Proposition 4.1.9 *The composition of T-duality r times gives a poset isomorphism between $L^r \text{Gr}_{k+r,n}^{\geq 0}$ and $CL^{-r} \text{Gr}_{k,n}^{\geq 0}$.*

4.1.2 T-Duality on Plabic Graphs

T-duality extends to an operation on plabic graphs.

Definition 4.1.10 A reduced plabic graph is called *black-trivalent* (resp. *white-trivalent*) if all of its interior black (resp. white) vertices are trivalent.

Note that in particular, black-trivalent (white-trivalent) graphs have no black (white) lollipops, so their trip permutations are loopless (coloopless).

Starting from a black-trivalent graph G with trip permutation π, we now give an explicit construction of a white-trivalent graph $\hat{G}$ with trip permutation $\hat{\pi}$. This construction streamlines the bijection of [8, Proposition 7.15] and [6, Proposition 8.3], and phrases the bijection entirely in terms of plabic graphs, rather than plabic and zonotopal tilings.

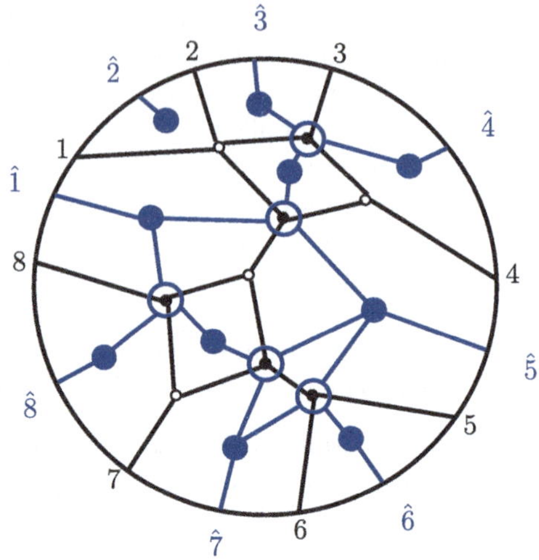

Fig. 4.1 In black: a plabic graph G of type $(4, 8)$ with trip permutation $(2, 4, 7, 1, 8, 5, 3, 6)$. In blue: the T-dual plabic graph $\hat{G}$ of type $(3, 8)$ with trip permutation $(6, 2, 4, 7, 1, 8, 5, 3)$, which is built using Definition 4.1.11

Definition 4.1.11 (*T-dual plabic graph*) Let G be a reduced black-trivalent plabic graph. The *T-dual* of G, denoted $\hat{G}$, is the graph obtained as follows:

1. In each face f of G, place a black vertex $\hat{b}(f)$.
2. "On top of" each black vertex b of G, place a white vertex $\hat{w}(b)$;
3. For each black vertex b of G in face f, put an edge $\hat{e}$ connecting $\hat{w}(b)$ and $\hat{b}(f)$;
4. Put $\hat{i}$ on the boundary of G between vertices $i-1$ and i and draw an edge from $\hat{i}$ to $\hat{b}(f)$, where f is the adjacent boundary face.

See (Fig. 4.1) for an example of a T-dual plabic graph.

Proposition 4.1.12 *Let G be a reduced black-trivalent plabic graph with trip permutation π. Then $\hat{G}$ is a reduced white-trivalent plabic graph with trip permutation $\hat{\pi}$.*

Proof We first show that if G has the trip $\gamma : i \to \pi(i)$, then $\hat{G}$ has the trip $\hat{\gamma} : \widehat{i+1} \to \widehat{\pi(i)}$.

Say γ starts at i. Let v be the first black vertex γ meets. By the rules of the road, there is one edge e attached to v at the left of γ (as G is black-trivalent). Note that vertex B is in the boundary face f containing boundary vertices i and $i+1$. This is because before meeting v, γ meets only white vertices, and by the rules of the road there are no edges involving these vertices lying to the left of γ. So $\hat{w}(v)$ is also connected to $\hat{b}(f)$. And by definition, $\widehat{i+1}$ is connected to $\hat{b}(f)$. Note that at the vertex $\hat{b}(f)$, if we start at the edge to $\widehat{i+1}$ and go counterclockwise, we see the edge to $\hat{w}(v)$. This means $\hat{\gamma}$ starts at $\widehat{i+1}$, goes to $\hat{b}(f)$, then to $\hat{w}(v)$ (see Fig. 4.2).

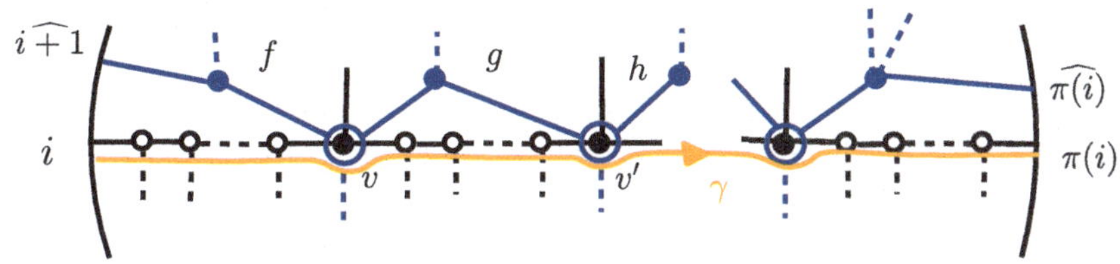

Fig. 4.2 Black edges and vertices are in G; blue are $\hat{G}$. In orange, the trip $\gamma : i \to \pi(i)$ in G. The trip $\hat{\gamma}$ follows the solid blue edges

Fig. 4.3 Black trivalent vertices of G correspond to white trivalent vertices of $\hat{G}$

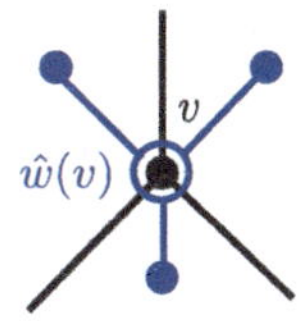

Now, let g be the face of G which contains e and the edge of γ following v. Clearly $\hat{b}(g)$ is connected to $\hat{w}(v)$. At the vertex $\hat{w}(v)$, if we start at the edge to $\hat{b}(f)$ and go clockwise, we see the edge to $\hat{b}(g)$. This means that γ goes from $\hat{w}(v)$ to $\hat{b}(g)$.

Now, let v' be the next black vertex γ meets. Again, the edges involving any white vertices on γ between v, v' must lie to the right of γ, and there is exactly one edge e' at v' to the left of γ. So the face g also contains v'. Let h be the face of G which contains e' and the edge of γ following v'. Then $\hat{\gamma}$ goes from $\hat{b}(g)$ to $\hat{w}(v')$ to $\hat{b}(h)$ (see Fig. 4.2). Continuing in this way, we see that if γ passes through a black vertex v, then $\hat{\gamma}$ passes through $\hat{w}(v)$ and then goes to $\hat{b}(f)$, where f is the face to the left of γ containing v and the edge of γ following v. If v is the last black vertex on γ, then f is the boundary face touching $\pi(i) - 1$ and $\pi(i)$. Note that at the vertex $\hat{b}(f)$, if we start at the edge to $\hat{w}(v)$ and go counterclockwise, we see the edge to $\widehat{\pi(i)}$. So γ will turn maximally right at $\hat{b}(f)$ to go to $\widehat{\pi(i)}$.

If γ meets no black vertices, there are no edges of G at the left of γ. This means $\pi(i) = i + 1$. The boundary face f between i and $\pi(i)$ contains only white vertices, so there will be a loop in $\hat{G}$ at boundary vertex $\widehat{i+1}$. Clearly $\hat{\pi}(i + 1) = i + 1 = \pi(i)$ as desired.

Now, that $\hat{G}$ is white-trivalent follows immediately from the fact that G is black-trivalent (see Fig. 4.3).

To show that $\hat{G}$ is reduced, it suffices to show that it has $\dim(S_{\hat{\pi}}) + 1$ many faces. Note that Definition 4.1.11 does not depend at all on the white vertices of G. So we may assume without loss of generality that G is bipartite and has a white vertex adjacent to every boundary vertex. With this assumption, it is not hard to see that the faces of $\hat{G}$ are in bijection with white vertices of G.

Let B, W, F, E denote the number of white vertices, black vertices, faces (excluding the infinite face), and edges (excluding edges between two boundary vertices) of G. Say that G is type $(k + 1, n)$. By [2, Proposition 5.17] (or the fact that T-duality preserves codimension), we have

$$\dim(S_{\hat{\pi}}) = \dim(S_{\pi}) - n + 2k + 1.$$

As G is reduced, $F = \dim(S_{\pi}) + 1$. So to show $W = \dim(S_{\hat{\pi}}) + 1$, it suffices to show that $W = F - n + 2k + 1$. This follows immediately from

$$E = 3B + n, \quad F = 1 - (W + B) + E, \quad W - B = k + 1.$$

The first equation holds because every edge between two internal vertices contains a unique black vertex, and all black vertices are trivalent. The second equation follows from Euler's formula for planar graphs. The third holds because G is type $(k + 1, n)$. □

Remark 4.1.13 It is straightforward to check that exchanging the roles of black and white vertices in Definition 4.1.11 gives a map from white-trivalent plabic graphs to black-trivalent graphs. This shows that T-duality is a bijection between black-trivalent graphs of type $(k + 1, n)$ and white trivalent graphs of type (k, n) (where we consider both sets of graphs up to edge contraction and bivalent vertex addition/removal).

The map $G \to \hat{G}$ can also be phrased in terms of plabic tilings [9], which are dual to plabic graphs. Note that our notion of plabic tiling is slightly looser than that in [9], where plabic tilings are defined with a particular planar embedding and are dual to bipartite plabic graphs with no bivalent vertices.

Definition 4.1.14 (*Plabic Tilings*) Let G be any connected reduced plabic graph with n boundary vertices, and let $\mathbf{P}_n$ be a convex n-gon, whose vertices are labelled from 1 to n in clockwise order. The *plabic tiling* $\mathcal{T}(G)$ dual to G is a tiling of $\mathbf{P}_n$ by coloured polygons (bigons allowed) such that: (i) it is the planar dual of G; (ii) each black (white) vertex of G is dual to a black (white) polygon in $\mathcal{T}(G)$; (iii) vertex i of $\mathbf{P}_n$ is dual to the face of G touching boundary vertices $i - 1$ and i. We consider two plabic tilings $\mathcal{T}(G)$ and $\mathcal{T}'(G')$ *equivalent* if G and G' are move-equivalent. We call the internal vertices of $\mathcal{T}(G)$ *punctures*.

Conversely, if $\mathcal{T}$ is a plabic tiling, then we denote the dual plabic graph by $G(\mathcal{T})$. Note that this is obtained from $\mathcal{T}$ by placing a black vertex in each black polygon, a white vertex in each white polygon, and connecting two vertices whenever they correspond to two polygons which share an edge.

Figure 4.4 shows three move-equivalent plabic graphs and the corresponding plabic tilings.

The construction of $\hat{G}$ from G of Proposition 4.1.12 can also easily be phrased in terms of plabic tilings as follows. This should be equivalent to the construction given in the proof of [6, Proposition 8.3], though the description there uses horizontal sections of fine zonotopal tilings (Fig. 4.5).

Proposition 4.1.15 (T-duality via Plabic Tilings) *Let G be a connected reduced black-trivalent plabic graph and let $\mathcal{T}$ be the dual plabic tiling. Then the plabic graph $\hat{G} = \hat{G}(\mathcal{T})$ which is T-dual to G is obtained as follows:*

1. *Place a black vertex at each vertex of each black triangle in $\mathcal{T}$.*
2. *Place a white vertex in the middle of each black triangle of $\mathcal{T}$ and connect it to the vertices of the triangle.*
3. *Add an edge of $\hat{G}$ from boundary vertex i on the disc to the black vertex on boundary vertex i of $\mathcal{T}$.*

See (Fig. 4.5) for an example of T-duality via plabic graphs.

Remark 4.1.16 The construction $\hat{G}(\mathcal{T})$ given in Proposition 4.1.15 is a simple generalization of the definition given for triangulated unpunctured plabic tilings in Definition 2.3.2. So Proposition 4.1.15 shows that the plabic graph $\hat{G}(\mathcal{T})$ from Definition 2.3.2 is T-dual to the plabic tree $G(\mathcal{T})$.

4.2 Positroid Polytopes Versus Grasstopes

In this section, we show T-duality gives a bijection between positroid tiles of $\Delta_{k+1,n}$ and positroid tiles for $\mathcal{A}_{n,k,2}(Z)$ (Corollary 4.2.1). We then investigate parallels between the inequalities cutting out positroid polytopes Γ_π and the T-dual Grasstopes

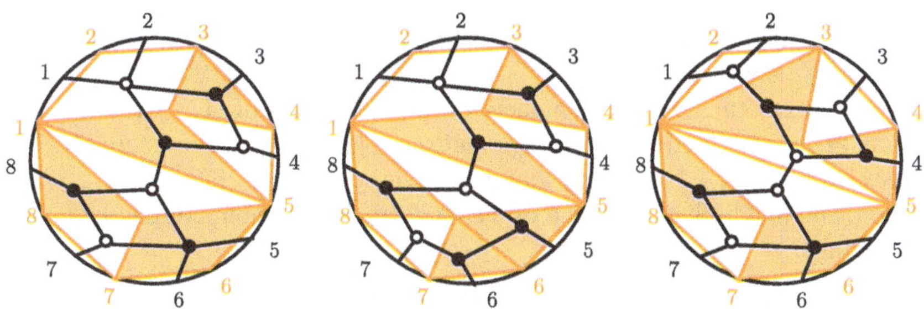

Fig. 4.4 Three equivalent plabic tilings $\mathcal{T}$ (in orange), and the corresponding dual plabic graphs $G(\mathcal{T})$ (in black). The center plabic tiling is dual to a black-trivalent plabic graph

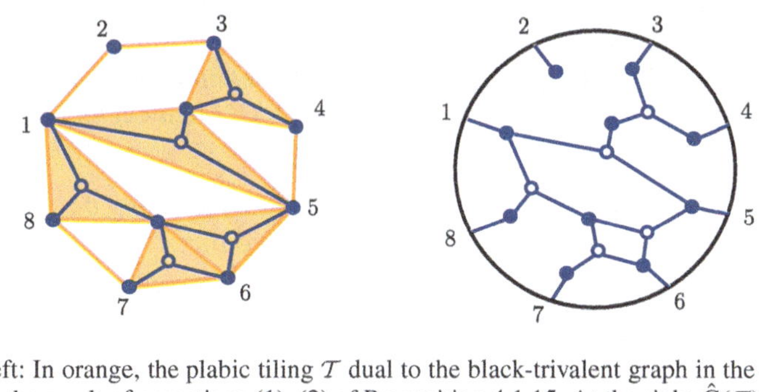

Fig. 4.5 Left: In orange, the plabic tiling $\mathcal{T}$ dual to the black-trivalent graph in the center of Fig. 4.4. In blue, the result of operations (1), (2) of Proposition 4.1.15. At the right $\hat{G}(\mathcal{T})$

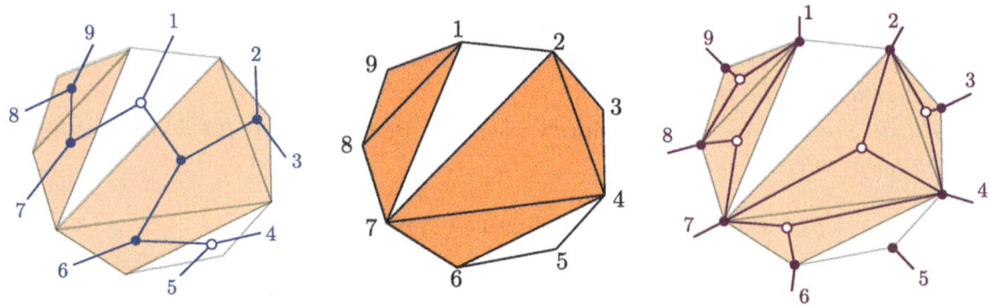

Fig. 4.6 A triangulation $\mathcal{T}'$ of a (5, 9)-unpunctured plabic tiling $\mathcal{T}$ (middle), its dual graph $G(\mathcal{T}')$ (left), and its T-dual graph $\hat{G}(\mathcal{T}')$ (right)

$Z_{\hat{\pi}}$; for positroid tiles, both pieces of data are encoded by the same plabic tiling (Theorem 4.2.2). We establish a similar parallel for facets of positroid tiles (Theorem 4.2.10), which we use later in Sect. 7.1.2 to prove the $m = 2$ *cluster adjacency* (Theorem 7.1.8).

4.2.1 T-Duality: Inequalities Versus Signs

In this subsection, we will see how unpunctured plabic tilings encode positroid tiles of both $\Delta_{k+1,n}$ and $\mathcal{A}_{n,k,2}(Z)$.

Theorems 2.3.25 and 3.1.23 characterize positroid tiles of $\mathcal{A}_{n,k,2}(Z)$ and $\Delta_{k+1,n}$ in terms of unpunctured plabic tilings and tree plabic graphs, respectively. These results with Remark 4.1.16 imply that positroid tiles of $\mathcal{A}_{n,k,2}(Z)$ and $\Delta_{k+1,n}$ are in bijection, and that both can be read off easily from (k, n)-unpunctured plabic tilings (see Fig. 4.6).

Corollary 4.2.1 (Positroid tiles in bijection via T-duality) *A positroid polytope Γ_G is a positroid tile of $\Delta_{k+1,n}$ if and only if the T-dual Grasstope $Z_{\hat{G}}$ is a positroid tile of $\mathcal{A}_{n,k,2}(Z)$. We read Γ_G and $Z_{\hat{G}}$ off of the same unpunctured plabic tiling $\mathcal{T}$ as follows:*

- *Choose any triangulation of the black polygons in $\mathcal{T}$.*
- *We let $G := G(\mathcal{T})$ be the dual plabic tree, as in Definition 4.1.14.*
- *We let $\hat{G} := \hat{G}(\mathcal{T})$ be the graph from Definition 2.3.2 (equivalently, in Proposition 4.1.15).*

From an unpunctured plabic tiling $\mathcal{T}$, we can obtain inequality descriptions of the positroid tile $\Gamma_{G(\mathcal{T})} \subset \Delta_{k+1,n}$ and the T-dual positroid tile $Z^{\circ}_{\hat{G}(\mathcal{T})} \subset \mathcal{A}_{n,k,2}(Z)$.

Given two positive numbers $a, b \in [n]$, the *cyclic interval* $[a, b]$ is defined to be

$$[a, b] := \begin{cases} \{a, a+1, \ldots, b-1, b\} & \text{if } a \leq b \\ \{a, a+1, \ldots, n, 1, \ldots, b\} & \text{otherwise.} \end{cases}$$

Theorem 4.2.2 (Inequalities and signs via T-duality) *Let $\mathcal{T}$ be an unpunctured plabic tiling and let $h \to j$ be a compatible arc, with $h < j$. Let $G(\mathcal{T})$ denote the tree plabic graph dual to $\mathcal{T}$, and $\hat{G}(\mathcal{T})$ the T-dual. Then:*

$$\begin{aligned} &(1) \qquad \operatorname{area}(h \to j) + 1 > x_{[h,j-1]} > \operatorname{area}(h \to j) \quad \text{for } x \in \Gamma^{\circ}_{G(\mathcal{T})} \\ &(2) \qquad \operatorname{sgn}\langle Yhj \rangle = (-1)^{\operatorname{area}(h \to j)} \quad \text{for } Y \in Z^{\circ}_{\hat{G}(\mathcal{T})}. \end{aligned}$$

The inequalities given by the arcs of any triangulation $\mathcal{T}'$ of $\mathcal{T}$ cut out $\Gamma^{\circ}_{G(\mathcal{T})}$ and $Z^{\circ}_{\hat{G}(\mathcal{T})}$.

Example 4.2.3 Consider the unpunctured plabic tiling $\mathcal{T}$ in Fig. 4.6. We have

$$\begin{aligned} &5 > x_{[1,7]} > 4, \quad 4 > x_{[1,6]} > 3, \quad 3 > x_{[2,5]} > 2, \quad \text{for } x \in \Gamma^{\circ}_{G(\mathcal{T})} \\ &\langle Y18 \rangle > 0, \quad \langle Y17 \rangle < 0, \quad \langle Y26 \rangle > 0, \quad \text{for } Y \in Z^{\circ}_{\hat{G}(\mathcal{T})}. \end{aligned}$$

◇

To prove Theorem 4.2.2, we need a few results on positroid polytopes Γ_G.

Lemma 4.2.4 *Let G be a bipartite plabic graph and let $\mathcal{T}$ be the dual plabic tiling. Let $W(G)$ and $B(G)$ denote the set of white and black vertices of G, respectively. Then*

$$|W(G)| - |B(G)| + |\{\text{bdry vt of } G \text{ adjacent to a black vt}\}| = \operatorname{area}(\mathcal{T}) - \operatorname{punc}(\mathcal{T}) + 1$$

where $\operatorname{area}(\mathcal{T})$ *is the number of black triangles in any triangulation of $\mathcal{T}$ and* $\operatorname{punc}(\mathcal{T})$ *is the number of internal vertices of $\mathcal{T}$.*

Proof Let E denote the edges of G involving at least one internal vertex. Each black vertex of G is dual to a black polygon of $\mathcal{T}$ with $\deg(v)$ many sides, so we have

$$\begin{aligned} \operatorname{area}(\mathcal{T}) &= \sum_{v \in B(G)} \deg(v) - 2 = \sum_{v \in B(G)} \deg(v) - 2|B(G)| \\ &= |E(G)| - |\{\text{bdry vt adjacent to a white vt}\}| - 2|B(G)| \end{aligned}$$

where the last equality follows from the fact that every edge of G contains a unique black vertex, except edges between a boundary vertex and a white vertex. The claim follows from this formula together with Euler's formula for planar graphs. □

Proposition 4.2.5 *Let $\mathcal{T}$ be a unpunctured plabic tiling and $G(\mathcal{T})$ the dual bipartite tree plabic graph. For all arcs $h \to j$ compatible with $\mathcal{T}$, points of $\Gamma_{G(\mathcal{T})}$ satisfy*

$$\operatorname{area}(h \to j) + 1 \geq x_{[h,j-1]} \geq \operatorname{area}(h \to j).$$

Proof Let G be the graph obtained from $G(\mathcal{T})$ by adding bivalent white vertices so that every boundary vertex is adjacent to a white vertex. Note that G is bipartite and represents the same positroid $\mathcal{M}$ as $G(\mathcal{T})$. In particular, the boundaries of matchings of G give the bases of $\mathcal{M}$. Let $W(G)$ and $B(G)$ denote the sets of white and black vertices of G, respectively.

Note that if $j = h + 1$, the inequality is clear.

We first deal with the case where $h \to j$ is an internal arc of $\mathcal{T}$. Let e be the edge of G which is dual to $h \to j$, and say the vertices of e are a white vertex w and black vertex b. If we remove the edge e, $G \setminus e$ has two connected components, G^w containing w and G^b containing b. Notice that both connected components are again bipartite plabic trees. Let I^w and I^b denote the boundary vertices of G^w and G^b, respectively. Because vertex i of $\mathcal{T}$ lies between boundary vertices $i-1, i$ of G, $\{I^w, I^b\} = \{[h, j-1], [j, h-1]\}$.

Now, we would like to compute the ranks of I_w, I_b. That is, for a matching M of G, we need to compute the maximum size of $\partial M \cap I^w$ and $\partial M \cap I^b$.

Let M be a matching of G. If M does not contain e, then M restricts to a matching of G^w and G^b. It is easy to see that

$$\begin{aligned}|\partial M \cap I^w| &= |W(G^w)| - |B(G^w)|\\ |\partial M \cap I^b| &= |W(G^b)| - |B(G^b)|.\end{aligned}$$

If M does contain e, then choose a path P from boundary to boundary which uses e and alternates between edges in M and edges not in M. Such a path can be constructed greedily because G is a tree. Orient P so it sees first w and then b. The edges of P in M are exactly the ones oriented from a white vertex to a black vertex. The first edge of P touches a boundary vertex in I^w and is oriented to a white vertex, so is not in M. The last edge of P touches a boundary vertex in I^b and is in M. Define a new matching N of G by $N := (M \setminus P) \cup (P \setminus M)$. The boundary ∂N contains one more element of I^w than ∂M, and one fewer element of I^b. The matching N does not contain e, so using the previous computation, we see that

$$\begin{aligned}|\partial M \cap I^w| &= |W(G^w)| - |B(G^w)| - 1\\ |\partial M \cap I^b| &= |W(G^b)| - |B(G^b)| + 1.\end{aligned}$$

We conclude that $\text{rank}(I^w) = |W(G^w)| - |B(G^w)|$ and $\text{rank}(I^b) = |W(G^b)| - |B(G^b)| + 1$.

From Proposition 3.1.10 and the fact that the rank of $\mathcal{M}$ is $|W(G)| - |B(G)| = |W(G^w)| - |B(G^w)| + |W(G^b)| - |B(G^b)|$, we see that the points of $\Gamma_{G(\mathcal{T})}$ satisfy

$$\begin{aligned}|W(G^w)| - |B(G^w)| - 1 \leq x_{I^w} &\leq |W(G^w)| - |B(G^w)|\\ |W(G^b)| - |B(G^b)| \leq x_{I^b} &\leq |W(G^b)| - |B(G^b)| + 1.\end{aligned}$$

All that remains is to rewrite the right hand sides of these inequalities in terms of area. Cut $\mathcal{T}$ along the arc $h \to j$, to get two smaller plabic tilings $\mathcal{T}^w$ and $\mathcal{T}^b$ containing the polygons dual to w and b, respectively. Notice that the graph $G(\mathcal{T}^w)$ dual to $\mathcal{T}^w$ can be obtained from G^w by adding a boundary vertex adjacent to w. Similarly, $G(\mathcal{T}^b)$ is obtained from G^b by adding a boundary vertex adjacent to b. So, using Lemma 4.2.4,

$$|W(G^w)| - |B(G^w)| = \text{area}(\mathcal{T}^w) + 1$$
$$|W(G^b)| - |B(G^b)| + 1 = \text{area}(\mathcal{T}^b) + 1.$$

Now, choose $v \in \{b, w\}$ so that $I^v = [h, j-1]$. Since $\mathcal{T}^v$ is exactly the part of $\mathcal{T}$ to the left of $h \to j$, the proposition now follows.

We now consider the case where $h \to j$ is not an arc of $\mathcal{T}$. In this case, let $\mathcal{T}'$ be the plabic tiling obtained from $\mathcal{T}$ by adding the arc $h \to j$. Let G be the tree plabic graph dual to $\mathcal{T}'$, which we make bipartite by adding an appropriately colored bivalent vertex v to the edge dual to $h \to j$. We also add bivalent white vertices to G to make all boundary vertices adjacent to a white vertex. Let e and f denote the edges containing v, and say e is to the left of $h \to j$. Similar to the first case, removing edges e, f and vertex v from G gives a graph with two connected components G^e, G^f which contain vertices adjacent to e and f, respectively. Notice that the boundary vertices I^e of G^e are exactly $[h, j-1]$. The rest of the argument is very similar to the first case. □

Recall that an arc $h \to j$ of an unpunctured plabic tiling is facet-defining if it bounds a black polygon on its left.

Proposition 4.2.6 *Let $\mathcal{T}$ be an unpunctured plabic tiling, and let G be the dual plabic tree of type $(k+1, n)$. Then Γ_G is cut out of $\mathbb{R}^n$ by the equality $x_{[n]} = k+1$ and the following inequalities, each of which defines a facet:*

1. *$x_i \geq 0$ for i a boundary vertex adjacent to a white vertex*
2. *$x_{[h,j-1]} \geq \text{area}(h \to j)$ for $h \to j$ a facet-defining arc of $\mathcal{T}$.*

Proof Recall from Theorem 3.1.23 that the moment map μ is a stratification-preserving homemorphism on the closure of S_G. So the facets of Γ_G are exactly the positroid polytopes $\Gamma_{G'}$ where $S_{G'}$ is a positroid cell contained in $\overline{S_G}$ with codimension 1. From [10, Corollary 18.10], each such cell is indexed by a reduced plabic graph G' obtained from G by removing a single edge (if the edge removed is between a boundary vertex and an internal vertex v, we also add a lollipop which is the opposite color of v).

Because G is a tree, $G' = G \setminus e$ is reduced for all edges e. If e is between a boundary vertex i and a white vertex, then G' has a black lollipop at i. Thus $S_{G'}$ has a loop at i, and $\Gamma_{G'}$ is contained in the hyperplane $x_i = 0$. Clearly Γ_G lies on the positive side of this hyperplane, which explains the facet inequalities of type 1.

If e is an edge between a boundary vertex i and a black vertex, then e is dual to the arc $(i+1) \to i$ of $\mathcal{T}$, which is a facet-defining arc. Then G' has a white lollipop at i, so $\Gamma_{G'}$ is contained in the hyperplane $x_i = 1$. Since we also have $x_{[n]} = k+1$, $\Gamma_{G'}$ is also contained in the hyperplane $x_{[i+1,i-1]} = k = \text{area}(i+1 \to i)$.

Now, consider the case when e is an edge between two internal vertices of G. The edge e is dual to the arc $h \to j$ of $\mathcal{T}$, which bounds a black polygon on the left. The proof of Lemma 4.2.5 shows that $\Gamma_{G'}$ is contained in the hyperplane $x_{[h,j-1]} = \text{area}(h \to j)$. This covers all edges of G, so we have described all facets. The directions of the facet inequalities follow immediately from Lemma 4.2.5. □

We can now prove Theorem 4.2.2.

Proof of Theorem 4.2.2 (1) follows from Lemma 4.2.5 and (2) follows from Theorem 2.3.14. The statement about inequalities cutting out $Z^{\circ}_{\hat{G}(\mathcal{T})}$ and $\Gamma^{\circ}_{G(\mathcal{T})}$ follow from Theorem 2.3.28, Proposition 4.2.6 and the fact that $x_{[n]} = \text{area}(h \to j) + \text{area}(j \to h) + 1$. □

We next generalize Theorem 4.2.2 by providing inequalities for full-dimensional positroid polytopes and Grasstopes from statistics of plabic tilings. We first generalize the definition of compatible arcs from Definition 2.3.6.

Definition 4.2.7 (*Statistics of plabic tilings*) Let $\mathcal{T}$ be a plabic tiling in a convex n-gon $\mathbf{P}_n$ and h, j a pair of vertices of $\mathbf{P}_n$. We say that the arc $h \to j$ is *compatible* with $\mathcal{T}$ if the arc either bounds or lies entirely inside a single polygon of $\mathcal{T}$. When $h \to j$ is compatible with $\mathcal{T}$, we let $\text{area}(h \to j) = \text{area}_{\mathcal{T}}(h \to j)$ denote the number of black triangles to the left of $h \to j$ in any triangulation of $\mathcal{T}$. We call the internal vertices of $\mathcal{T}$ *punctures*. We also let $\text{punc}(h \to j) = \text{punc}_{\mathcal{T}}(h \to j)$ denote the number of punctures of $\mathcal{T}$ to the left of the arc $h \to j$. Note that black bigons do not contribute to the area.

For example, the tiling $\mathcal{T}$ in Fig. 4.4 has two punctures, and $1 \to 3$, $1 \to 5$, $5 \to 7$ are compatible arcs. We have $\text{area}(1 \to 3) = 0$, $\text{area}(1 \to 5) = 2$, $\text{area}(5 \to 7) = 1$, $\text{punc}(1 \to 3) = \text{punc}(5 \to 7) = 0$, $\text{punc}(1 \to 5) = 1$.

Theorem 4.2.8 *Let $\mathcal{T}$ be a plabic tiling and let $h \to j$ be a compatible arc, with $h < j$. Let $G(\mathcal{T})$ denote the plabic graph dual to $\mathcal{T}$, and $\hat{G}(\mathcal{T})$ the T-dual. Then:*

(1) $\text{area}(h \to j) - \text{punc}(h \to j) + 1 > x_{[h,j-1]} > \text{area}(h \to j) - \text{punc}(h \to j)$ *for* $x \in \Gamma^{\circ}_{G(\mathcal{T})}$

(2) $\text{sgn}\langle Yhj \rangle = (-1)^{\text{area}(h\to j)-\text{punc}(h\to j)}$ *for* $Y \in Z^{\circ}_{\hat{G}(\mathcal{T})}$.

Note that compatible arcs depend on the tiling $\mathcal{T}$, while $\Gamma^{\circ}_{G(\mathcal{T})}$ and $Z^{\circ}_{\hat{G}(\mathcal{T})}$ depend only on $\overline{\mathcal{T}}$. Any arc compatible with any tiling equivalent to $\mathcal{T}$ gives inequalities for $\Gamma^{\circ}_{G(\mathcal{T})}$ and $Z^{\circ}_{\hat{G}(\mathcal{T})}$ via Proposition 4.2.8.

Proof The proof of (1) proceeds similarly as in Lemma 4.2.5, where we compute the rank of $[h, j-1]$. The arc $h \to j$ is dual to an edge e of $G(\mathcal{T})$ (or a graph which differs from $G(\mathcal{T})$ only by uncontracting an edge and adding a bivalent vertex). Removing e gives two connected components, the boundary vertices of which are

$[h, j-1]$ and $[j, h-1]$. Again, any matching of $G(\mathcal{T})$ will either use e or differs from a matching using e by a "swivel" (see [11, Appendix B]) along one of the two boundary faces containing e (which changes the boundary's intersection with $[h, j-1]$ by precisely 1) followed by swivels at faces contained in one of the connected components (which do not change the boundary's intersection with $[h, j-1]$). In this way we compute the rank of $[h, j-1]$ and $[j, h-1]$. One must apply Lemma 4.2.4 to obtain the ranks in terms of area and punc.

The proof of (2) proceeds similarly as in Theorem 2.3.14. Any almost-perfect matching M of $\hat{G}(\mathcal{T})$ which does not have h or j in ∂M will have $|\partial M \cap [h+1, j-1]| = \text{area}(h \to j) - \text{punc}(h \to j)$. Indeed, there are exactly $\text{area}(h \to j)$ internal white vertices to the left of $h \to j$, which must be covered by an edge of M, and exactly $\text{punc}(h \to j)$ many internal black vertices, which also must be covered. This leaves $\text{area}(h \to j) - \text{punc}(h \to j)$ edges of M which cover a boundary vertex. □

Example 4.2.9 For the plabic tiling $\mathcal{T}$ in Fig. 4.4, Proposition 4.2.8 tells us that

$$\begin{array}{llll} 1 > x_{[1,2]} > 0, & 2 > x_{[1,4]} > 1, & 2 > x_{[5,6]} > 1, & \text{for } x \in \Gamma_{G(\mathcal{T})}; \\ \langle Y13\rangle > 0, & \langle Y15\rangle < 0, & \langle Y57\rangle < 0, & \text{for } Y \in Z^{\circ}_{\hat{G}(\mathcal{T})}. \end{array}$$

◇

4.2.2 T-Duality and Facets

From an unpunctured plabic tiling $\mathcal{T}$, we can also read off the facets of both $\Gamma_{G(\mathcal{T})}$ and $Z_{\hat{G}(\mathcal{T})}$ (see Definition 2.1.8).

Theorem 4.2.10 (Facets via T-duality) *Let $\mathcal{T}$ be a triangulated unpunctured plabic tiling and let $h \to j$ be a facet-defining arc of $\mathcal{T}$. Let $G := G(\mathcal{T})$ be the plabic tree dual to $\mathcal{T}$ and let G' be the plabic forest obtained from G by deleting the edge dual to $h \to j$. Let $\hat{G}$ and $\hat{G}'$ denote their T-duals.*

1. *The positroid polytope $\Gamma_{G'}$ is a facet of Γ_G, and lies on the hyperplane*

$$x_{[h,j-1]} = \text{area}(h \to j).$$

2. *The Grasstope $Z_{\hat{G}'}$ is a facet of $Z_{\hat{G}}$, and lies on the hypersurface*

$$\langle Yhj\rangle = 0.$$

Moreover, if we let $h \to j$ range over the facet-defining arcs of $\mathcal{T}$ which are not on the boundary of $\mathbf{P}_n$, we obtain all facets of Γ_G and $Z_{\hat{G}}$ in the interior of $\Delta_{k+1,n}$ and $\mathcal{A}_{n,k,2}(Z)$.

Proof (1) follows immediately from Proposition 4.2.6 and its proof.

For (2), we first show that $Z_{\hat{G}'}$ is contained in the hypersurface $\{\langle Yhj\rangle = 0\}$. The arc $h \to j$ is in a unique triangle T_r of $\mathcal{T}$; say its third vertex is i. Using Proposition 4.1.12, it is not hard to see that $\hat{G}'$ is obtained from $\hat{G}$ by deleting the edge e from B_i to W_r. This means that every almost perfect matching of $\hat{G}'$ must use either the edge from B_h to W_r or the edge from B_j to W_r, so h or j is in the boundary. From Lemma 2.2.6, we immediately conclude that $\langle Yhj\rangle$ is identically zero on $Z^{\circ}_{\hat{G}'}$ and thus on $Z_{\hat{G}'}$.

Now, we show that $\tilde{Z}$ is injective on $Z_{\hat{G}'}$, by showing Proposition 2.3.19 holds for $Z_{\hat{G}'}$ and then applying the first paragraph in the proof of Theorem 2.3.25. Consider $Y \in Z_{\hat{G}'}$, let $C := C^{\text{tw}}_{\mathcal{T}}(Y)$ be the twistor coordinate matrix of Y and let $Y' := CZ$. We would like to show that $C \in S_{\hat{G}'}$ and that rowspan $Y' =$ rowspan Y; by Proposition 2.3.30, it suffices to show the former. Note that the Kasteleyn matrix K' for $\hat{G}'$ is obtained from the Kasteleyn matrix K for $\hat{G}$ by setting the parameter in row r and column i to 0. So we only need to show that for all arcs $a \to b$ of $\mathcal{T}$ with $\{a, b\} \neq \{h, j\}$, $\langle Y'ab\rangle$ is nonzero.

Pick such an arc $a \to b$ of $\mathcal{T}$. It suffices to show that there is a matching of $\hat{G}$ which does not use e and does not have a or b in its boundary. We will argue by induction on the number of black triangles of $\mathcal{T}$. The base case, with 1 triangle, is clear by inspection. The arc $a \to b$ bounds some black triangle T_s of $\mathcal{T}$, with third vertex c. Cut $\mathcal{T}$ along the arcs $a \to b$, $b \to c$ and $c \to a$ to obtain smaller triangulated unpunctured tilings $\mathcal{T}_1, \mathcal{T}_2, \mathcal{T}_3$ (one of which may be empty), which each contain a single edge of T_s. One will have $h \to j$ as a facet-defining arc. By induction, $\hat{G}(\mathcal{T}_i)$ has an almost-perfect matching M_i whose boundary avoids the appropriate vertices of T_s and does not use the edge e. Take the union of these matchings, together with the edge f from W_s to B_c. This gives a matching of $\hat{G}$ whose boundary avoids a, b. Note that since $\{a, b\} \neq \{h, j\}$, the edge f is different from e, so this matching does not use e.

Now we check that $Z_{\hat{G}'}$ is a facet of $Z_{\hat{G}}$. We first show that the hypersurface $H := \{\langle Yhj\rangle = 0\}$ intersects $Z_{\hat{G}}$ only on the boundary of $Z_{\hat{G}}$, which shows $Z_{\hat{G}'}$ is contained in the boundary as well. Recall that the open positroid tile $Z^{\circ}_{\hat{G}}$ is dense in $Z_{\hat{G}}$ and moreover, $(-1)^{\text{area}(h\to j)}\langle Yhj\rangle$ is positive on $Z^{\circ}_{\hat{G}}$ (Theorem 2.3.14). This implies that $(-1)^{\text{area}(h\to j)}\langle Yhj\rangle$ is positive on the interior of $Z_{\hat{G}}$. Indeed, if the hypersurface H intersected the interior of $Z_{\hat{G}}$, one could find an open set in the interior where $(-1)^{\text{area}(h\to j)}\langle Yhj\rangle$ is negative. (This is because $\langle Yhj\rangle$ is linear in the Plücker coordinates, so $\langle Yhj\rangle$ takes both positive and negative values on any open set in $\text{Gr}_{k,k+2}$ containing a point of H). But such a set cannot be in $\overline{Z^{\circ}_{\hat{G}}}$.

Now we verify that $Z_{\hat{G}'}$ has the correct codimension. From the proof of Proposition 4.2.6, $S_{G'}$ is codimension 1 in $\overline{S_G}$. Since T-duality is a rank-preserving poset isomorphism, we also have that $S_{\hat{G}'}$ is contained in $\overline{S_{\hat{G}'}}$ and has codimension 1; that is, $S_{\hat{G}'}$ has dimension $2k-1$. Because $\tilde{Z}$ is injective on $S_{\hat{G}'}$, $Z^{\circ}_{\hat{G}'}$ and $Z_{\hat{G}'}$ also have dimension $2k-1$.

To see the last statement of the proposition, note that any codimension 1 cell $S_H \subset \overline{S_{\hat{G}}}$ with a coloop q will have Z_H contained in the hypersurface $\{\langle Yq(q+1)\rangle = 0\}$. So the facets Z_H avoiding the amplituhedron boundaries must come from coloopless cells S_H. These coloopless cells are T-dual to loopless codimension 1 cells contained in $\overline{S_G}$. As $h \to j$ varies over all facet-defining arcs of $\mathcal{T}$, $S_{G'}$ varies over all such loopless cells, by Proposition 4.2.6. So the facets Z_H avoiding the amplituhedron boundary are of the form $Z_{\hat{G}'}$ for some arc $h \to j$. From the proof above, we see that $Z_{\hat{G}'}$ is not contained in an amplituhedron boundary precisely when $h \to j$ is not a boundary arc of $\mathcal{T}$. □

Example 4.2.11 Consider the unpunctured plabic tiling $\mathcal{T}$ in Fig. 2.3. The facet-defining arcs not on the boundary of $\mathbf{P}_9$ are $1 \to 7, 2 \to 7$ and $4 \to 6$, with area$(1 \to 7) =$ area$(2 \to 7) = 3$, area$(4 \to 6) = 0$. The corresponding internal facets lie on the following hyperplanes:

$$\langle Y17\rangle = 0, \qquad \langle Y27\rangle = 0, \qquad \langle Y46\rangle = 0, \qquad \text{for } Z_{\hat{G}(\mathcal{T})};$$
$$x_{[1,6]} = 3, \qquad x_{[2,6]} = 3, \qquad x_{[4,5]} = 0, \qquad \text{for } \Gamma_{G(\mathcal{T})}.$$

One facet-defining arc at the boundary of $\mathbf{P}_9$ is $2 \to 3$, with area$(2 \to 3) = 0$. This gives an external facet lying on $\langle Y23\rangle = 0$ for $Z_{\hat{G}(\mathcal{T})}$ and $x_2 = 0$ for $\Gamma_{G(\mathcal{T})}$. ◇

4.2.3 Positroid Tiles: Separable Permutations and Schröder Numbers

Recall from Corollary 4.2.1 that positroid tiles for both $\mathcal{A}_{n,k,2}(Z)$ and $\Delta_{k+1,n}$ are in bijection with (k, n)-unpunctured plabic tilings and tree positroids in $\mathrm{Gr}^{\geq 0}_{k+1,n}$. In this section we prove that the number $R_{k,n-2}$ of positroid tiles for $\mathcal{A}_{n,k,2}$ is given by [12, A175124]—a refinement of the *large Schröder numbers* (see Table 4.2). We will do this by giving a bijection between tree positroids in $\mathrm{Gr}^{\geq 0}_{k+1,n}$ and *separable* permutations on $[n-1]$ with k descents (enumerated by $R_{k,n-2}$).

Definition 4.2.12 A permutation $w = w_1 \dots w_n$ (in one-line notation) is *separable* if it is 3142- and 2413-avoiding, i.e. there are not four indices $i_1 < i_2 < i_3 < i_4$ such that $w_{i_3} < w_{i_1} < w_{i_4} < w_{i_2}$ or $w_{i_2} < w_{i_4} < w_{i_1} < w_{i_3}$.

Definition 4.2.13 Let π and ν be permutations on $[k]$ and $[l]$, respectively. The *direct sum* $\pi \oplus \nu$ and the *skew sum* $\pi \ominus \nu$ of π and ν are permutations on $[k + l]$ defined by:

$$(\pi \oplus \nu)_i = \begin{cases} \pi_i, & i \in [1, k] \\ \nu_{i-k} + k, & i \in [k+1, k+l] \end{cases}, \qquad (\pi \ominus \nu)_i = \begin{cases} \pi_i + l, & i \in [1, k] \\ \nu_{i-k}, & i \in [k+1, k+l] \end{cases}.$$

For example, $123 \oplus 21 = 12354$ and $123 \ominus 21 = 34521$.

Proposition 4.2.14 ([13]) *A permutation is separable if and only if w can be built from the permutation 1 by repeatedly applying $\oplus$ and $\ominus$ repeatedly.*

For example, the permutation $w = 231654$ can be written as

$$((1 \oplus 1) \ominus 1) \oplus ((1 \ominus 1) \ominus 1) = (12 \ominus 1) \oplus (21 \ominus 1) = 231 \oplus 321 = 231654.$$

Proposition 4.2.15 *Let β be the map sending a permutation $w = w_1 \dots w_{n-1}$ in one-line notation to the permutation $\beta(w) = (w_1, \dots, w_{n-1}, n)$ in cycle notation. Then β is a bijection between separable permutations on $[n-1]$ with k descents and trip permutations of tree positroids in $\mathrm{Gr}_{k+1,n}^{\geq 0}$.*

Proof We use strong induction on n; the base case $n = 2$ is trivial. It is enough to show that β is well-defined and surjective. Suppose that $w \in S_{n-1}$ is separable. Then either $w = u \oplus v$ or $w = u \ominus v$, for some $u \in S_{\ell-1}$, $v \in S_{r-1}$ separable, with $\ell - 1 + r - 1 = n - 1$. By the induction hypothesis, $\beta(u) \in S_\ell$ and $\beta(v) \in S_r$ are the trip permutations of tree plabic graphs S and T. We now "glue" together S and T in order to obtain a tree plabic graph with boundary vertices $\{1, 2, \dots, n\}$ with trip permutation $\beta(w) \in S_n$ (see Fig. 4.7).

It is straighforward to check that the trip permutation of the resulting tree is $\beta(w)$. This shows that β is well-defined.

For surjectivity, consider a trivalent tree plabic graph G on $[n]$. Let v be the internal vertex adjacent to the boundary vertex n. Then deleting v gives two trees: S on $[\ell]$ and T on $[\ell + 1, n - 1]$. Let π be the trip permutation of S. Subtract ℓ from the boundary labels of T to get a tree T' on $[1, n - \ell - 1]$ and let ν be its trip permutation. Then define w to be either $\beta^{-1}(\pi) \oplus \beta^{-1}(\nu)$ or $\beta^{-1}(\pi) \ominus \beta^{-1}(\nu)$, based on whether v is white or black. By the argument used above to show well-definedness, $\beta(w)$ is the trip permutation of G. □

Remark 4.2.16 If S and T are tree plabic graphs, the positroids associated to $S \oplus T$ and $S \ominus T$ are the *parallel-connection* and *series-connection* of the matroids associated to S, T.

Table 4.2 Large Schröder numbers R_{n-2} and their refinement $R_{k,n-2}$ which count the number of (k, n)-unpunctured plabic tilings

n \ k	0	1	2	3	4	5	R_{n-2}
2	1						1
3	1	1					2
4	1	4	1				6
5	1	10	10	1			22
6	1	20	48	20	1		90
7	1	35	161	161	35	1	394

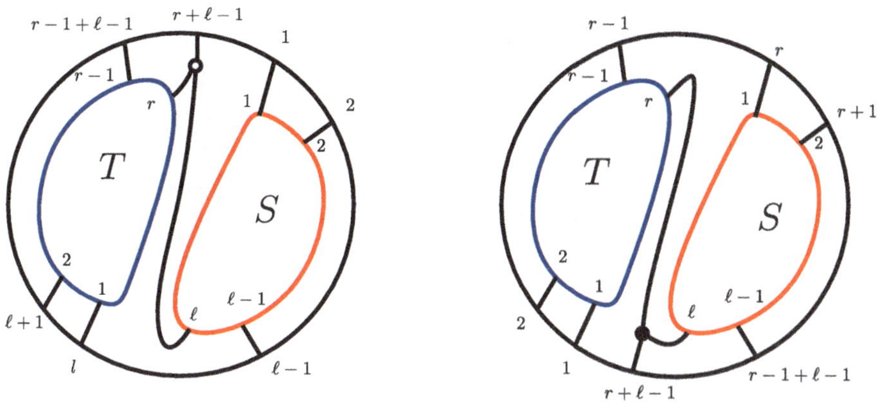

Fig. 4.7 How to glue S, T together when $w = u \oplus v$ (on the left) and when $w = u \ominus v$ (on the right)

The large Schröder number R_{n-2} counts separable permutations on $[n-1]$ [14] and $R_{k,n-2}$counts separable permutations on $[n-1]$ with k descents [15, Theorem 1.1].

Corollary 4.2.17 *Positroid tiles of Δ_{k+1} and $\mathcal{A}_{n,k,2}(Z)$ are in bijection with separable permutations on $[n-1]$ with k descents. They are enumerated by $R_{k,n-2}$ from [12, A175124].*

4.3 w-Simplices Versus w-Chambers: Eulerian Numbers

In this section we study the amplituhedron chambers of $\mathcal{A}_{n,k,2}(Z)$. Separately, the hypersimplex $\Delta_{k+1,n}$ has a well-known decomposition into simplices which refines every positroid tiling. Both decompositions have chambers/maximal simplices which are naturally indexed by permutations of $n-1$ with k descents. We use this correspondence in Sect. 5.1 to establish results on positroid tilings.

We begin by reviewing the decomposition of the hypersimplex $\Delta_{k+1,n}$. It is well-known that the volume of the hypersimplex $\Delta_{k+1,n}$ is the *Eulerian number* $E_{k+1,n-1}$ [16], which counts the permutations on $n-1$ letters with k descents. A triangulation[1] of $\Delta_{k+1,n}$ into unit simplices indexed by such permutations was first discovered by Stanley [17]. Sturmfels [18] later gave an *a priori* different triangulation of $\Delta_{k+1,n}$. Lam and Postnikov [19] then gave two other triangulations, and showed that all four triangulations coincide. After defining some permutation statistics, we will define this triangulation.

[1] Where we mean here a polytopal subdivision into simplices.

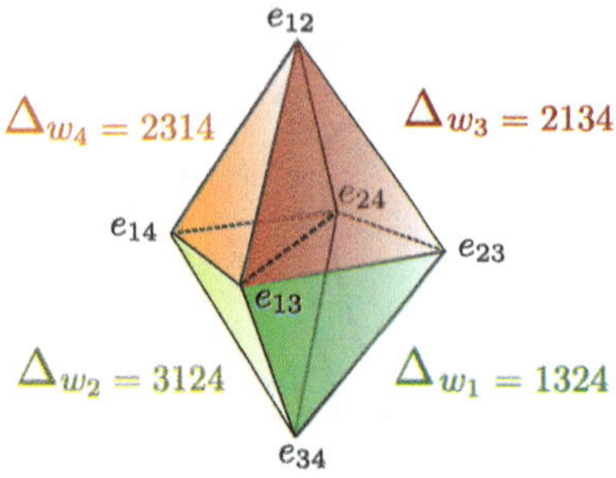

Fig. 4.8 w-simplices for $\Delta_{2,4}$

Definition 4.3.1 Let $w \in S_n$. We call a letter $i \geq 2$ in w a *left descent* (or a *left descent top*) if i occurs to the left of $i-1$ in w. In other words, $w^{-1}(i) < w^{-1}(i-1)$. And we say that $i \in [n]$ in w is a *cyclic left descent* if either $i \geq 2$ is a left descent of w or if $i = 1$ and 1 occurs to the left of n in w, that is, $w^{-1}(1) < w^{-1}(n)$. We let $\mathrm{cDes}_{\mathrm{L}}(w)$ denote the set of cyclic left descents of w, and $\mathrm{Des}_{\mathrm{L}}(w)$ the set of left descents. We frequently refer to cyclic left descents as simply *cyclic descents*.

Remark 4.3.2 Left and right descents and descent sets are discussed extensively in [20, Chap. 1]. Left descents are sometimes called *recoils* in the literature.

Let $D_{k+1,n}$ be the set of permutations $w \in S_n$ with $k+1$ cyclic descents and $w_n = n$. Note that $|D_{k+1,n}|$ equals the *Eulerian number*

$$E_{k,n-1} := \sum_{\ell=0}^{k+1} (-1)^{\ell} \binom{n}{\ell} (k+1-\ell)^{n-1}.$$

Example 4.3.3 $D_{2,4}$ contains the following $E_{2,3} = 4$ permutations: $w_1 = 1324$, $w_2 = 3124$, $w_3 = 2134$, $w_4 = 2314$. Moreover, we have: $\mathrm{cDes}_{\mathrm{L}}(w_1) = \{1, 3\}$, $\mathrm{cDes}_{\mathrm{L}}(w_2) = \{1, 3\}$, $\mathrm{cDes}_{\mathrm{L}}(w_3) = \{1, 2\}$ and $\mathrm{cDes}_{\mathrm{L}}(w_4) = \{1, 2\}$. ◇

Definition 4.3.4 (w-*simplices*) For $w \in D_{k+1,n}$, let $w^{(a)}$ denote the cyclic rotation of w ending at a. We define

$$I_r = I_r(w) := \mathrm{cDes}_{\mathrm{L}}(w^{(r-1)}).$$

The w-*simplex* $\Delta_w \subseteq \Delta_{k+1,n}$ is the simplex with vertices $e_{I_1}, \ldots, e_{I_n}$.

Example 4.3.5 From Example 4.3.3, we have 4 w-simplices for $\Delta_{2,4}$, see Fig. 4.8. For example, in order to determine the vertices of Δ_{w_1}, we consider the cyclic descents of rotations of w_1: $\mathrm{cDes}_{\mathrm{L}}(w_1) = \{1, 3\} = I_1$, $\mathrm{cDes}_{\mathrm{L}}(w_1^{(1)}) = \mathrm{cDes}_{\mathrm{L}}(3241) = \{2, 3\} = I_2$, $\mathrm{cDes}_{\mathrm{L}}(w_1^{(2)}) = \mathrm{cDes}_{\mathrm{L}}(4132) = \{3, 4\} = I_3$, $\mathrm{cDes}_{\mathrm{L}}(w_1^{(3)}) = \mathrm{cDes}_{\mathrm{L}}(2413) = \{2, 4\} = I_4$. Therefore, Δ_{w_1} is the convex hull of $e_{13}, e_{23}, e_{34}, e_{24}$, i.e. it is the green simplex in Fig. 4.8. ◇

Notice that r is always in I_r and $r-1$ is never in I_r.

The following triangulation of the hypersimplex first appeared in [17], though the description there was slightly different.

Proposition 4.3.6 ($\Delta_{k+1,n}$ is the union of w-simplices [17]) *The w-simplices $\{\Delta_w : w \in D_{k+1,n}\}$ are the maximal simplices of a triangulation of the hypersimplex $\Delta_{k+1,n}$. Moreover, projecting $\{\Delta_w : w \in S_n\}$ into $\mathbb{R}^{n-1}$ (see Remark 3.1.2), we obtain the maximal simplices in a triangulation of the hypercube $\square_{n-1}$ which refines the subdivision of the hypercube into hypersimplices.*

Remark 4.3.7 The w-simplex Δ_w as defined above agrees with the simplex denoted $\Delta_{(w)}$ in [19, Sect. 2.4]. In particular, the directed circuit the authors use to define $\Delta_{(w)}$ is given by $e_{I_1} \to e_{I_{w_1+1}} \to e_{I_{w_2+1}} \to \cdots \to e_{I_{w_{n-1}+1}} \to e_{I_1}$. Another way to say this is I_{w_i+1} is equal to $(I_{w_{i-1}+1} \setminus \{w_i\}) \cup \{w_i + 1\}$.

It follows from the results of [19] that every full-dimensional positroid polytope also has a triangulation into w-simplices. Indeed, the triangulation of $\Delta_{k+1,n}$ given by w-simplices is the simultaneous refinement of all positroid subdivisions of $\Delta_{k+1,n}$.

We now turn to the amplituhedron side. We define some special chambers in $\mathcal{A}_{n,k,2}(Z)$ whose sign vectors are obtained from cyclic descents of permutations. We later will show that these are precisely the realizable sign chambers (Theorems 4.3.13, 4.4.12).

Recall that for $v \in \mathbb{R}^n$, Flip(v) records where coordinates of v change sign (Definition 2.2.15).

Definition 4.3.8 (*w-chambers*) Let $w \in D_{k+1,n}$ and $I_1, \ldots, I_n$ as in Definition 4.3.4. We define $\hat{\Delta}^\circ_w(Z)$ to be the open amplituhedron chamber consisting of $Y \in \mathcal{A}_{n,k,2}(Z)$ such that for $a = 1, \ldots, n$,

$$\text{Flip}(\langle Ya\hat{1}\rangle, \langle Ya\hat{2}\rangle, \ldots, \langle Ya\widehat{a-1}\rangle, \langle Yaa\rangle, \langle Yaa+1\rangle, \ldots, \langle Yan\rangle) = I_a \setminus \{a\}. \tag{4.3.9}$$

Equivalently, $\hat{\Delta}^\circ_w(Z)$ consists of $Y \in \mathcal{A}_{n,k,2}(Z)$ such that

$$\begin{aligned} \text{sgn}\langle Yaj\rangle &= (-1)^{|I_a \cap [a,j-1]|-1} \quad \text{for } j > a \\ \text{sgn}\langle Ya\hat{j}\rangle &= (-1)^{|I_a \cap [a,j-1]|-1} \quad \text{for } j < a. \end{aligned}$$

We refer to $\hat{\Delta}^\circ_w(Z)$ and $\hat{\Delta}_w(Z) := \overline{\hat{\Delta}^\circ_w(Z)}$ as open and closed *amplituhedron w-chambers*. We will often refer to closed amplituhedron w-chambers as simply *w-chambers*.

Example 4.3.10 From Example 4.3.3, we have 4 w-chambers for $\mathcal{A}_{4,1,2}$, see Fig. 4.8. For example, in order to determine $\hat{\Delta}_{w_1}(Z)$, we need to consider the sign of twistor coordinates. From Eq. 4.3.9 and $I_1(w_1) = \{1, 3\}$:

$$\text{Flip}(0, \langle Y12\rangle > 0, \langle Y13\rangle, \langle Y14\rangle < 0) = \{3\} \Rightarrow \langle Y13\rangle > 0.$$

From $I_2(w_1) = \{2, 3\}$:

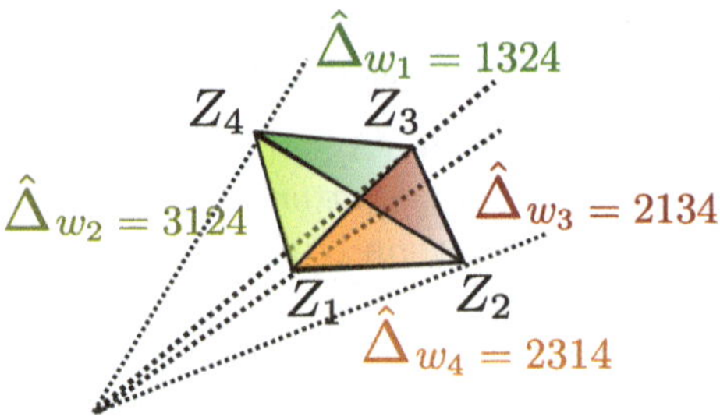

Fig. 4.9 w-chambers for $\mathcal{A}_{4,1,2}$

$$\text{Flip}(\langle Y2\hat{1}\rangle < 0, 0, \langle Y23\rangle > 0, \langle Y24\rangle) = \{3\} \Rightarrow \langle Y24\rangle < 0.$$

Therefore, $\hat{\Delta}_{w_1}$ is the (closure of the) region in $\text{Gr}_{1,3} \cong \mathbb{P}^2$ where $\langle Y14\rangle$, $\langle Y24\rangle$ are negative and all the remaining twistor coordinates are positive. This corresponds to the green triangle in Fig. 4.9 with vertices Z_3, Z_4 and the point where the two diagonals intersect. ◇

Remark 4.3.11 One might hope that the structure of $\hat{\Delta}_w(Z)$ does not depend on the choice of $Z \in \text{Mat}^{>0}_{n,k+2}$. Each $\hat{\Delta}_w(Z)$ is nonempty for some choice of Z (Theorem 4.4.12), but may be empty for other choices of Z (see Sect. 5.1.2).

Because the positroid tiles of $\mathcal{A}_{n,k,2}(Z)$ can be described entirely in terms of signs of twistor coordinates and the signs of twistor coordinates in $\hat{\Delta}_w(Z)$ are constant, we have the following lemma. It is the analogue of the fact that for a tree positroid polytope Γ_π, either $\Delta_w \cap \Gamma^\circ_\pi = \emptyset$ or $\Delta_w \subseteq \Gamma_\pi$.

Lemma 4.3.12 *Let Z°_π be a positroid tile for $\mathcal{A}_{n,k,2}(Z)$ and let $\hat{\Delta}_w(Z)$ be a non-empty w-chamber. Then either $\hat{\Delta}_w(Z) \cap Z^\circ_\pi = \emptyset$ or $\hat{\Delta}_w(Z) \subset Z_\pi$.*

Despite the subtleties regarding the nonemptiness of $\hat{\Delta}_w(Z)$, the closed w-chambers always cover $\mathcal{A}_{n,k,2}(Z)$, in direct analogy to w-simplices in $\Delta_{k+1,n}$.

Theorem 4.3.13 ($\mathcal{A}_{n,k,2}$ is the union of w-chambers) *Fix $k < n$ and $Z \in \text{Mat}^{>0}_{n,k+2}$. Then*

$$\mathcal{A}_{n,k,2}(Z) = \bigcup_{w \in D_{k+1,n}} \hat{\Delta}_w(Z).$$

To prove Theorem 4.3.13, we use a characterization of the simplices Δ_w given by Sturmfels [18], involving sorted collections. We follow the presentation of [19, Sect. 2.2].

Definition 4.3.14 Let $(J_1, \ldots, J_t)$ be a tuple of distinct elements of $\binom{[n]}{k+1}$, where we write $J_s = \{j_{s1} < j_{s2} < \cdots < j_{s(k+1)}\}$. We call $(J_1, \ldots, J_t)$ a *sorted collection* if $j_{11} \le j_{21} \le \cdots \le j_{t1} \le j_{12} \le j_{22} \le \cdots \le j_{t(k+1)}$. If (J_1, J_2) is a sorted collection, we call them a *sorted pair*.

The w-simplices of $\Delta_{k+1,n}$ are exactly the simplices with vertices $e_{J_1}, \ldots, e_{J_n}$ for $(J_1, \ldots, J_n)$ a sorted collection. To see if a collection is sorted, one need only check pairs of elements.

Lemma 4.3.15 *Given* $\{J_1, \ldots, J_t\} \subset \binom{[n]}{k+1}$*, suppose that for all* $a \neq b$*, either* (J_a, J_b) *or* (J_b, J_a) *is a sorted pair. Then* $J_1, \ldots, J_t$ *can be ordered to give a sorted collection.*

Proof First, notice that if (J_a, J_b) is a sorted pair and (J_b, J_c) is a sorted pair, then (J_a, J_c) is a sorted pair. Indeed, if $a \neq b$, then there exists i such that $j_{ai} < j_{bi} \leq j_{ci}$. It follows that (J_c, J_a) is not a sorted pair, so (J_a, J_c) must be. So on $\{J_1, \ldots, J_t\}$, the property of being a sorted pair is reflexive, antisymmetric, and transitive, which means it is a partial order. We've assumed every pair is comparable, so we have a total order. The result follows. □

Proof of Theorem 4.3.13 Let $Y \in \mathcal{A}_{n,k,2}$ be a point whose twistor coordinates are all nonzero. We will show that Y lies in $\hat{\Delta}_w(Z)$ for some w. The points with nonzero twistor coordinates form a dense subset of $\mathcal{A}_{n,k,2}$ (their complement, a union of hypersurfaces, has codimension 1), so this will show the desired equality.

$$\text{Set } L_a := \text{Flip}(\langle Ya\hat{1}\rangle, \langle Ya\hat{2}\rangle, \ldots, \langle Ya\,\widehat{a-1}\rangle, \langle Ya\; a\rangle, \langle Ya\; a+1\rangle, \ldots, \langle Ya\; n\rangle).$$

By Corollary 2.4.2, we have $|L_a| = k$. Choose $a < b$. We will show that $I_a := L_a \cup \{a\}$ and $I_b := L_b \cup \{b\}$ are distinct and, for some ordering, form a sorted pair. We temporarily abuse notation by omitting the Y's and hats from our notation; if $a > i$, we write $\langle ai\rangle$ for $\langle Ya\hat{i}\rangle$.

Certain 3-term Plücker relations constrain sign flips, as noted in [21, Sect. 5]. For $j \in [a-2] \cup [b+1, n]$, we have the relation

$$\langle j\; j+1\rangle\langle a\; b\rangle = \langle a\; j\rangle\langle b\; j+1\rangle - \langle b\; j\rangle\langle a\; j+1\rangle \tag{4.3.16}$$

and for $j \in [a+1, b-2]$ we have

$$\langle j\; j+1\rangle\langle b\; a\rangle = \langle b\; j\rangle\langle a\; j+1\rangle - \langle a\; j\rangle\langle b\; j+1\rangle. \tag{4.3.17}$$

Because $\text{sgn}\langle j\; j+1\rangle = +$ for all j, the sign of the left hand sides of Eq. (4.3.16) and (4.3.17) does not depend on j. This means that if $\text{sgn}\langle a\; b\rangle = +$, then for $j \in [a-2] \cup [b+1, n]$

$$\begin{pmatrix} \text{sgn}\langle a\; j\rangle & \text{sgn}\langle a\; j+1\rangle \\ \text{sgn}\langle b\; j\rangle & \text{sgn}\langle b\; j+1\rangle \end{pmatrix} \neq \begin{pmatrix} \delta & \epsilon \\ \epsilon & -\delta \end{pmatrix} \tag{4.3.18}$$

for any $\delta, \epsilon \in \{+, -\}$. Similarly, if $\text{sgn}\langle a\; b\rangle = -$, then for $j \in [a-2] \cup [b+1, n]$

$$\begin{pmatrix} \text{sgn}\langle a\; j\rangle & \text{sgn}\langle a\; j+1\rangle \\ \text{sgn}\langle b\; j\rangle & \text{sgn}\langle b\; j+1\rangle \end{pmatrix} \neq \begin{pmatrix} \delta & -\epsilon \\ \epsilon & \delta \end{pmatrix}. \tag{4.3.19}$$

If $\text{sgn}\langle b\ a\rangle = +$ (respectively, $-$), then for $j \in [a+1, b-1]$, the sign pattern in Eq. (4.3.19) (respectively, Eq. (4.3.18)) never occurs.

Suppose j is a value where the sign pattern in Eq. (4.3.18) is forbidden. If there is a sign flip after $\langle a\ j\rangle$ and not after $\langle b\ j\rangle$, then $\text{sgn}\langle a\ j\rangle = \text{sgn}\langle b\ j\rangle$; if there is a sign flip after $\langle b\ j\rangle$ and not after $\langle a\ j\rangle$, then $\text{sgn}\langle a\ j\rangle \neq \text{sgn}\langle b\ j\rangle$. When the sign pattern in Eq. (4.3.19) is forbidden, there are analogous statements with conclustions swapped. This means that for any interval $I \subset [n]$ where one of the patterns is forbidden, $|L_a \cap I|$ and $|L_b \cap I|$ differ by at most one and either $(L_a \cap I, L_b \cap I)$ or $(L_b \cap I, L_a \cap I)$ is sorted.[2]

Which one of $(L_a \cap I, L_b \cap I)$ and $(L_b \cap I, L_a \cap I)$ is sorted gives us additional information.

Let $P := L_a \cap [a-2]$, $Q := L_b \cap [a-2]$. Suppose $P \neq Q$, and consider the smallest j so that there is a sign flip after exactly one of $\langle a\ j\rangle$ and $\langle b\ j\rangle$. Clearly, (P, Q) is sorted if and only if there is a sign flip after $\langle a\ j\rangle$ and not after $\langle b\ j\rangle$. If the latter occurs, then $\text{sgn}\langle ab\rangle = +$ and $\text{sgn}\langle a\ 1\rangle = \text{sgn}\langle b\ 1\rangle$, or $\text{sgn}\langle ab\rangle = -$ and $\text{sgn}\langle a\ 1\rangle \neq \text{sgn}\langle b\ 1\rangle$; in short, $\text{sgn}\langle a\ b\rangle \cdot \text{sgn}\langle a\ 1\rangle = \text{sgn}\langle b\ 1\rangle$. Analogously, if (Q, P) is sorted, then $\text{sgn}\langle a\ b\rangle \cdot \text{sgn}\langle a\ 1\rangle \neq \text{sgn}\langle b\ 1\rangle$.

Let $T := L_a \cap [a+1, b-2]$ and $U = L_b \cap [a+1, b-2]$, and suppose $T \neq U$. By essentially identical reasoning as in the previous paragraph, if (T, U) is sorted, then $\text{sgn}\langle b\ a\rangle \cdot \text{sgn}\langle a\ a+1\rangle \neq \text{sgn}\langle b\ a+1\rangle$. Since $\langle a\ a+1\rangle > 0$ by assumption, the latter condition implies we have a sign flip between $\langle b\ a\rangle$ and $\langle b\ a+1\rangle$, so $a \in L_b$. Similar reasoning gives that if (U, T) is sorted, then $a \notin L_b$.

Let $V := L_a \cap [b+1, n]$ and $W := L_b \cap [b+1, n]$ and suppose that $V \neq W$. Repeating the arguments of the previous paragraphs gives that if (V, W) is sorted, then $b \notin L_a$, and if (W, V) is sorted, then $b \in L_a$.

Now, there are two cases: $|L_a \cap [b, n]|$ and $|L_b \cap [b, n]|$ have the same parity or they have opposite parity. They are similar, so we will assume we are in the first case, and leave the second to the reader.

Suppose $|L_a \cap [b, n]|$ and $|L_b \cap [b, n]|$ have the same parity. Note that $(-1)^{|L_i \cap [i+1, j-1]|}$ is $\text{sgn}\langle i\ j\rangle$, and, since $b \notin L_b$, $L_b \cap [b, n]$ and $L_b \cap [b+1, n]$ are equal. So

$$\begin{aligned}\text{sgn}\langle b\ n\rangle &= (-1)^{|L_a \cap [b,n]|}\\ &= (-1)^{|L_a \cap [a+1,b-1]|}(-1)^{|L_a \cap [a+1,n]|}\\ &= \text{sgn}\langle a\ b\rangle \cdot \text{sgn}\langle a\ 1\rangle\end{aligned}$$

and thus (P, Q) is sorted. We will show that (I_a, I_b) is sorted and $I_a \neq I_b$.

If $b \in L_a$, then $|V|$ and $|W|$ have different parity. In particular, V and W are not equal, so (W, V) is sorted and $|W| = |V| + 1$. The two sets interweave like

$$w_1 \leq v_1 \leq w_2 \leq v_2 \leq \cdots \leq w_r \leq v_r \leq w_{r+1}.$$

---

[2] Extending the definition of sorted in the obvious way to sets whose sizes differ by at most one.

Since $V = I_a \cap [b+1, n]$ and $W = I_b \cap [b+1, n]$, we also have that I_a and I_b are distinct. Note that $I_b \cap [b, n] = W \cup \{b\}$ and $I_a \cap [b, n] = V \cup \{b\}$ and we have

$$b \le b < w_1 \le v_1 \le w_2 \le v_2 \le \cdots \le w_r \le v_r \le w_{r+1},$$

so $(I_b \cap [b, n], I_a \cap [b, n])$ form a sorted pair. If $b \notin L_a$, then $|V|$ and $|W|$ have the same parity. The pair (V, W) is sorted, and $I_a \cap [b, n] = V$, while $I_b \cap [b, n] = W \cup \{b\}$. So $(I_b \cap [b, n], I_a \cap [b, n])$ are a sorted pair in this case as well, since we have

$$b < v_1 \le w_1 \le v_2 \le w_2 \le \cdots \le v_r \le w_r.$$

Note that b is in I_b but not in I_a, so we also have that I_a and I_b are distinct in this case.

Now we turn to the sets T and U. Because $\operatorname{sgn}\langle a\ b\rangle = (-1)^k \operatorname{sgn}\langle b\ a\rangle$, we have

$$\begin{aligned}(-1)^{|L_a \cap [a, b-1]|} &= (-1)^k (-1)^{|L_b \cap [b, a-1]|} \\ &= (-1)^{|L_b \cap [1, n]| + |L_b \cap [b, a-1]|} \\ &= (-1)^{|L_b \cap [a, b-1]|}.\end{aligned}$$

Note that $|T| \le |L_a \cap [a, b-1]| \le 1 + |T|$, since $a \notin L_a$. If $a \in L_b$, then (T, U) is sorted and $|L_b \cap [a, b-1]| = 1 + |U|$, since $b - 1 \notin L_b$. If T and U have the same cardinality, then $|L_a \cap [a, b-1]|$ must be equal to $1 + |T|$ in order to have the same parity as $1 + |U|$. Thus $b - 1 \in L_a$. This means that $(I_a \cap [a, n], I_b \cap [a, n])$ is a sorted pair, as we have

$$a \le a < t_1 \le u_1 \le \cdots \le t_j \le u_j < b - 1 < b \le \cdots \le v_r \le w_r.$$

If $|T| = |U| + 1$, we conclude by similar reasoning that $b - 1 \notin L_a$, and again $(I_a \cap [a, n], I_b \cap [a, n])$ is a sorted pair, as we have

$$a \le a < t_1 \le u_1 \le \cdots \le t_j \le u_j \le t_{q+1} < b \le \cdots \le v_r \le w_r.$$

If $a \notin L_b$, then (U, T) is sorted and $L_b \cap [a, b-1] = U$, since a and $b - 1$ are not in L_b. A parity argument as in the last paragraph shows that if $|U| = |T|$, then $b - 1 \notin L_a$; if $|U| = |T| + 1$, then $b - 1 \in L_a$. Either way, $(I_a \cap [a, n], I_b \cap [a, n])$ is a sorted pair; we see

$$a \le u_1 \le t_1 \le \cdots \le u_j \le t_j < b \le \cdots \le v_r \le w_r$$

in the first case and

$$a \le u_1 \le t_1 \le \cdots \le u_j \le t_j \le u_{q+1} < b - 1 < b \le \cdots \le v_r \le w_r$$

in the second.

Finally, we deal with P and Q. Recall that (P, Q) are sorted. Since $|L_a \cap [b, n]|$ and $|L_b \cap [b, n]|$ have the same parity and $|L_a \cap [a, b-1]|$ and $|L_b \cap [a, b-1]|$ have the same parity, $|L_a \cap [1, a-1]|$ and $|L_b \cap [1, a-1]|$ have the same parity. Since $a - 1 \notin L_a$, we have that $P = L_a \cap [1, a-1]$. On the other hand $|Q| \leq |L_b \cap [1, a-1]| \leq |Q| + 1$. If $|P| = |Q|$, then for parity reasons $Q = L_b \cap [1, a-1]$ and thus $a - 1 \notin L_b$. So $(I_a \cap [1, a-1], I_b \cap [1, a-1])$ are a sorted pair, as we have

$$r_1 \leq s_1 \leq \cdots \leq r_i \leq s_i.$$

Similarly, if $|P| = |Q| + 1$, then $a - 1 \in L_b$ and $(I_a \cap [1, a-1], I_b \cap [1, a-1])$ again are a sorted pair, since we have

$$r_1 \leq s_1 \leq \cdots \leq r_i \leq s_i \leq r_{i+1} \leq a - 1.$$

Since $(I_a \cap [1, a-1], I_b \cap [1, a-1])$ is a sorted pair ending in an element of I_b and $(I_a \cap [a, n], I_b \cap [a, n])$ is a sorted pair, it follows that (I_a, I_b) is a sorted pair. □

Using Theorem 4.3.13, we can conclude that positroid tiles are unions of amplituhedron w-chambers, just as tree positroid polytopes are unions of hypersimplex w-simplices. More precisely, we have the following corollary, which we sharpen further in Proposition 5.1.4.

Corollary 4.3.20 (Positroid tiles are unions of w-chambers) *Let Z_π be a positroid tile for $\mathcal{A}_{n,k,2}(Z)$. Then*

$$Z_\pi = \bigcup_{\substack{\hat{\Delta}_w(Z): \\ \hat{\Delta}^\circ_w(Z) \cap Z^\circ_\pi \neq \emptyset}} \hat{\Delta}_w(Z).$$

4.4 The Hypercube Versus the Total Amplituhedron

In this subsection, we embed $\mathcal{A}_{n,k,2}(Z)$ into a full-dimensional subset of $\mathrm{Gr}_{2,n}$—the '$\mathcal{G}$-amplituhedron' $\mathcal{G}_{n,k,2}$– which does not depend on Z. We use sign chambers in the $\mathcal{G}$-amplituhedron to prove that all w-chambers of $\mathcal{A}_{n,k,2}$ are realizable (Theorem 4.4.12). We also draw another parallel between the hypersimplex and the ampliltuhedron. In Remark 3.1.2 we saw that the union of the (projected) hypersimplices $\tilde{\Delta}_{k+1,n}$ is the hypercube $\square_{n-1}$. Analogously, we take the union of $\mathcal{G}$-amplituhedra varying over all k to obtain the *total amplituhedron* $\mathcal{G}^{(2)}_n$, which is the amplituhedron-analogue of $\square_{n-1}$.

The following definition is inspired by Corollary 2.4.3. The only difference is that we drop the dependence on $Z \in \mathrm{Mat}^{>0}_{n,k+2}$ (or equivalently, its column span $W \in \mathrm{Gr}^{>0}_{k+2,n}$).

Definition 4.4.1 (*The $\mathcal{G}$-Amplituhedron*) Fix $k < n$ and let

$$\mathcal{G}^\circ_{n,k,2} := \{z \in \mathrm{Gr}_{2,n} \mid p_{i,i+1}(z) > 0 \text{ for } 1 \leq i \leq n-1, \text{ and } p_{n,\hat{1}}(z) > 0, \\ \text{and } \mathrm{var}((p_{12}(z), p_{13}(z), \ldots p_{1n}(z)) = k\},$$

The closure $\mathcal{G}_{n,k,2} := \overline{\mathcal{G}^\circ_{n,k,2}}$ in $\mathrm{Gr}_{2,n}$ is the *$\mathcal{G}$-amplituhedron.*

Remark 4.4.2 Following the sign-flip descriptions from [21, 22], one can generalise most of the definitions in this section for any m. We leave this to future work.

Comparing with Corollary 2.4.3 we have:

Proposition 4.4.3 *Fix $k < n$, and $W \in \mathrm{Gr}^{>0}_{k+2,n}$. Then*

$$\mathcal{G}^\circ_{n,k,2}(W) = \{z \in \mathcal{G}^\circ_{n,k,2} \mid z \subset W\} = \mathcal{G}^\circ_{n,k,2} \cap \mathrm{Gr}_2(W) \quad \textit{and} \quad \mathcal{B}_{n,k,2}(W) = \overline{\mathcal{G}^\circ_{n,k,2} \cap \mathrm{Gr}_2(W)}.$$

Remark 4.4.4 Note that $\mathcal{G}_{n,k,2}$ is full-dimensional in $\mathrm{Gr}_{2,n}$, i.e. it has dimension $2(n-2)$, whereas $\mathcal{G}^\circ_{n,k,2}(W)$ and $\mathcal{B}_{n,k,2}(W)$ are full-dimensional in $\mathrm{Gr}_2(W)$, i.e. have dimension $2k$.

Motivated by the decomposition of $\mathcal{A}_{n,k,2}(Z)$ into w-chambers, we analogously define w-chambers for $\mathcal{G}_{n,k,2}$.

Definition 4.4.5 Let $w \in D_{k+1,n}$ and let $I_a := \mathrm{cDes_L}(w^{(a-1)})$. Then the *(open) $\mathcal{G}$-amplituhedron w-chamber* $\hat{\Delta}^\circ_w(\mathcal{G})$ consists of $z \in \mathcal{G}_{n,k,2}$ such that for $a = 1, \ldots, n$,

$$\mathrm{Flip}(p_{a\hat{1}}(z), p_{a\hat{2}}(z), \ldots, p_{a\widehat{a-1}}(z), p_{aa}(z), p_{aa+1}(z), \ldots, p_{an}(z)) = I_a \setminus \{a\}.$$

Equivalently, $\Delta^\circ_w(\mathcal{G})$ consists of $z \in \mathrm{Gr}_{2,n}$ such that

$$\mathrm{sgn}\, p_{aj}(z) = (-1)^{|I_a \cap [a,j-1]|-1} \text{ for } j > a \quad \text{and} \quad \mathrm{sgn}\, p_{a\hat{j}}(z) = (-1)^{|I_a \cap [a,j-1]|-1} \text{ for } j < a. \tag{4.4.6}$$

The *closed $\mathcal{G}$-amplituhedron w-chamber* is the closure $\hat{\Delta}_w(\mathcal{G}) := \overline{\hat{\Delta}^\circ_w(\mathcal{G})}$. Abusing notation, we will often omit 'closed' when referring to closed $\mathcal{G}$-amplituhedron w-chambers.

The situation for $\mathcal{G}$-amplituhedron w-chambers is quite straightforward. We will see that the second part of (4.4.6) follows from the first part, so each $\hat{\Delta}^\circ_w(\mathcal{G})$ is an oriented matroid stratum, whose underlying matroid is the rank 2 uniform matroid on $[n]$.

Proposition 4.4.7 *Let $w \in D_{k+1,n}$. Then $\hat{\Delta}^\circ_w(\mathcal{G})$ is nonempty and is contractible.*

Proof Consider n vectors $v_1, v_2, \ldots, v_n$ in $\mathbb{R}^2$ so that the matrix

$$\begin{bmatrix} v_1 & v_{w_1+1} & v_{w_2+1} & \cdots & v_{w_{n-1}+1} \end{bmatrix}$$

has all maximal minors positive. In particular, drawing the vectors in the plane and going counterclockwise, we see $v_1, v_{w_1+1}, v_{w_2+1}, \ldots, v_{w_{n-1}+1}$ in that order.

Now, set $z_1 := v_1$ and $z_b := (-1)^{|I_1 \cap [1,b-1]|-1} v_b$ for $b > 2$. We claim that

$$z = \left[z_1\ z_2\ z_3\ \ldots\ z_n\right]$$

represents a point in $\hat{\Delta}^{\circ}_{w}(\mathcal{G})$.

Clearly $p_{1b}(z)$ has the correct sign. Consider $1 \neq a < j$. We will assume $\det[v_a v_j] > 0$; the other case is similar. Note that $p_{aj}(z)$ has sign $(-1)^{|I_1 \cap [a,j-1]|}$; we would like to show that this is equal to $(-1)^{|I_a \cap [a,j-1]|-1}$. Because $\det[v_a v_j] > 0$, $a-1$ occurs before $j-1$ in w, written in one-line notation. Recall from Remark 4.3.7 that $I_{w_i+1} = I_{w_{i-1}+1} \setminus \{w_i\} \cup \{w_i + 1\}$. That is, I_a can be obtained from I_1 by removing w_1 and adding $w_1 + 1$, then removing w_2 and adding $w_2 + 1$, and so on until one removes $w_q = a - 1$ and adds a. Note that for $c = w_1, \ldots, w_{q-1}$, the numbers c and $c+1$ are either both in $[a, j-1]$ or both not in $[a, j-1]$, so $|I_1 \cap [a, j-1]| = |I_{c+1} \cap [a, j-1]|$. Removing $a-1$ from $I_{w_{q-1}+1}$ and adding a increases the size of the intersection with $[a, j-1]$ by one, so $|I_1 \cap [a, j-1]| = |I_a \cap [a, j-1]| - 1$. This shows $p_{aj}(z)$ has the correct sign for $a < j$; a similar argument shows that for $a > j$, $p_{a\hat{j}}(z)$ has the desired sign so long as $p_{ja}(z)$ does.

So $\hat{\Delta}^{\circ}_{w}(\mathcal{G})$ is an oriented matroid stratum for a rank 2 oriented matroid. By [23, Corollary 8.2.3], all rank 2 oriented matroid strata are contractible. □

Example 4.4.8 Let $w = (2, 6, 1, 4, 5, 3, 7) \in D_{k+1,n}$ with $k = 3$ and $n = 7$. We have $I_1 = \{1, 2, 4, 6\}$. Following the proof of Proposition 4.4.7, we can choose

$$\begin{aligned}(v_1, v_{w_1+1}, v_{w_2+1}, v_{w_3+1}, v_{w_4+1}, v_{w_5+1}, v_{w_6+1}) &= (v_1, v_3, v_7, v_2, v_5, v_6, v_4)\\ &= \begin{pmatrix} 1\ 1\ 1\ 1\ 1\ 1\ 1 \\ 1\ 2\ 3\ 4\ 5\ 6\ 7 \end{pmatrix}.\end{aligned}$$

We then get

$$z = \begin{pmatrix} 1\ 1\ -1\ -1\ 1\ 1\ -1 \\ 1\ 4\ -2\ -7\ 5\ 6\ -3 \end{pmatrix}.$$

One can check that z lies in $\hat{\Delta}^{\circ}_{w}(\mathcal{G})$. Also note that both row vectors $z^{(1)}$ and $z^{(2)}$ of z have $\mathrm{var}(z^{(1)}) = \mathrm{var}(z^{(2)}) = k$ by construction. ◇

Remark 4.4.9 The w-chambers of the $\mathcal{G}$-amplituhedron do *not* depend on Z. Roughly speaking, the amplituhedron w-chambers are linear slices of $\mathcal{G}$-amplituhedron w-chambers. More precisely, for $Z \in \mathrm{Mat}^{>0}_{n,k+2}$ with column span $W \in \mathrm{Gr}^{>0}_{k+2,n}$, we have

$$f_Z(\hat{\Delta}^{\circ}_{w}(\mathcal{G}) \cap \mathrm{Gr}_2(W)) = \hat{\Delta}^{\circ}_{w}(Z),$$

where f_Z is the homeomorphism from Proposition 2.2.3.

Our next goal is to use Proposition 4.4.7 and the connection with the $\mathcal{B}$-amplituhedron from Proposition 4.4.3 to deduce Theorem 4.4.12 on realizability of w-chambers. We start by proving the following lemma.

Lemma 4.4.10 *Given a* $2 \times n$ *matrix* z *as constructed in the proof of Proposition 4.4.7, we can construct a* $(k+2) \times n$ *matrix* A' *representing a point* $W \in \mathrm{Gr}_{k+2,n}^{\geq 0}$ *which contains* $\mathrm{rowspan}(z)$ *as a subspace.*

Proof Let $z^{(1)} = (z_1^{(1)}, \dots, z_n^{(1)})$ and $z^{(2)} = (z_1^{(2)}, \dots, z_n^{(2)})$ denote the rows of z. By construction, $\mathrm{var}(z^{(1)}) = \mathrm{var}(z^{(2)}) = k$ and moreover we can partition $[n]$ into disjoint consecutive intervals $H_1 \sqcup \cdots \sqcup H_{k+1}$ such that the entries of $z^{(1)}$ and $z^{(2)}$ in positions H_i are positive if i is odd and negative if i is even.

By [24, Lemma 4.1], since $\mathrm{var}(z^{(2)}) = k$, we can construct a $(k+1) \times n$ matrix A with maximal minors nonnegative whose row sum is $z^{(2)}$. More explicitly, we define the ith row of A to be the vector $(a_{i1}, \dots, a_{i2})$ such that $a_{ij} = z_j^{(2)}$ for $j \in H_i$ and $a_{ij} = 0$ for $j \notin H_i$. Therefore the nonvanishing Plücker coordinates of A are precisely the $p_B(A)$ such that $B = \{b_1 < b_2 < \cdots < b_{k+1}\}$ with $b_i \in H_i$.

Let A' be the matrix obtained from A by adding $z^{(1)}$ as a new top (0th) row. We will label the rows of A' from 0 to $k+1$. The nonvanishing Plücker coordinates of A' are precisely the $p_{B'}(A')$ where $B' = \{b_1 < b_2 < \cdots < b_{k+1}\} \cup \{b'_j\}$ with $b_i \in H_i$ and both b_j, b'_j lie in H_j.

Now we can compute the Plücker coordinates of A' in terms of Plücker coordinates of z and minors of A. Let $B' = \{b_1 < b_2 < \cdots < b_{k+1}\} \cup \{b'_j\}$ as above. Then we have

$$
\begin{aligned}
p_{B'}(A') &= (-1)^{j-1} \Delta_{0j, b_j b'_j}(A') \cdot \Delta_{[k+1] \setminus j, B' \setminus \{b_j, b'_j\}}(A') \\
&= (-1)^{j-1} p_{b_j b'_j}(z) \prod_{i \neq j} z_{b_i}^{(2)}
\end{aligned}
$$

where $\Delta_{R,C}(A')$ denotes the minor of A' on rows R and columns C. Now it follows from the construction of z that since both b_j, b'_j lie in H_j, we have $p_{b_j b'_j}(z) > 0$. Additionally, we have that the sign of $\prod_{i \neq j} z_{b_i}^{(2)}$ is $(-1)^{j+1}$. Therefore $p_{B'}(A')$ is positive, as desired. □

Example 4.4.11 We illustrate the proof of Lemma 4.4.10 using our running example from Example 4.4.8. We have

$$
z = \begin{pmatrix} 1 & 1 & -1 & -1 & 1 & 1 & -1 \\ 1 & 4 & -2 & -7 & 5 & 6 & -3 \end{pmatrix}
$$

so

$$
A = \begin{pmatrix} 1 & 4 & 0 & 0 & 0 & 0 & 0 \\ 0 & 0 & -2 & -7 & 0 & 0 & 0 \\ 0 & 0 & 0 & 0 & 5 & 6 & 0 \\ 0 & 0 & 0 & 0 & 0 & 0 & -3 \end{pmatrix} \text{ and } A' = \begin{pmatrix} 1 & 1 & -1 & -1 & 1 & 1 & 1 \\ 1 & 4 & 0 & 0 & 0 & 0 & 0 \\ 0 & 0 & -2 & -7 & 0 & 0 & 0 \\ 0 & 0 & 0 & 0 & 5 & 6 & 0 \\ 0 & 0 & 0 & 0 & 0 & 0 & -3 \end{pmatrix}.
$$

Both matrices have maximal minors nonnegative. If $B' = \{2, 3, 5, 6, 7\}$ then $2 \in H_1, 3 \in H_2, 5, 6 \in H_3, 7 \in H_4$ and we have

$$p_{B'}(A') = \Delta_{03,56}(A')\Delta_{124,237}(A') = p_{56}(z) \cdot (4 \cdot (-2) \cdot (-3)).$$

◇

Theorem 4.4.12 (All w-chambers are realisable) *For each $w \in D_{k+1,n}$, there exists some $Z \in \mathrm{Mat}_{n,k+2}^{>0}$ such that the amplituhedron w-chamber $\hat{\Delta}_w(Z)$ in $\mathcal{A}_{n,k,2}(Z)$ is nonempty.*

Proof By Proposition 2.2.3, we know that $\mathcal{B}_{n,k,2}(W)$ is homeomorphic to $\mathcal{A}_{n,k,2}(Z)$, where $W \in \mathrm{Gr}_{k+2,n}^{>0}$ is the column span of Z. Moreover the Plücker coordinates of the former agree with the twistor coordinates of the latter. Proposition 4.4.7 gives an explicit construction of a $2 \times n$ matrix z representing a point in $\hat{\Delta}_w^{\circ}(\mathcal{G})$, and by Proposition 4.4.3 we have $\mathcal{B}_{n,k,2}(W) = \overline{\mathcal{G}_{n,k,2}^{\circ} \cap \mathrm{Gr}_2(W)}$, so to prove the theorem, we just need to realize z as a two-dimensional subspace contained in some $(k+2)$-plane $W \in \mathrm{Gr}_{k+2,n}^{>0}$.

By Lemma 4.4.10, we can realize z as a two-dimensional subspace contained in a $(k+2)$-plane $W \in \mathrm{Gr}_{k+2,n}^{\geq 0}$. (Here $W = \mathrm{rowspan}(A')$.) We want to now slightly deform A' to make it totally positive.

We claim that $A' \in \mathrm{Gr}_{k+2,n}^{\geq 0}$ is the limit of a sequence of points $\{\tilde{A}_t\} \in \mathrm{Gr}_{k+2,n}^{>0}$ where $\mathrm{rowspan}(\tilde{A}_t)$ contains a 2-plane $z(t)$ which lies in the same sign-chamber as z. To see this, we use the fact that $\mathrm{Gr}_{k+2,n}^{\geq 0} = \overline{\mathrm{Gr}_{k+2,n}^{>0}}$ (see Remark 2.1.3). We can therefore write A' as the limit of a sequence of matrices of the form $A' + (\epsilon_{ij}(t)) \in \mathrm{Gr}_{k+2,n}^{>0}$, where $(\epsilon_{ij}(t))$ is a $(k+2) \times n$ matrix, and each $\epsilon_{ij}(t)$ is a function of t with small absolute value and $\epsilon_{ij}(t) \to 0$ as $t \to 0$.

We denote the rows of $A' + (\epsilon_{ij}(t))$ by $r_i(t)$ for $0 \leq i \leq k+1$. Let $z^{(1)}(t) := r_0(t)$, let $z^{(2)}(t) := r_1(t) + r_2(t) + \cdots + r_{k+1}(t)$, and let $z(t)$ be the matrix with rows $z^{(1)}(t)$ and $z^{(2)}(t)$.

Then when $t = 0$, we have $z = z(t)$. Moreover for small t, the Plücker coordinates of $z(t)$ have the same signs as the Plücker coordinates of z, so $z(t)$ lies in the same w-simplex $\hat{\Delta}_w^{\circ}(\mathcal{G})$ as z. But now by construction, $\mathrm{rowspan}(z(t))$ lies in the positive $(k+2)$-plane $W = \mathrm{rowspan}(A' + (\epsilon_{ij}(t)))$. This completes the proof of the theorem. □

Corollary 4.4.13 (Amplituhedron chambers and Eulerian numbers) *The realizable amplituhedron chambers $\mathcal{A}_{n,k,2}^{\sigma}(Z)$ are exactly the w-chambers $\hat{\Delta}_w^{\circ}(Z)$ where $w \in D_{k+1,n}$.*

Proof Theorem 4.4.12 shows that each w-chamber is realizable. Theorem 4.3.13 shows that no other sign chambers are realizable. □

We now turn to the structure of the $\mathcal{G}$-amplituhedron. The proof of Theorem 4.3.13 implies the following.

Theorem 4.4.14 ($\mathcal{G}_{n,k,2}$ is the union of w-chambers) *Fix $k < n$, then*

$$\mathcal{G}_{n,k,2} = \bigcup_{w \in D_{k+1,n}} \hat{\Delta}_w(\mathcal{G}).$$

Definition 4.4.15 (*Total Amplituhedron*) The *total amplituhedron* $\mathcal{G}_n^{(2)}$ is

$$\mathcal{G}_n^{(2)} := \bigcup_{k=0}^{n-2} \mathcal{G}_{n,k,2}.$$

Note that $\mathcal{G}_n^{(2)}$ has top dimension $2(n-2)$ in $\mathrm{Gr}_{2,n}$, and it does *not* depend on Z.

Recall that the hypercube $\square_{n-1} \subset \mathbb{R}^{n-1}$ can be decomposed into $(n-1)!$ w-simplices in a way which is compatible with its slicing into (projected) hypersimplices $\tilde{\Delta}_{1,n}, \tilde{\Delta}_{2,n}, \ldots, \tilde{\Delta}_{n-1,n}$. Each $\tilde{\Delta}_{k+1,n}$ is a union of exactly $E_{k,n-1}$ simplices, where $E_{k,n-1}$ is the Eulerian number.

Analogously, by Theorem 4.4.14, the total amplituhedron $\mathcal{G}_n^{(2)} \subset \mathrm{Gr}_{2,n}$ can be decomposed into $(n-1)!$ w-chambers in a way which is compatible with its decomposition into the $\mathcal{G}$-amplituhedra $\mathcal{G}_{n,0,2}, \mathcal{G}_{n,1,2}, \ldots, \mathcal{G}_{n,n-2,2}$. Each $\mathcal{G}_{n,k,2}$ is a union of exactly $E_{k,n-1}$ w-chambers. This is the '$m = 2$' equivalent of encoding all helicity sectors at once for tree-level scattering amplitudes of $\mathcal{N} = 4$ SYM for $m = 4$. A related space was discussed in the context of the $\mathcal{B}$-amplituhedron [22, Sect. 3.4].

References

1. M. Parisi, M. Sherman-Bennett, L. Williams, The $m = 2$ amplituhedron and the hypersimplex: signs, clusters, triangulations, Eulerian numbers, 2023. Accepted and to be published in *Communications of the American Mathematical Society*
2. T. Łukowski, M. Parisi, L.K. Williams, The positive tropical Grassmannian, the hypersimplex, and the $m = 2$ amplituhedron. Int. Math. Res. Not. 03 2023. rnad010
3. S.N. Karp, L.K. Williams, Y.X. Zhang, Decompositions of amplituhedra (2017)
4. N. Arkani-Hamed, J. Bourjaily, F. Cachazo, A. Goncharov, A. Postnikov, J. Trnka, *Grassmannian Geometry of Scattering Amplitudes* (Cambridge University Press, 2016)
5. C. Benedetti, A. Chavez, D. Tamayo, Quotients of uniform positroids (2019)
6. P. Galashin, Critical varieties in the grassmannian (2021)
7. A. Knutson, T. Lam, D.E. Speyer, Positroid varieties: juggling and geometry. Compositio Mathematica **149**(10), 1710–1752 (2013)
8. P. Galashin, A. Postnikov, L. Williams, Higher secondary polytopes and regular plabic graphs (2019)
9. S. Oh, A. Postnikov, D.E. Speyer, Weak separation and plabic graphs. Proc. Lond. Math. Soc. (3) **110**(3), 721–754 (2015)
10. A. Postnikov, Total positivity, Grassmannians, and networks. Preprint, http://math.mit.edu/~apost/papers/tpgrass.pdf
11. G. Muller, D.E. Speyer, The twist for positroid varieties. Proc. Lond. Math. Soc. (3) **115**(5), 1014–1071 (2017)
12. N.J. Sloane et al., The on-line encyclopedia of integer sequences (2003–)
13. S. Kitaev, *Patterns in Permutations and Words* (Springer, 2011)

14. J. West, Generating trees and the Catalan and Schröder numbers. Discrete Math. **146**(1), 247–262 (1995)
15. S. Fu, Z. Lin, J. Zeng, On two unimodal descent polynomials. Discrete Math. **341**(9), 2616–2626 (2018)
16. R.P. Stanley, *Enumerative Combinatorics. Volume 1*, Cambridge Studies in Advanced Mathematics, vol. 49, 2nd edn. (Cambridge University Press, Cambridge, 2012)
17. R. Stanley, Eulerian partitions of a unit hypercube, in *Higher combinatorics: Proceedings of the NATO Advanced Study Institute held in Berlin, 1–10 Sept 1976* ed. by M. Aigner, p. 49 (D. Reidel Publishing Co., Dordrecht-Boston, Mass., 1977). NATO Advanced Study Institute Series. Ser. C: Mathematical and Physical Sciences, 31
18. B. Sturmfels, *Gröbner Bases and Convex Polytopes*, University Lecture Series, vol. 8 (American Mathematical Society, Providence, RI, 1996)
19. T. Lam, A. Postnikov, Alcoved polytopes. I. Discrete Comput. Geom. **38**(3), 453–478 (2007)
20. A. Björner, F. Brenti, *Combinatorics of Coxeter Groups*, Graduate Texts in Mathematics, vol. 231 (Springer, New York, 2005)
21. N. Arkani-Hamed, H. Thomas, J. Trnka, Unwinding in binary. Preprint, http://arxiv.org/abs/1704.05069
22. S.N. Karp, L.K. Williams, The m = 1 amplituhedron and cyclic hyperplane arrangements. Int. Math. Res. Not, IMRN (to appear) (2017)
23. A. Björner, M. Las Vergnas, B. Sturmfels, N. White, G.M. Ziegler, *Oriented Matroids*, *Encyclopedia of Mathematics and its Applications*, vol. 46, 2nd edn. (Cambridge University Press, Cambridge, 1999)
24. S.N. Karp, Sign variation, the Grassmannian, and total positivity. J. Combin. Theory Ser. A **145**, 308–339 (2017)

Chapter 5
Positroid Tilings

The combinatorics of T-duality is realised via a bijection between loopless positroid cells S_π of $\mathrm{Gr}_{k+1,n}^{\geq 0}$ and coloopless positroid cells $S_{\hat{\pi}}$ of $\mathrm{Gr}_{k,n}^{\geq 0}$, see Sect. 4.1. This is a simple operation on decorated permutations—which induces a poset isomorphism too (Proposition 4.1.6). Remarkably, T-duality sends *positroid tiles* of the hypersimplex (images of cells where the moment map is injective) to positroid tiles of the amplituhedron (images of cells where $\widetilde{Z}$ is injective), see Corollary 4.2.1. Properties of positroid polytopes are in general related to the ones of their T-dual Grasstopes (Proposition 4.2.8 and Theorem 4.2.10). Moreover, the decomposition of $\Delta_{k+1,n}$ into w-simplicies is T-dual to the decomposition of the amplituhedron $\mathcal{A}_{n,k,2}$ into w-chambers—an Eulerian number of regions where all twistor coordinates have fixed signs. These results allow us to prove another crucial feature of T-duality about *positroid tilings* of $\Delta_{k+1,n}$ and $\mathcal{A}_{n,k,2}$.

Given any surjective map $\phi : \mathrm{Gr}_{r,n}^{\geq 0} \to X$ where $\dim X = d$, it is natural to try to decompose X using images of positroid cells under ϕ. This leads to the following definition.

Definition 5.0.1 (*Positroid tilings/dissections*) Let $\phi : \mathrm{Gr}_{r,n}^{\geq 0} \to X$ be a continuous surjective map where $\dim X = d$. A *positroid tiling* of X (with respect to ϕ) is a collection $\{\overline{\phi(S_\pi)}\}$ of images of d-dimensional positroid cells such that

- ϕ is injective on each S_π from the collection
- pairs of distinct images $\phi(S_\pi)$ and $\phi(S_{\pi'})$ are disjoint
- $\cup\overline{\phi(S_\pi)} = X$.

If we relax the injectivity assumption (e.g. we exclude the first property), then we call $\{\overline{\phi(S_\pi)}\}$ a *positroid dissection*. When ϕ is the moment map, the (closures of) the images of the positroid cells S_π are the *positroid polytopes* Γ_π [1], so a positroid tiling of the hypersimplex is a decomposition into positroid polytopes (cf. Chap. 3). When ϕ is the amplituhedron map $\tilde{Z}$, the (closures of) the images of the positroid

M. Parisi, *Combinatorial Aspects of Scattering Amplitudes*, Springer Theses,
https://doi.org/10.1007/978-3-031-41069-7_5

cells S_π are *Grasstopes* Z_π. These were first studied in [2] as the building blocks of conjectural positroid tilings of the amplituhedron—hence providing expressions for $\mathcal{N} = 4$ SYM scattering amplitudes.

Remark 5.0.2 Note that *positroid tilings* in general are *not* the same as *polyhedral triangulations*. In this work, we introduced the word 'tiling' to avoid the confusion. However, in the literature they have often been referred to as *(positroid) triangulations*.

Remark 5.0.3 In the following, if the collection $\{\overline{\phi(S_\pi)}\}$ is a positroid tiling (dissection), we may say for brevity that the collection of positroid cells $\{S_\pi\}$ *gives* a positroid tiling (dissection).

Note that neither the amplituhedron nor the Grasstopes are polytopes. Nevertheless, the main result of this chapter is showing that positroid tilings of the hypersimplex $\Delta_{k+1,n}$ and of the ampliuthedron $\mathcal{A}_{n,k,2}(Z)$ are in bijection via T-duality (Theorem 5.1.9). Considering the various relations via T-duality, should there be a map from $\Delta_{k+1,n}$ to $\mathcal{A}_{n,k,2}$ or vice-versa? We have dim $\Delta_{k+1,n} = n - 1$ and dim $\mathcal{A}_{n,k,2} = 2k$, with no relation between $n - 1$ and $2k$ (apart from $k \leq n$) so it is not obvious that such map should exist.

T-duality reveals to be an important tool for studying the amplituhedron $\mathcal{A}_{n,k,2}$: we can try to understand properties of the amplituhedron by studying the hypersimplex and applying T-duality. For example, we can obtain a whole class of 'regular' tilings of the amplituhedron from correspondingly regular positroid tilings of the hypersimplex. By Proposition 3.2.14, the regular positroid tilings (dissections) of $\Delta_{k+1,n}$ come precisely from the positive tropical Grassmannian $\text{Trop}^+ \text{Gr}_{k+1,n}$. Hence using T-duality we can get positroid tilings (dissections) of the amplituhedron $\mathcal{A}_{n,k,2}$, see Corollary 5.2.22. We speculate that $\text{Trop}^+ \text{Gr}_{k+1,n}$ plays the role of secondary fan for the regular positroid dissections of $\mathcal{A}_{n,k,2}$.

Summary of the Chapter. This chapter is based on the following works from the author: [3] and [4, Sect. 11].

In this chapter we introduce and discuss about positroid tilings and dissections—which arise from subdividing a space using images of positroid cells of a totally non-negative Grassmannian. In Sect. 5.1 we show how T-duality relates tilings and dissections of the hypersimplex $\Delta_{k+1,n}$ and the amplituhedron $\mathcal{A}_{n,k,2}(Z)$. In Sect. 5.1.1 we state one of the main results of our work: a collection of positroid polytopes is a positroid tiling of $\Delta_{k+1,n}$ if and only if the collection of T-dual Grasstopes is a positroid tiling of $\mathcal{A}_{n,k,2}(Z)$ for all Z. We also conjecture the same holds true if we replace "tilings" with "dissections". In Sect. 5.1.2 we discuss algorithms to find positroid tilings using w-simplices of $\Delta_{k+1,n}$ and (possibly empty) w-chambers of $\mathcal{A}_{n,k,2}(Z)$. In Sect. 5.2 we show how to obtain different types of positroid tilings (and dissections) of $\Delta_{k+1,n}$ and $\mathcal{A}_{n,k,2}(Z)$ from: BCFW recursion relations (Sect. 5.2.1), the positive tropical Grassmannian $\text{Trop}^+ \text{Gr}_{k+1,n}$ (Sect. 5.2.2), and positions of descents/sign flips (Sect. 5.2.3). Finally, in Sect. 5.3 we discuss the relation of T-duality with cyclic symmetry and *parity duality*—which relates the ampliuthedron $\mathcal{A}_{n,k,m}$ with $\mathcal{A}_{n,n-m-k,m}$.

5.1 Positroid Tilings of $\Delta_{k+1,n}$ and $\mathcal{A}_{n,k,2}$

In this section we prove that T-duality is a bijection between positroid tilings of $\Delta_{k+1,n}$ and positroid tilings of $\mathcal{A}_{n,k,2}(Z)$. For this, we use that realisable amplituhedron chambers are exactly labelled by Eulerian numbers Corollary 4.4.13. Furthermore, we explore the phenomena that some w-chambers can be empty and describe algorithms to obtain positroid tilings exploiting w-simplices and w-chambers.

Recall the definition of positroid tiling from Definition 5.0.1. Specializing to $\mathcal{A}_{n,k,2}(Z)$, we get the following.

Definition 5.1.1 (*Positroid Tilings/Dissections of $\mathcal{A}_{n,k,2}$*) Let $\mathcal{C} = \{Z_\pi\}$ be a collection of Grasstopes, with $\{S_\pi\}$ positroid cells of $\mathrm{Gr}^{\geq 0}_{k,n}$. We say that $\mathcal{C}$ is a *positroid tiling* of $\mathcal{A}_{n,k,2}(Z)$ if we have that:

1. each Grasstope Z_π is a positroid tile (i.e. $\widetilde{Z}$ is injective on S_π and $\dim Z_\pi = 2k$);
2. pairs of distinct open Grasstopes Z°_π and $Z^\circ_{\pi'}$ in the collection are disjoint;
3. $\bigcup_\pi Z_\pi = \mathcal{A}_{n,k,2}(Z)$.

We say that $\mathcal{C}$ is a *positroid dissection* of $\mathcal{A}_{n,k,2}(Z)$ if each Grasstope Z_π is full-dimensional (i.e. $\dim Z_\pi = 2k$) and (2) and (3) are satisfied.

Remark 5.1.2 Alternatively, one could define a positroid tiling/dissections as coming from a collection $\{S_\pi\}$ of cells such that $\{Z_\pi\}$ is a tiling/dissection (as above) for *all* choices of Z in $\mathrm{Mat}^{>0}_{n,k+2}$. We use Definition 5.1.1 here since some objects we define will be sensitive to the choice of Z.

Because positroid tiles in $\mathcal{A}_{n,k,2}(Z)$ are defined by sign conditions, the decomposition of $\mathcal{A}_{n,k,2}(Z)$ into chambers refines every positroid tiling.

Example 5.1.3 (*Positroid tiling of $\mathcal{A}_{4,1,2}$*) By Example 2.1.7 positroid tiles of $\mathcal{A}_{4,1,2}$ are just (projective) triangles. Then positroid tilings of $\mathcal{A}_{4,1,2}$ are just standard (polyhedral) triangulations of a 4-gon. Figure 5.1 shows a triangulation $\{Z_{\nu_1}, Z_{\nu_2}\}$, where Z_{ν_1} is the triangle with vertices Z_1, Z_3, Z_4 considered in Example 2.1.7, and Z_{ν_2} is the triangle obtained from the positroid cell S_{ν_2}. Elements of the cell can be represented by a row vector $C = (c_1, c_2, c_3, c_4)$, with $p_4(C) = c_4 = 0$ and $c_1, c_2, c_3 > 0$. Therefore Z_{ν_2} is the triangle with vertices Z_1, Z_2, Z_3. ◇

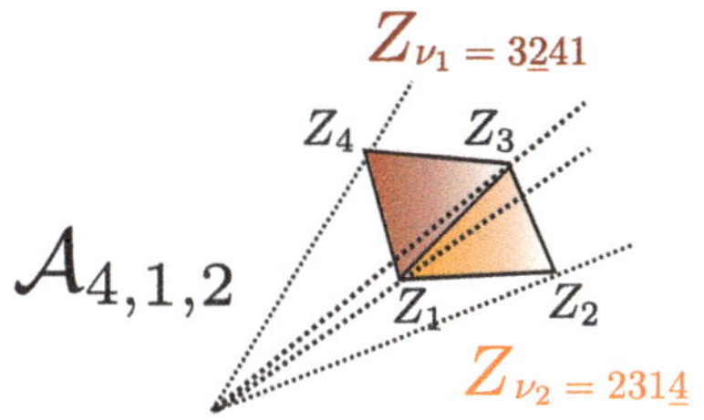

Fig. 5.1 A positroid tiling of $\mathcal{A}_{4,1,2}$

5.1.1 Tilings in Bijection via T-Duality

Recall that w-simplices in $\Delta_{k+1,n}$ are indexed by $D_{k+1,n}$. One main tool is the following.

Proposition 5.1.4 *Fix* $k < n$ *and* $Z \in \mathrm{Mat}_{n,k+2}^{>0}$. *Suppose* $w \in D_{k+1,n}$ *and* $\hat{\Delta}_w(Z) \neq \emptyset$. *For any tree positroid polytope* Γ_π *we have*

$$\Delta_w \subset \Gamma_\pi \quad \Leftrightarrow \quad \hat{\Delta}_w(Z) \subset Z_{\hat{\pi}}. \tag{5.1.5}$$

Proof Fix a triangulated unpunctured plabic tiling $\mathcal{T}$ so that $G(\mathcal{T})$ is a plabic tree with trip permutation π and $\hat{G}(\mathcal{T})$ has trip permutation $\hat{\pi}$. From Theorem 4.2.2, $Z^\circ_{\hat{\pi}}$ consists of $Y \in \mathrm{Gr}_{k,k+2}$ such that for all arcs $a \to b$ of $\mathcal{T}$

$$\begin{cases} \mathrm{sgn}\langle Yab\rangle = (-1)^{\mathrm{area}(a\to b)} & \text{if } a < b \\ \mathrm{sgn}\langle Ya\hat{b}\rangle = (-1)^{\mathrm{area}(a\to b)} & \text{if } a > b \end{cases}$$

and Γ_π consists of the points $x \in \mathbb{R}^n$ satisfying

$$\mathrm{area}(a \to b) \leq x_{[a,b-1]} \leq \mathrm{area}(a \to b) + 1$$

for all arcs $a \to b$ of $\mathcal{T}$. (In fact, to cut out $Z^\circ_{\hat{\pi}}$, it suffices to consider arcs with $a < b$.)

Suppose $\Delta_w \subset \Gamma_\pi$. Then the vertices $e_{I_1}, \ldots, e_{I_n}$ of Δ_w satisfy the defining inequalities of Γ_π. In particular, for each arc $a \to b$ of $\mathcal{T}$,

$$\mathrm{area}(a \to b) \leq |I_a \cap [a, b-1]| \leq \mathrm{area}(a \to b) + 1.$$

By Remark 4.3.7, there is another vertex e_{I_r} of Δ_w satisfying $I_r = I_a \setminus \{a\} \cup \{a-1\}$. This vertex also satisfies the defining inequalities of Γ_π. Moreover, $|I_r \cap [a, b-1]|$ is 1 smaller than $|I_a \cap [a, b-1]|$, so we must have

$$|I_a \cap [a, b-1]| = \mathrm{area}(a \to b) + 1.$$

Consider $Y \in \hat{\Delta}^\circ_w(Z)$. By definition, for $a < b$, $\mathrm{sgn}\langle Yab\rangle = (-1)^{|I_a\cap[a,b-1]|-1}$. By the above computation, $\mathrm{sgn}\langle Yab\rangle = (-1)^{\mathrm{area}(a\to b)}$ for every arc $a \to b$, so we have shown $\hat{\Delta}^\circ_w(Z) \subset Z^\circ_{\hat{\pi}}$. Taking closures gives the desired containment.

Now, suppose $\hat{\Delta}_w(Z) \subset Z_{\hat{\pi}}$. This means that for all arcs $a \to b$ of $\mathcal{T}$, area $(a \to b) + 1$ is the same parity as $|I_a \cap [a, b-1]|$. We will show that for all q,

$$\mathrm{area}(a \to b) \leq |I_q \cap [a, b-1]| \leq \mathrm{area}(a \to b) + 1.$$

From the alcove description of w-simplices in [5, Sect. 2.3], there is some d so that Δ_w lies between the hyperplanes $\{x_{[a,b-1]} = d - 1\}$ and $\{x_{[a,b-1]} = d\}$.

As noted above, there is a vertex e_{I_r} of Δ_w satisfying $I_r = I_a \setminus \{a\} \cup \{a-1\}$. Since $|I_r \cap [a, b-1]| = |I_a \cap [a, b-1]| - 1$, we conclude that d is $|I_a \cap [a, b-1]|$. Thus, it suffices to show that

$$\text{area}(a \to b) + 1 = |I_a \cap [a, b-1]|. \tag{5.1.6}$$

This is proved in the following lemma. □

Lemma 5.1.7 *Let $\mathcal{T}$ be a triangulated (k, n)-unpunctured plabic tiling and let $\Delta_w \in \Delta_{k+1,n}$ be a w-simplex with vertices $e_{I_1}, \dots, e_{I_n}$. Suppose for all arcs $a \to b$ of $\mathcal{T}$,*

$$\text{area}(a \to b) + 1 \equiv |I_a \cap [a, b-1]| \pmod 2.$$

Then $\text{area}(a \to b) + 1 = |I_a \cap [a, b-1]|$ *for all arcs $a \to b$ of $\mathcal{T}$.*

Proof We use induction on n. The base cases are $n = 3$ and $k = 0, 1$, which are clear.

Without loss of generality, we may assume that $\mathcal{T}$ contains the arc $1 \to (n-1)$. Indeed, $\mathcal{T}$ contains some arc $(r+1) \to (r-1)$. We can rotate $\mathcal{T}$ by r to obtain a new triangulation with an arc $1 \to (n-1)$. We can also apply the corresponding cyclic shift $e_i \mapsto e_{i-r}$ to Δ_w to obtain a new simplex Δ_u. The vertex e_{I_p} of Δ_w is mapped to vertex $e_{J_{p-r}}$ of Δ_u, where $J_{p-r} = \{i - r : i \in I_p\}$. If the proposition is true for the new triangulation and Δ_u, it is easy to see (by shifting back) that it is true for $\mathcal{T}$ and Δ_w.

Let $\mathcal{T}'$ be the $(k', n-1)$-unpunctured triangulated plabic tiling obtained by chopping the triangle with vertices $1, n-1, n$ off of $\mathcal{T}$. Note that $k' = k$ if this triangle is white, and $k' = k-1$ otherwise. Let $v \in S_{n-1}$ be the permutation obtained from w by deleting $w_n = n$ and moving $n-1$ to the end.

Case I: Suppose the triangle deleted from $\mathcal{T}$ is white, so $k' = k$. Then $\text{area}_{\mathcal{T}}(1 \to (n-1))$ is equal to $\text{area}_{\mathcal{T}}(1 \to n)$, so the assumption on parities means that $I_1 \cap [1, n-1]$ has the same size as $I_1 \cap [1, n-2]$. That is, $n-1 \notin I_1$, which means that $n-1$ appears to the right of $n-2$ in w. Deleting w_n and moving $n-1$ to the end results in a permutation with the same number of cyclic descents as w, meaning that $\Delta_v \subset \Delta_{k',n-1}$.

The vertices of Δ_v are $e_{J_1}, \dots, e_{J_{n-1}}$, where

$$J_a = \begin{cases} I_a & \text{if } n \notin I_a \\ I_a \setminus \{n\} \cup \{n-1\} & \text{if } n \in I_a. \end{cases}$$

For the moment, we will denote cyclic intervals in $[n-1]$ by $[a, b]'$.

Let $a \to b$ be an arc of $\mathcal{T}'$. Because $b \neq n$, $[a, b-1]$ either contains both $n-1$ and n, or neither. So $J_a \cap [a, b-1]'$ and $I_a \cap [a, b-1]$ have the same cardinality. Also, $\text{area}_{\mathcal{T}'}(a \to b)$ is equal to $\text{area}_{\mathcal{T}}(a \to b)$, so $\mathcal{T}'$ and $\hat{\Delta}_v(Z)$ satisfy the assumptions of the proposition. By induction, we can conclude that $|J_a \cap [a, b-1]'| = \text{area}_{\mathcal{T}'}(a \to b) + 1$. In light of the equalities in this paragraph, this means that for all arcs $a \to b$ of $\mathcal{T}$ where a, b are not n, we have $|I_a \cap [a, b-1]| = \text{area}_{\mathcal{T}}(a \to b) + 1$. It remains to check that a similar equality for the arcs $1 \to n$, $(n-1) \to n$ and their reverses, which are trivial.

Case II: Suppose the triangle deleted from $\mathcal{T}$ is black, so $k' = k - 1$. Then $\text{area}_{\mathcal{T}}(1 \to (n-1))$ is equal to $\text{area}_{\mathcal{T}}(1 \to n) - 1$. The assumption on parities implies that $I_1 \cap [1, n-1]$ and $I_1 \cap [1, n-2]$ are different sizes, so $n - 1 \in I_1$. This means that $n-1$ appears to the left of $n-2$ in w, and v has one fewer left descent than w. So $\Delta_v \subset \Delta_{k',n-1}$ as desired.

The vertices of Δ_v are $e_{J_1}, \ldots, e_{J_{n-1}}$, where

$$J_a = \begin{cases} I_a \setminus \{n\} & \text{if } n \in I_a \\ I_a \setminus \{n-1\} & \text{if } n-1 \in I_a, n \notin I_a. \end{cases}$$

Let $a \to b$ be an arc of $\mathcal{T}'$. Again, the cyclic interval $[a, b-1]$ either contains both $n-1$ and n, or contains neither. If $[a, b-1]$ contains neither, then clearly $|J_a \cap [a, b-1]'| = |I_a \cap [a, b-1]|$; in this case, $\text{area}_{\mathcal{T}'}(a \to b) = \text{area}_{\mathcal{T}}(a \to b)$ as well. If $[a, b-1]$ contains both, then $|J_a \cap [a, b-1]'| = |I_a \cap [a, b-1]| - 1$ and $\text{area}_{\mathcal{T}'}(a \to b) = \text{area}_{\mathcal{T}}(a \to b) - 1$. So again, $\mathcal{T}'$ and Δ_v satisfy the assumptions of the proposition. As in Case I, we can conclude that for all arcs $a \to b$ of $\mathcal{T}$ where a, b are not n, we have $|I_a \cap [a, b-1]| = \text{area}_{\mathcal{T}}(a \to b) + 1$. The equalities for the arcs $1 \to n$, $(n-1) \to n$, and their reverses are clear. □

Remark 5.1.8 Proposition 5.1.4 motivates the intuition that the w-simplex $\Delta_w \subset \Delta_{k+1,n}$ and the w-chamber $\hat{\Delta}_w(Z) \subset \mathcal{A}_{n,k,2}(Z)$ are 'T-dual' to each other. In Proposition 5.2.40 we will show that any w-simplex is the intersection of n distinguished positroid polytopes $\{\Gamma_\pi\}$, and the corresponding w-chamber is the intersection of the n T-dual Grasstopes $\{Z_{\hat{\pi}}\}$.

We can now show the main result of this section.

Theorem 5.1.9 (Tilings of $\Delta_{k+1,n}$ and $\mathcal{A}_{n,k,2}$ are T-dual) *The collection $\mathcal{C} = \{\Gamma_\pi\}$ is a positroid tiling of $\Delta_{k+1,n}$ if and only if for all $Z \in \text{Mat}_{n,k+2}^{>0}$, the collection of T-dual Grasstopes $\hat{\mathcal{C}} = \{Z_{\hat{\pi}}\}$ is a positroid tiling of $\mathcal{A}_{n,k,2}(Z)$.*

Proof ($\Longrightarrow$) : Suppose $\mathcal{C}$ is a positroid tiling of $\Delta_{k+1,n}$ and choose $Z \in \text{Mat}_{n,k+2}^{>0}$. We already know that $Z_{\hat{\pi}}$ is a positroid tile from Corollary 4.2.1.

We first show that the Grasstopes in $\hat{\mathcal{C}}$ are dense in the amplituhedron. Consider a nonempty amplituhedron w-chamber $\hat{\Delta}_w(Z)$. Since $\mathcal{C}$ is a positroid tiling, there exists a tree positroid polytope $\Gamma_\pi \in \mathcal{C}$ which contains Δ_w. By Proposition 5.1.4, $\hat{\Delta}_w^\circ(Z) \subset Z_{\hat{\pi}}^\circ$, where the latter is by definition in $\hat{\mathcal{C}}$. So we have

$$\bigcup_w \hat{\Delta}_w^\circ(Z) \subseteq \bigcup_{\hat{\mathcal{C}}} Z_{\hat{\pi}}^\circ \subseteq \mathcal{A}_{n,k,2}(Z).$$

By Theorem 4.3.13, the closure of the left-most set is equal to the right, so the closure of the middle set is $\mathcal{A}_{n,k,2}(Z)$, as desired.

Now, suppose for the sake of contradiction that two distinct $Z^\circ_{\hat{\pi}}, Z^\circ_{\hat{\mu}} \in \hat{\mathcal{C}}$ are not disjoint. They are open, so their intersection is open, and thus their intersection contains a point in $\hat{\Delta}^\circ_w(Z)$ for some w. Lemma 4.3.12 implies that in fact the entire w-simplex $\hat{\Delta}^\circ_w(Z)$ is contained in their intersection. But then by Proposition 5.1.4, Δ_w is contained in $\Gamma_\pi \cap \Gamma_\mu$, a contradiction (Fig. 5.2).

($\Longleftarrow$): Suppose that for all $Z \in \text{Mat}^{>0}_{n,k+2}$, $\hat{\mathcal{C}}$ is a positroid tiling of $\mathcal{A}_{n,k,2}(Z)$. By Theorem 4.4.12, for all $w \in D_{k+1,n}$, we can choose Z so that $\hat{\Delta}_w(Z)$ is nonempty. In particular, $\hat{\Delta}^\circ_w(Z)$ must intersect one of the positroid tiles $Z^\circ_{\hat{\pi}}$ and thus by Lemma 4.3.12, $\hat{\Delta}_w(Z) \subset Z_{\hat{\pi}}$. Because $\hat{\mathcal{C}}$ is a positroid tiling, $\hat{\Delta}_w(Z)$ is not contained in any other positroid tile in $\hat{\mathcal{C}}$. Using Proposition 5.1.4, we see that every w-simplex is contained in precisely one positroid polytope in $\mathcal{C}$, and thus $\mathcal{C}$ is a positroid tiling of $\Delta_{k+1,n}$. □

Example 5.1.10 (*Positroid tilings of $\Delta_{2,4}$ and $\mathcal{A}_{4,1,2}$ are T-dual*) By Example 2.1.7 $\mathcal{A}_{4,1,2}$ is a 4-gon whose positroid tilings are just standard (polyhedral) triangulations. There are just 2 of them given by the pairs of triangles in the right side of Fig. 5.2. In turn, the hypersimplex $\Delta_{2,4}$ also has only 2 positroid tilings given by the pairs of positroid tiles in the left side of Fig. 5.2. The bijection in Corollary 4.2.1 between positroid tiles of $\Delta_{2,4}$ and $\mathcal{A}_{4,1,2}$ via the T-duality map on permutation induces a bijection between positroid tilings. For example, the positroid tiling $\{\Gamma_{\pi_1}, \Gamma_{\pi_2}\}$ of $\Delta_{2,4}$ corresponds to the positroid tiling $\{Z_{\hat{\pi}_1}, Z_{\hat{\pi}_2}\}$ of $\mathcal{A}_{4,1,2}$. ◇

Example 5.1.11 (*w-simplices and w-chambers are T-dual*) We observe that the decomposition $\Delta_{2,4} = \Delta_{w_1} \cup \Delta_{w_2} \cup \Delta_{w_3} \cup \Delta_{w_4}$ refines both positroid tilings of $\Delta_{2,4}$ in Fig. 5.3. Analogously, the decomposition $\mathcal{A}_{4,1,2}(Z) = \hat{\Delta}_{w_1}(Z) \cup \hat{\Delta}_{w_2}(Z) \cup \hat{\Delta}_{w_3}(Z) \cup \hat{\Delta}_{w_4}(Z)$ refines both positroid tilings of $\mathcal{A}_{4,1,2}(Z)$. Moreover, one can see that T-duality preserves the property of inclusion of a w-simplex in a positroid tile. For example, $\Delta_{w_1} \subset \Gamma_{\nu_1}, \Gamma_{\pi_2}$ (and even more: $\Delta_{w_1} = \Gamma_{\nu_1} \cap \Gamma_{\pi_2}$) and $\hat{\Delta}_{w_1} \subset Z_{\hat{\nu}_1}, Z_{\hat{\pi}_2}$ (and even more: $\hat{\Delta}_{w_1} = Z_{\hat{\nu}_1} \cap Z_{\hat{\pi}_2}$).

Using the sign characterization of a positroid tile $Z_{\mathcal{T}}$ of $\mathcal{A}_{n,k,2}(Z)$ (Theorem 2.3.28), one can define a *positroid tile* $\mathcal{G}_{\mathcal{T}}$ in $\mathcal{G}_{n,k,2}$ as (the closure of) the region in $\text{Gr}_{2,n}$ whose Plücker coordinates satisfy the same sign conditions as the twistor coordinates of $Z_{\mathcal{T}}$:

Definition 5.1.12 (*Positroid tiles of $\mathcal{G}_{n,k,2}$*) Let $\mathcal{T}$ be a triangulated (k, n)-unpunctured plabic tiling. Then we define

$$\mathcal{G}^\circ_{\mathcal{T}} := \{z \in \text{Gr}_{2,n} \mid \text{sgn}\, p_{ij}(z) = (-1)^{\text{area}(i \to j)} \text{ for all black arcs } i \to j \text{ of } \mathcal{T} \text{ with } i < j\}.$$

We denote the closure of $\mathcal{G}^\circ_{\mathcal{T}}$ as $\mathcal{G}_{\mathcal{T}}$ and refer to it as a *positroid tile* of $\mathcal{G}_{n,k,2}$.

Analogously to Corollary 4.3.20, $\mathcal{G}_{\mathcal{T}}$ is a union of $\mathcal{G}$-amplituhedron w-chambers. Moreover, $Z_{\mathcal{T}}$ is a linear slice of $\mathcal{G}_{\mathcal{T}}$ (analogously to Remark 4.4.9). We say a *positroid tiling* of $\mathcal{G}_{n,k,2}$ is a collection of positroid tiles which cover $\mathcal{G}_{n,k,2}$ and

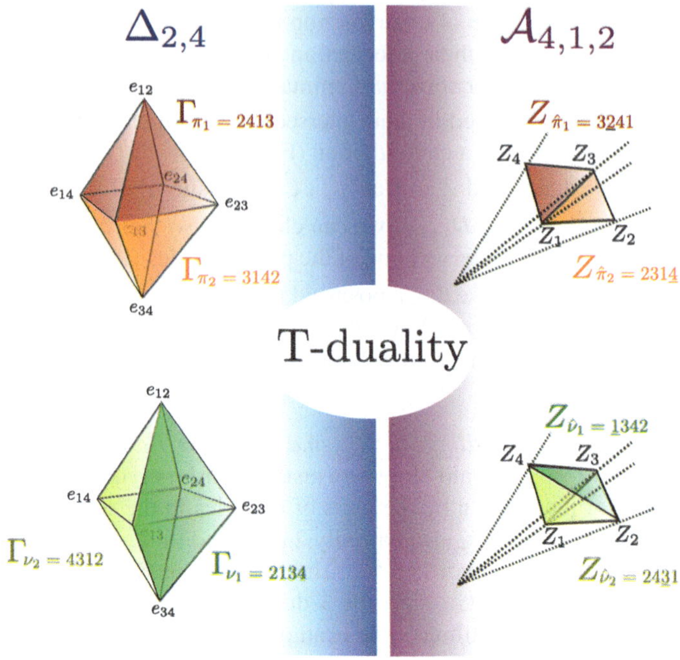

Fig. 5.2 Positroid tilings of $\Delta_{2,4}$ and $\mathcal{A}_{4,1,2}$ in bijection via T-duality

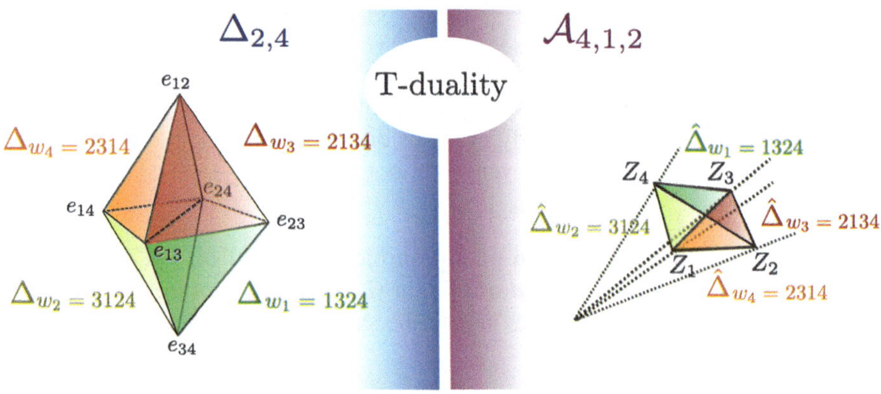

Fig. 5.3 w-simplices of $\Delta_{2,4}$ and w-chambers of $\mathcal{A}_{4,1,2}$ are T-dual

have disjoint interiors. Since all $\mathcal{G}$-amplituhedron w-chambers are non-empty, the $\mathcal{G}$-amplituhedron analogue of Theorem 5.1.9 holds:

Theorem 5.1.13 (Tilings of $\Delta_{k+1,n}$ and $\mathcal{G}_{n,k,2}$ are T-dual) *The collection $\mathcal{C} = \{\Gamma_{\mathcal{T}}\}$ is a positroid tiling of $\Delta_{k+1,n}$ if and only if the collection of T-dual positroid tiles $\hat{\mathcal{C}} = \{\mathcal{G}_{\mathcal{T}}\}$ is a positroid tiling of $\mathcal{G}_{n,k,2}$.*

We end this section by stating a conjecture about dissections. This is motivated by Sect. 5.2.1 about BCFW-type dissections, discussions in Sect. 5.2.2 about the tropical positive Grassmannian $\mathrm{Trop}^+ \mathrm{Gr}_{k+1,n}$, and experimental data in Sect. 5.2.2.

Conjecture 5.1.14 (Dissections of $\Delta_{k+1,n}$ and $\mathcal{A}_{n,k,2}$ are T-dual) *The collection $\mathcal{C} = \{\Gamma_\pi\}$ is a positroid dissection of $\Delta_{k+1,n}$ if and only if for all $Z \in \mathrm{Mat}^{>0}_{n,k+2}$, the collection of T-dual Grasstopes $\hat{\mathcal{C}} = \{Z_{\hat{\pi}}\}$ is a positroid dissection of $\mathcal{A}_{n,k,2}(Z)$.*

5.1.2 Tilings from w-Simplices and w-Chambers

In this section we provide algorithms to find *all* positroid tilings of the hypersimplex and the amplituhedron using w-simplices and w-chambers. As mentioned in Remark 4.3.11, $\hat{\Delta}_w(Z)$ may be empty for some choice of Z. We take a closer look at this phenomenon and give some examples.

Remark 5.1.15 It is a priori possible for an amplituhedron $\mathcal{A}_{n,k,2}(Z)$ to have a tiling $\hat{\mathcal{C}}$ which is not T-dual to an hypersimplex tiling. However, Theorem 5.1.9 tells us that the collection of Grasstopes $\hat{\mathcal{C}}$ will fail to be a tiling for some other amplituhedron $\mathcal{A}_{n,k,2}(Z')$. We have not found any instances of such "sporadic" tilings in the examples we considered.

Proposition 5.1.16 (Algorithm for tilings of $\Delta_{k+1,n}$) *In order to find all tilings of $\Delta_{k+1,n}$ proceed as follows. Call two positroid tiles Γ_{π_1} and Γ_{π_2}* compatible *if they do not contain any common w-simplex.*

Step 1. Define a graph $\mathcal{G}$ whose vertices are positroid tiles of $\Delta_{k+1,n}$ and edges connect compatible positroid tiles.
Step 2. Compute the set $Cl(\mathcal{G})$ of all maximal cliques of $\mathcal{G}$;
Step 3. For each clique $C \in Cl(\mathcal{G})$, compute the list $\mathcal{L}_C$ of all w-simplices contained in any positroid tile $\Gamma_\pi \in C$;
Step 4. If $\mathcal{L}_C$ consists of all w-simplices of $\Delta_{k+1,n}$, then C is a tiling of $\Delta_{k+1,n}$, otherwise it is not.

Proposition 5.1.17 (Algorithm for tilings of $\mathcal{A}_{n,k,2}$) *In order to find all tilings of $\mathcal{A}_{n,k,2}(Z)$ proceed as follows. Let $\mathcal{E}_Z$ be the list of all w-simplices Δ_w in $\Delta_{k+1,n}$ such that $\hat{\Delta}_w(Z) = \emptyset$. Call two positroid tiles $Z_{\hat{\pi}_1}$, $Z_{\hat{\pi}_2}$* compatible *if and only if $\Gamma_{\pi_1} \cap \Gamma_{\pi_2}$ is empty or contains w-simplices which are in $\mathcal{E}_Z$.*

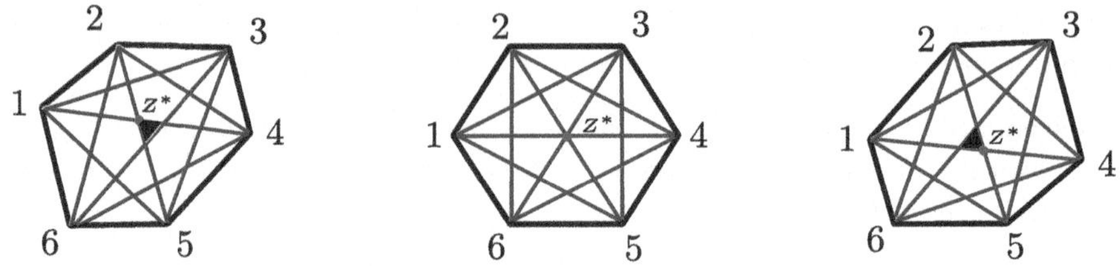

Fig. 5.4 From left to right: For $\langle Z^*, Z_3, Z_6\rangle > 0$, $\hat{\Delta}_{w^{(+)}}$ is nonempty (in black) but $\hat{\Delta}_{w^{(-)}}$ is empty; if $\langle Z^*, Z_3, Z_6\rangle = 0$ then $\hat{\Delta}_{w^{(+)}}$ and $\hat{\Delta}_{w^{(-)}}$ are both empty; for $\langle Z^*, Z_3, Z_6\rangle < 0$ then $\hat{\Delta}_{w^{(+)}}$ is empty but $\hat{\Delta}_{w^{(-)}}$ is not (shown in black)

Step 1. *Make a graph $\hat{\mathcal{G}}$ whose vertices are positroid tiles of $\mathcal{A}_{n,k,2}(Z)$ and edges connect compatible positroid tiles;*
Step 2. *Compute the set $Cl(\hat{\mathcal{G}})$ of all maximal cliques of $\hat{\mathcal{G}}$;*
Step 3. *For each clique $\hat{\mathcal{C}} \in Cl(\hat{\mathcal{G}})$, consider the collection $\mathcal{C}$ of T-dual positroid tiles in $\Delta_{k+1,n}$. Compute the list $\mathcal{L}_{\mathcal{C}}$ of all w-simplices in $\Delta_{k+1,n}$ contained in any positroid tile $\Gamma_\pi \in \mathcal{C}$;*
Step 4. *If the (possibly empty) complement of $\mathcal{L}_{\mathcal{C}}$ is contained in $\mathcal{E}_Z$, then $\hat{\mathcal{C}}$ is a tiling of $\mathcal{A}_{n,k,2}(Z)$, otherwise it is not.*

Remark 5.1.18 If we would like to find a positroid tiling $\hat{\mathcal{C}}$ of the amplituhedron $\mathcal{A}_{n,k,2}(Z)$ which is *not* a positroid tiling of $\Delta_{k+1,n}$, then after Step 3 we need check that either: i) the complement of $\mathcal{L}_{\mathcal{C}}$ is *nonempty* and contained in $\mathcal{E}_Z$; or ii) $\mathcal{L}_{\mathcal{C}}$ is the set of all w-simplices of $\Delta_{k+1,n}$ and there is a pair of positroid tiles $\Gamma_{\pi_1}, \Gamma_{\pi_2}$ in $\mathcal{C}$ which both contain a w-simplex in $\mathcal{E}_Z$.

Below, we report some results on empty w-chambers in the cases $k = 1, 2$.

$k = 1$ **Case.** The amplituhedron $\mathcal{A}_{n,1,2}(Z)$ is just an n-gon $\mathbf{P}_n(Z)$ in $\mathbb{P}^2$ with vertices $Z_1, \ldots, Z_n$ going clockwise. Let $i \to j$ be a side or a diagonal of $\mathbf{P}_n(Z)$, with $i < j$. The twistor coordinate $\langle Yij\rangle$ is positive, negative or zero if Y lies to the right, left, or on the diagonal $i \to j$ respectively. Then the nonempty w-chambers $\hat{\Delta}_w(Z)$ are the connected components of the complement of all diagonals of $\mathbf{P}_n(Z)$ (see Fig. 5.4). If no three diagonals of $\mathbf{P}_n(Z)$ intersect at a point in the interior, it is well known the number of connected components is given by:

$$N_n = \sum_{r=2}^{4} \binom{n-1}{r} = \binom{n}{4} + \binom{n-1}{2}.$$

The number of empty w-chambers in this case is show in Table 5.1.

If three diagonals of $\mathbf{P}_n(Z)$ intersect at a point in its interior, then the number of empty w-chambers is larger (as the number of regions realised is smaller).

Example 5.1.19 The easiest example with an empty w-chamber is for $\mathcal{A}_{6,1,2}(Z)$, which is an hexagon. Let us consider the permutations $w^{(+)} = 145236$ and $w^{(-)} =$

Table 5.1 Empty w-chambers versus Eulerian numbers for $k = 1$

n	3	4	5	6	7	8	9
N_n	1	4	11	25	50	91	154
$E_{1,n-1}$	1	4	11	26	57	120	247
# Empty $\hat{\Delta}_w$	0	0	0	1	7	29	93

341256. Using Definition 4.3.8, it is easy to see that points in $\hat{\Delta}_{w^{(+)}}(Z)$ and $\hat{\Delta}_{w^{(-)}}(Z)$ have all twistor coordinates with the same sign, except for $\{\langle Y14\rangle, \langle Y25\rangle, \langle Y36\rangle\}$, whose signs are $\{+\ -\ +\}$ and $\{-\ +\ -\}$, respectively. Let Z^* be the intersection of the diagonals $(1, 4)$ and $(2, 5)$. Then $\hat{\Delta}_{w^{(+)}}(Z)$ (respectively, $\hat{\Delta}_{w^{(-)}}(Z)$) is non-empty if and only if Z^* is to the right (respectively, left) of the diagonal $3 \to 6$. This happens when

$$\langle Z_1, Z_2, Z_5\rangle\langle Z_4, Z_3, Z_6\rangle - \langle Z_1, Z_3, Z_6\rangle\langle Z_4, Z_2, Z_5\rangle \tag{5.1.20}$$

is positive (respectively, negative), see Fig. 5.4. Therefore we conclude that for any choice of Z either $\hat{\Delta}_{w^{(+)}}(Z) = \emptyset$ or $\hat{\Delta}_{w^{(-)}}(Z) = \emptyset$. Moreover, both are empty if (5.1.20) vanishes. Similar phenomena occur for higher n as well. ◇

In the following, we show an example of overlapping positroid polytopes $\{\Gamma_\pi\}$ which do *not* cover $\Delta_{2,n}$, but whose T-dual Grasstopes $\{Z_{\hat{\pi}}\}$ cover $\mathcal{A}_{n,1,2}(Z)$, for some Z.

Example 5.1.21 Consider $\mathcal{A}_{6,1,2}(Z_0)$, with Z_0 vertices of the regular hexagon in $\mathbb{P}^2$ (i.e. Z_0 is the point in $Gr_{3,6}^{>0}$ invariant under cyclic symmetry). Consider the positroid tiles (actual triangles) $Z_{\hat{\pi}_1}, \ldots, Z_{\hat{\pi}_6}$ below.

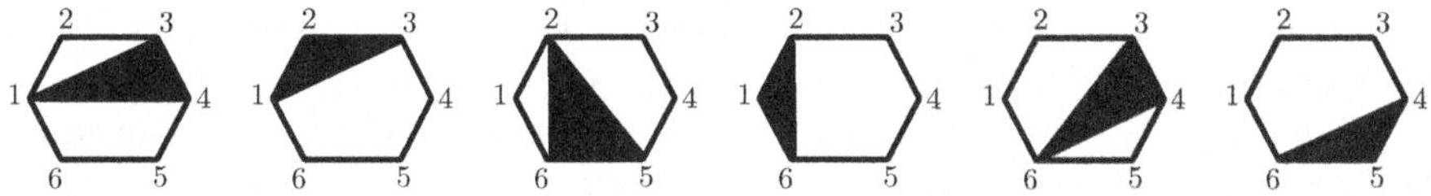

Clearly, they *do* cover $\mathcal{A}_{6,1,2}(Z_0)$ (and overlap). However, $\Delta_{w^{(+)}}$ is not included in any of the positroid tiles $\Gamma_{\pi_1}, \ldots, \Gamma_{\pi_6}$ of $\Delta_{2,6}$. Therefore they do *not* cover $\Delta_{2,6}$. ◇

Despite the presence of empty w-chambers, for *any* Z in $\mathrm{Mat}_{k+2,n}^{>0}$, tilings of $\Delta_{2,n}$ and $\mathcal{A}_{n,1,2}(Z)$ are still in bijection:

Proposition 5.1.22 *A collection of tree positroid polytopes $\{\Gamma_\pi\}$ is a tiling of $\Delta_{2,n}$ if and only if $\{Z_{\hat{\pi}}\}$ is a tiling of $\mathcal{A}_{n,1,2}(Z)$. All such tilings are regular.*

Proof The forward direction comes from Theorem 5.1.9. The other direction comes from the fact that $\mathcal{A}_{n,1,2}(Z)$ is just an n-gon. Its tilings are in bijection with the regular tilings of $\Delta_{2,n}$ described in [3, Proposition 10.7] (of 'Catalan' type). □

$k = 2$ **Case.** We used Mathematica and the package 'positroid' [6].

For $n = 6$, there are choices of Z such that all w-chambers of $\mathcal{A}_{6,2,2}(Z)$ are nonempty.

For $n = 7$ and some choices of Z, there are empty w-chambers $\hat{\Delta}_w$ for which Δ_w is the intersection of just 2 positroid tiles of $\Delta_{3,7}$. This implies that in general the compatibility graph $\mathcal{G}$ of positroid tiles of $\Delta_{k+1,n}$ differs from the one $\hat{\mathcal{G}}$ of positroid tiles of $\mathcal{A}_{n,k,2}(Z)$ (cf. Proposition 5.1.17). For example, let $w = 1645237$, then $\Delta_w = \Gamma_{\pi_1} \cap \Gamma_{\pi_2}$, with $\pi_1 = (2, 3, 7, 8, 6, 11, 12)$, $\pi_2 = (6, 7, 4, 5, 9, 10, 8)$. The positroid polytopes $\{\Gamma_{\pi_1}, \Gamma_{\pi_2}\}$ are *not* compatible in $\Delta_{3,7}$, but the T-dual Grasstopes $\{Z_{\pi_1}, Z_{\pi_2}\}$ are compatible in $\mathcal{A}_{7,2,2}(Z)$, as $Z_{\pi_1} \cap Z_{\pi_2} = \hat{\Delta}_w = \emptyset$. Nevertheless, the 3073 tilings of $\mathcal{A}_{7,2,2}(Z)$ are still in bijection with the 3073 tilings of $\Delta_{3,7}$.

For $n = 8$, we checked only a few choices of Z, but found that there are more than 100 w-chambers which can be empty depending on Z. As for $n = 7$, the compatibility graph of $\Delta_{3,8}$ differs from that of $\mathcal{A}_{8,2,2}(Z)$: e.g. we found multiple collections $\{Z_\pi\}$ of 13 Grasstopes in $\mathcal{A}_{8,2,2}(Z)$ which are mutually compatible (but don't form a tiling), whose T-duals are *not* mutually compatible positroid polytopes. Nevertheless, the 6443460 tilings of $\mathcal{A}_{8,2,2}(Z)$ (each of which has size 15) are in bijection with the tilings of $\Delta_{3,8}$.

5.2 Different Types of Tilings and Dissections

Is it possible to classify all positroid tilings (dissections) of $\Delta_{k+1,n}$ and $\mathcal{A}_{n,k,2}$? In particular, it would be desirable to have a combinatorial characterization of when a given collection of positroid tiles gives a positroid tiling. While this still remains an open question, in this section we present broad classes of positroid tilings. Those are built from *BCFW recursions* (Sect. 5.2.1)—involving tilings with lower n and k, and the positive tropical Grassmannian $\mathrm{Trop}^+ \mathrm{Gr}_{k+1,n}$ (Sect. 5.2.2)—governing the *regular* tilings. We will also explore particular positroid tilings (both regular and of BCFW type) obtained from the combinatorics of certain descent/sign-flip configurations (Sect. 5.2.3).

5.2.1 Tilings and Dissections from BCFW Recursions

In scattering amplitudes, *BCFW* recursion relations allowed to compute amplitudes very effectively. In this section we show that an analogous relation holds for the hypersimplex too. Moreover, we show that these recursions are T-dual with recursions for the $m = 2$ amplituhedron proved by Bao and He [7]. They have two types of terms[1]: one involving tilings with lower n, and one involving tilings with lower n

[1] In physics, the first would be equivalent to a 'soft limit'; the second would be a 'collinear limit'.

and k. We note that, thanks to Theorem 5.1.9, Bao and He's results are now an immediate corollary.

BCFW recursion for the Hypersimplex Δ_{k+1}.

Definition 5.2.1 Consider the set of reduced plabic graphs with $n-1$ boundary vertices, associated to cells of $\mathrm{Gr}^{\geq 0}_{k+1,n-1}$ (respectively, $\mathrm{Gr}^{\geq 0}_{k,n-1}$), which do not have a loop at vertex $n-1$. Let $\mathfrak{i}_{\mathrm{pre}}$ (respectively, $\mathfrak{i}_{\mathrm{inc}}$) be the map on these graphs, which takes a reduced plabic graph G and replaces the $(n-1)$st boundary vertex with a trivalent internal white (resp. black) vertex attached to boundary vertices $n-1$ and n, as in the middle (resp. rightmost) graph of Fig. 5.5. Using Appendix A, it is straightforward to verify that both $\mathfrak{i}_{\mathrm{pre}}(G)$ and $\mathfrak{i}_{\mathrm{inc}}(G)$ are reduced plabic graphs for cells of $\mathrm{Gr}^{\geq 0}_{k+1,n}$.[2]

Abusing notation slightly, we also use $\mathfrak{i}_{\mathrm{pre}}$ and $\mathfrak{i}_{\mathrm{inc}}$ to denote the corresponding maps on positroid cells and decorated permutations associated with the plabic graph (see Definition A.0.4).

Lemma 5.2.2 *Let $\pi = (a_1, a_2, \ldots, a_{n-1})$ be a decorated permutation on $n-1$ letters; assume that $(n-1) \mapsto a_{n-1}$ is not a black fixed point. Then $\mathfrak{i}_{\mathrm{pre}}(\pi) = (a_1, a_2, \ldots, a_{n-2}, n, a_{n-1})$.*

Let us now assume that $(n-1) \mapsto a_{n-1}$ is not a white fixed point. Let $j = \pi^{-1}(n-1)$. Then $\mathfrak{i}_{\mathrm{inc}}(\pi) = (a_1, a_2, \ldots, a_{j-1}, n, a_{j+1}, \ldots, a_{n-1}, n-1)$.

Theorem 5.2.3 (BCFW recursions for $\Delta_{k+1,n}$) *Let $\mathcal{C}_{k+1,n-1}$ (respectively $\mathcal{C}_{k,n-1}$) be a collection of cells in $\mathrm{Gr}^{\geq 0}_{k+1,n-1}$ (resp. $\mathrm{Gr}^{\geq 0}_{k,n-1}$) which gives a dissection of the hypersimplex $\Delta_{k+1,n-1}$ (resp. $\Delta_{k,n-1}$). Then*

$$\mathcal{C}_{k+1,n} = \mathfrak{i}_{\mathrm{pre}}(\mathcal{C}_{k+1,n-1}) \cup \mathfrak{i}_{\mathrm{inc}}(\mathcal{C}_{k,n-1})$$

gives a dissection $\Delta_{k+1,n}$.

Diagrammatically, the main theorem reads as (Fig. 5.5).

Remark 5.2.4 Because of the cyclic symmetry of the positive Grassmannian and the hypersimplex (see e.g. Theorem 5.3.4) there are $n-1$ other versions of Theorem 5.2.3 (and Fig. 5.5) in which all plabic graph labels get shifted by i modulo n (for $1 \leq i \leq n-1$).

Proof The hypersimplex $\Delta_{k+1,n}$ is cut out by the inequalities $0 \leq x_i \leq 1$, as well as the equality $\sum_i x_i = k+1$. We will show that Fig. 5.5 represents the partition of $\Delta_{k+1,n}$ into two pieces, with the middle graph representing the piece cut out by $x_{n-1} + x_n \leq 1$, and the rightmost graph representing the piece cut out by $x_{n-1} + x_n \geq 1$.

[2] We can in fact define $\mathfrak{i}_{\mathrm{pre}}(G)$ (resp. $\mathfrak{i}_{\mathrm{inc}}(G)$) on any reduced plabic graph for $\mathrm{Gr}^{\geq 0}_{k+1,n-1}$ (resp. $\mathrm{Gr}^{\geq 0}_{k,n-1}$) which does not have a black (resp. white) lollipop at vertex $n-1$, and will again have that $\mathfrak{i}_{\mathrm{pre}}(G)$ and $\mathfrak{i}_{\mathrm{inc}}(G)$ represent cells of $\mathrm{Gr}^{\geq 0}_{k+1,n}$.

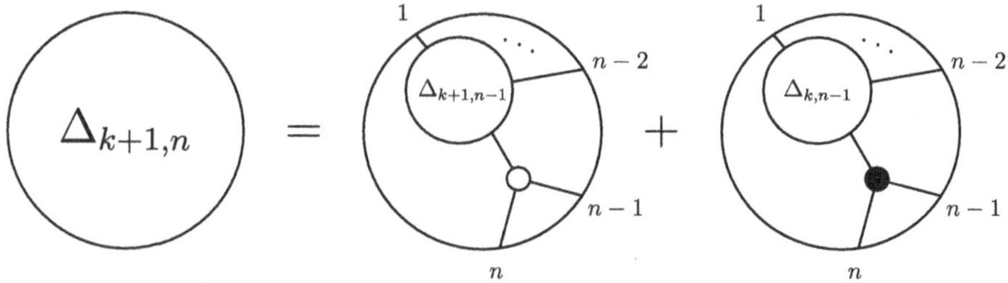

Fig. 5.5 A recursion for dissecting the hypersimplex. There is a parallel recursion obtained from this one by cyclically shifting all boundary vertices of the plabic graphs by i (modulo n)

Towards this end, it follows from Proposition A.0.6 that if G is a reduced plabic graph representing a cell of $\mathrm{Gr}_{k+1,n-1}^{\geq 0}$, such that the positroid M_G has bases $\mathcal{B}$, then the bases of $M_{\mathfrak{i}_{\mathrm{pre}}(G)}$ are precisely $\mathcal{B} \sqcup \{(B \setminus \{n-1\}) \cup \{n\} \mid B \in \mathcal{B}, n-1 \in B\}$. In particular, each basis of $M_{\mathfrak{i}_{\mathrm{pre}}(G)}$ may contain at most one element of $\{n-1, n\}$.

Meanwhile, it follows from Proposition A.0.6 that if G is a reduced plabic graph representing a cell of $\mathrm{Gr}_{k,n-1}^{\geq 0}$, such that the positroid M_G has bases $\mathcal{B}$, then the bases of $M_{\mathfrak{i}_{\mathrm{inc}}(G)}$ are precisely $\{B \cup \{n\} \mid B \in \mathcal{B}\} \sqcup \{B \cup \{n-1\} \mid B \in \mathcal{B}, n-1 \notin B\}$. In particular, each basis of $M_{\mathfrak{i}_{\mathrm{inc}}(G)}$ must contain at least one element of $\{n-1, n\}$.

It is now a straightforward exercise (using e.g. [8, Proposition 5.6]) to determine that if $\mathcal{C}_{k+1,n-1}$ is a collection of cells in $\mathrm{Gr}_{k+1,n-1}^{\geq 0}$ which dissects $\Delta_{k+1,n-1}$ then $\mathfrak{i}_{\mathrm{pre}}(\mathcal{C}_{k+1,n-1})$ dissects the subset of $\Delta_{k+1,n}$ cut out by the inequality $x_{n-1} + x_n \leq 1$. Similarly for $\mathfrak{i}_{\mathrm{inc}}(\mathcal{C}_{k,n-1})$ and the subset of $\Delta_{k+1,n}$ cut out by $x_{n-1} + x_n \geq 1$. □

Example 5.2.5 Let $n = 5$ and $k = 2$. We will use Theorem 5.2.3 to obtain a dissection of $\Delta_{k+1,n} = \Delta_{3,5}$. We start with a dissection of $\Delta_{3,4}$ coming from the plabic graph shown below (corresponding to the decorated permutation $(4, 1, 2, 3)$), and a dissection of $\Delta_{2,4}$ (corresponding to the permutations $(2, 4, 1, 3)$ and $(3, 1, 4, 2)$). Applying the theorem leads to the three plabic graphs in the bottom line, which correspond to the permutations $(4, 1, 2, 5, 3)$, $(2, 5, 1, 3, 4)$, $(3, 1, 5, 2, 4)$. ◇

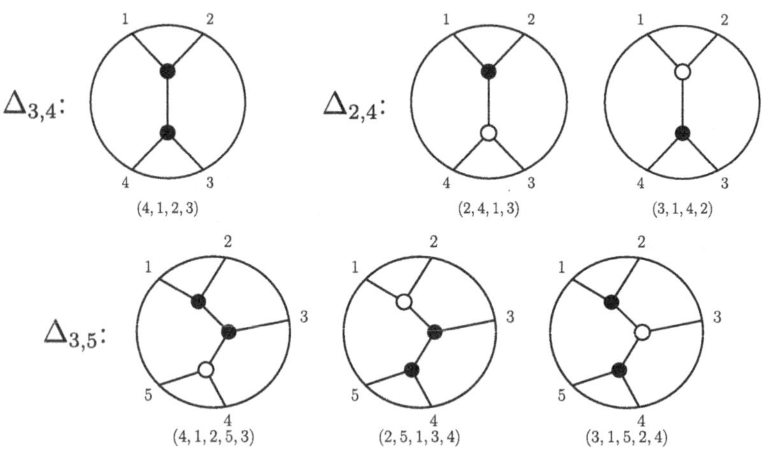

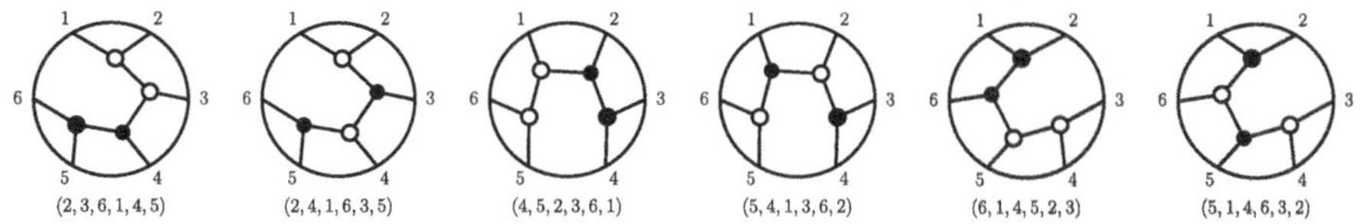

Fig. 5.6 An example of dissection of $\Delta_{3,6}$ that cannot be obtained from the recursion in Theorem 5.2.3

Remark 5.2.6 It is worth pointing out that the recursion does not provide all possible dissections of the hypersimplex. This comes from the fact that in each step of the recursion we divide the hypersimplex into two pieces, while there are some dissections coming from 3-splits. The simplest example of a dissection which cannot be obtained from the recursion can be found already for $\Delta_{3,6}$ and is depicted in Fig. 5.6.

BCFW recursion for the Amplituhedron $\mathcal{A}_{n,k,2}$. We now introduce some maps on plabic graphs, and recall a result of Bao and He [7].

Definition 5.2.7 Consider the set of reduced plabic graphs with $n-1$ boundary vertices, associated to cells of $\mathrm{Gr}_{k,n-1}^{\geq 0}$. Let ι_{pre} be the map which takes such a graph G and adds a black lollipop at a new boundary vertex n, as shown in the middle graph of Fig. 5.7. Clearly the resulting graph $\iota_{\mathrm{pre}}(G)$ is a reduced plabic graph for a cell of $\mathrm{Gr}_{k,n}^{\geq 0}$.

Similarly, let ι_{inc} be the map on plabic graphs for $\mathrm{Gr}_{k-1,n-1}^{\geq 0}$ which modifies such a graph G, changing the graph locally around vertices $1, n, n-1$, as shown at the right of of Fig. 5.7 (in physics lore it corresponds to attaching an inverse soft factor). It is not hard to show that (as long as G does not have white fixed points at vertices 1 or $n-1$) then $\iota_{\mathrm{inc}}(G)$ is a reduced plabic graph for a cell of $\mathrm{Gr}_{k,n}^{\geq 0}$.

Abusing notation slightly, we also use ι_{pre} and ι_{inc} to denote the corresponding maps on positroid cells and decorated permutations (see Definition A.0.4).

Lemma 5.2.8 *Let $\pi = (a_1, a_2, \ldots, a_{n-1})$ be a decorated permutation on $n-1$ letters. Then $\iota_{\mathrm{pre}}(\pi) = (a_1, a_2, \ldots, a_{n-2}, a_{n-1}, n)$, where n is a black fixed point.*

Let us now assume that neither positions 1 nor $n-1$ are white fixed points of π. Let $h = \pi^{-1}(n-1)$. Then $\iota_{\mathrm{inc}}(\pi)$ is the permutation such that $1 \mapsto n-1$, $h \mapsto n$, $n \mapsto a_1$, and $j \mapsto a_j$ for all $j \neq 1, h, n$.

The construction below is closely related to the recursion from [9, Definition 4.4], which is a sort of $m = 2$ version of the BCFW recurrence.

Theorem 5.2.9 (BCFW recursions for $\mathcal{A}_{n,k,2}$) *[7, Theorem A] Let $\mathcal{C}_{n-1,k,2}$ (respectively $\mathcal{C}_{n-1,k-1,2}$) be a collection of cells in $\mathrm{Gr}_{k,n-1}^{\geq 0}$ (resp. $\mathrm{Gr}_{k-1,n-1}^{\geq 0}$) which gives a dissection of the amplituhedron $\mathcal{A}_{n-1,k,2}$ (resp. $\mathcal{A}_{n-1,k-1,2}$). Then*

$$\mathcal{C}_{n,k,2} = \iota_{\mathrm{pre}}(\mathcal{C}_{n-1,k,2}) \cup \iota_{\mathrm{inc}}(\mathcal{C}_{n-1,k-1,2})$$

gives a dissection of $\mathcal{A}_{n,k,2}$.

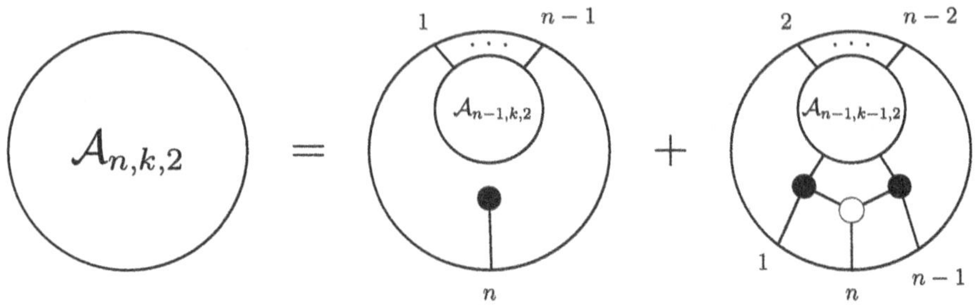

Fig. 5.7 A recursion for dissecting the amplituhedron. There is a parallel recursion obtained from this one by cyclically shifting all boundary vertices of the plabic graphs by i (modulo n)

Diagrammatically, the main theorem reads as in (Fig. 5.4):

Remark 5.2.10 Because of the cyclic symmetry of the positive Grassmannian and the amplituhedron (see e.g. Theorem 5.3.5) there are $n-1$ other versions of Theorem 5.2.3 (and Fig. 5.5) in which all plabic graph labels get shifted by i modulo n (for $1 \leq i \leq n-1$).

Note that [7] worked in the setting of *tilings*—i.e. they were only considering collections of cells that map injectively from the positive Grassmannian to the amplituhedron—but Theorem 5.2.9 holds in the more general setting of dissections.

Example 5.2.11 Let $n = 5$ and $k = 2$. We will use Theorem 5.2.9 to obtain a dissection of $\mathcal{A}_{n,k,2} = \mathcal{A}_{5,2,2}$. We start with a dissection of $\mathcal{A}_{4,2,2}$ coming from the plabic graph shown below (corresponding to the decorated permutation $(3, 4, 1, 2)$), and a dissection of $\mathcal{A}_{4,1,2}$ (corresponding to the permutations $(3, \underline{2}, 4, 1)$ and $(2, 3, 1, \underline{4})$). Applying the theorem leads to the three plabic graphs in the bottom line, which correspond to the permutations $(3, 4, 1, 2, \underline{5})$, $(4, \underline{2}, 5, 1, 3)$, $(4, 3, 1, 5, 2)$.

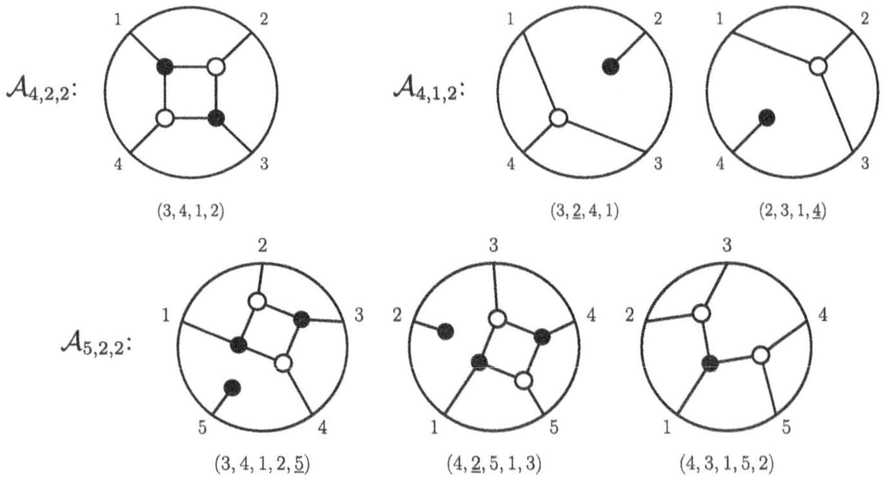

◇

T-duality for BCFW recursions. We now show that the BCFW recursions for the hypersimplex $\Delta_{k+1,n}$ Theorem 5.2.3 and for the amplituhedron $\mathcal{A}_{n,k,2}$ Theorem 5.2.9 are T-dual to each other.

Theorem 5.2.12 (BCFW recursions for $\Delta_{k+1,n}$ and $\mathcal{A}_{n,k,2}$ are T-dual) *The T-duality map provides a bijection between the tilings of the hypersimplex $\Delta_{k+1,n}$ constructed in Theorem 5.2.3, and the tilings of the amplituhedron $\mathcal{A}_{n,k,2}$ constructed in Theorem 5.2.9. More specifically, let $\{S_\pi\}$ be a collection of cells of $\mathrm{Gr}_{k+1,n}^{\geq 0}$ constructed from Theorem 5.2.3. Then this collection gives a tiling of $\Delta_{k+1,n}$ if and only if the collection $\{S_{\hat{\pi}}\}$ of cells of $\mathrm{Gr}_{k,n}^{\geq 0}$ gives a tiling of $\mathcal{A}_{n,k,2}$.*

The same statement holds if we replace the word "tiling" everywhere with "dissection."

Proof We prove this by induction on $k+n$, using Theorems 5.2.3 and 5.2.9. It suffices to show:

- if $\{S_\pi\}_{\pi\in\mathcal{C}}$ dissects $\Delta_{k+1,n-1}$ and $\{S_{\hat{\pi}}\}_{\pi\in\hat{\mathcal{C}}}$ dissects $\mathcal{A}_{n-1,k,2}$ then for any $\pi\in\mathcal{C}$, $\widehat{\mathrm{i}_{\mathrm{pre}}(\pi)} = \iota_{\mathrm{pre}}(\hat{\pi})$.
- if $\{S_\pi\}_{\pi\in\mathcal{C}}$ dissects $\Delta_{k,n-1}$ and $\{S_{\hat{\pi}}\}_{\pi\in\hat{\mathcal{C}}}$ dissects $\mathcal{A}_{n-1,k-1,2}$ then for any $\pi\in\mathcal{C}$, $\widehat{\mathrm{i}_{\mathrm{inc}}(\pi)} = \iota_{\mathrm{inc}}(\hat{\pi})$.

Let $\pi=(a_1,\ldots,a_{n-1})$ be a decorated permutation. We first verify the first statement. Then $\mathrm{i}_{\mathrm{pre}}(\pi)=(a_1,a_2,\ldots,a_{n-2},n,a_{n-1})$, so $\widehat{\mathrm{i}_{\mathrm{pre}}(\pi)}=(a_{n-1},a_1,a_2,\ldots,a_{n-2},n)$, where n is a black fixed point. Meanwhile, $\hat{\pi}=(a_{n-1},a_1,a_2,\ldots,a_{n-2})$, so $\iota_{\mathrm{pre}}(\hat{\pi})=(a_{n-1},a_1,a_2,\ldots,a_{n-2},n)$, where n is a black fixed point.

We now verify the second statement. Let $j=\pi^{-1}(n-1)$. Then we have that $\mathrm{i}_{\mathrm{inc}}(\pi)=(a_1,a_2,\ldots,a_{j-1},n,a_{j+1},\ldots,a_{n-1},n-1)$, and $\widehat{\mathrm{i}_{\mathrm{inc}}(\pi)}=(n-1,a_1,a_2,\ldots,a_{j-1},n,a_{j+1},\ldots,a_{n-1})$. Meanwhile $\hat{\pi}=(a_{n-1},a_1,a_2,\ldots,a_{n-2})$. Then it is straightforward to verify that $\iota_{\mathrm{inc}}(\hat{\pi})$ is exactly the permutation $\widehat{\mathrm{i}_{\mathrm{inc}}(\pi)}=(n-1,a_1,a_2,\ldots,a_{j-1},n,a_{j+1},\ldots,a_{n-1})$, as desired. □

Remark 5.2.13 Theorem 5.2.12 can also be easily proved by observing how T-duality maps plabic graphs in Proposition 4.1.15. Moreover, Bao and He's Theorem 5.2.9 for tilings of $\mathcal{A}_{n,k,2}$ is an immediate consequence of our Theorem 5.2.3 and Theorem 5.1.9.

5.2.2 Regular Tilings and Dissections from Trop$^+$ *$Gr_{k+1,n}$*

Good Tilings and Dissections. Among all possible positroid dissections, there are some with particularly nice features, which we will call *good*, as well as others with rather unpleasant properties. In particular, one can consider dissections where the boundaries of the strata interact nicely.

Definition 5.2.14 (*Good Dissections of* $\Delta_{k+1,n}$) Let $\{\Gamma_\pi\}$ be a dissection of $\Delta_{k+1,n}$. $\{\Gamma_\pi\}$ is a *good dissection* if positrod polytopes either intersect in common facets or in higher codimension, i.e. if $\Gamma_\pi \cap \Gamma_\nu$ has codimension one, then $\Gamma_\pi \cap \Gamma_\nu$ equals $\Gamma_{\pi'}$, where $\Gamma_{\pi'}$ is a facet of both Γ_π and Γ_ν.

Analogously, we define good dissections of amplituhedron.

Definition 5.2.15 (*Good Dissections of* $\mathcal{A}_{n,k,m}$) Let $\{Z_\pi\}$ be a dissection of $\mathcal{A}_{n,k,m}$. $\{Z_\pi\}$ is a *good dissection* if Grasstopes either intersect in common facets or in higher codimension, i.e. if $Z_\pi \cap Z_\nu$ has codimension one, then $Z_\pi \cap Z_\nu$ equals $Z_{\pi'}$, where $Z_{\pi'}$ is a facet of both Z_π and Z_ν.

Example 5.2.16 Let us study the following tiling of $\mathcal{A}_{6,2,2}$:

$$\mathcal{C}_1 = \{Z_{\pi^{(1)}}, Z_{\pi^{(2)}}, Z_{\pi^{(3)}}, Z_{\pi^{(4)}}, Z_{\pi^{(5)}}, Z_{\pi^{(6)}}\}$$

with

$$\pi^{(1)} = (\underline{1}, \underline{2}, 5, 6, 3, 4), \quad \pi^{(2)} = (\underline{1}, 3, 6, 5, 2, 4), \quad \pi^{(3)} = (\underline{1}, 4, 6, 2, \underline{5}, 3)\,,$$
$$\pi^{(4)} = (2, 6, \underline{3}, 5, 1, 4), \quad \pi^{(5)} = (2, 6, 4, 1, \underline{5}, 3), \quad \pi^{(6)} = (3, 6, 1, 4, \underline{5}, 2)\,.$$

All elements of $\mathcal{C}_1$ are positroid tiles and their images through $\tilde{Z}$-map are 4-dimensional. The tiling $\mathcal{C}_1$ is a refinement of the following dissection

$$\mathcal{C}_2 = \{Z_{\pi^{(1)}}, Z_{\pi^{(7)}}, Z_{\pi^{(8)}}, Z_{\pi^{(6)}}\}$$

with

$$\pi^{(7)} = (1, 4, 6, 5, 2, 3)\,, \qquad \pi^{(8)} = (2, 6, 4, 5, 1, 3)\,.$$

The dissection $\mathcal{C}_2$ has the property that if a pair of cell images under $\tilde{Z}$-map intersect along a 3-dimensional surface then this surface is an image of another positroid cell in $\mathrm{Gr}_{2,6}^{\geq 0}$:

$$Z_{\pi^{(1)}} \cap Z_{\pi^{(7)}} = Z_{(\underline{1},\underline{2},6,5,3,4)}$$
$$Z_{\pi^{(7)}} \cap Z_{\pi^{(8)}} = Z_{(\underline{1},6,4,5,2,3)}$$
$$Z_{\pi^{(8)}} \cap Z_{\pi^{(6)}} = Z_{(2,6,1,\underline{4},\underline{5},3)}$$

and all remaining pairs of images intersect along lower dimensional surfaces. We consider the dissection $\mathcal{C}_2$ "good" because all its elements have compatible codimension one boundaries. However, the dissection $\mathcal{C}_1$ does not have this property. Let us observe that

$$Z_{\pi^{(2)}} \cup Z_{\pi^{(3)}} = Z_{\pi^{(7)}}$$
$$Z_{\pi^{(4)}} \cup Z_{\pi^{(5)}} = Z_{\pi^{(8)}}$$

We expect that, after we subdivide $Z_{\pi^{(7)}}$ and $Z_{\pi^{(8)}}$, the boundary $Z_{(\underline{1},6,4,5,2,3)}$ which they share will also get subdivided. This however happens in two different ways and we do not get compatible facets for the dissection $\mathcal{C}_1$. It is a similar picture to the one we get when we consider polyhedral subdivisions of a double square pyramid: it is possible to subdivide it into two pieces along its equator, and then further subdivide each pyramid into two simplices. However, in order to get a polyhedral triangulation of the double square pyramid, we need to do it in a compatible way, along the same diagonal of the equatorial square. ◇

In the following, we will conjecture that good dissections of the hypersimplex are in bijection via T-duality with good dissections of the amplituhedron. Towards this goal, we start by providing a characterization of good intersections of positroid polytopes.

Proposition 5.2.17 *Let $\Gamma_{\pi^{(1)}}$ and $\Gamma_{\pi^{(2)}}$ be two $(n-1)$-dimensional positroid polytopes whose intersection $\Gamma_{\pi^{(1)}} \cap \Gamma_{\pi^{(2)}}$ is a polytope of dimension $n-2$. Then $\Gamma_{\pi^{(1)}} \cap \Gamma_{\pi^{(2)}}$ is a positroid polytope of the form $\Gamma_{\pi^{(3)}}$, where $\pi^{(3)}$ is a loopless permutation.*

Proof By Theorem 3.1.12, $\Gamma_{\pi^{(1)}} \cap \Gamma_{\pi^{(2)}}$ is a positroid polytope and hence has the form $\Gamma_{\pi^{(3)}}$, for some decorated permutation $\pi^{(3)}$. (Using Proposition 3.1.9, the fact that $\dim(\Gamma_{\pi^{(3)}}) = n-2$ implies that the positroid associated to $\pi^{(3)}$ has precisely two connected components.)

Now we claim that the positroid associated to $\pi^{(3)}$ is loopless. In general there is an easy geometric way of recognizing when a matroid M is loopless from the polytope Γ_M: M is loopless if and only if Γ_M is not contained in any of the n facets of the hypersimplex of the type $x_i = 0$ for $1 \leq i \leq n$. Since $\Gamma_{\pi^{(3)}}$ arises as the codimension 1 intersection of two full-dimensional matroid polytopes contained in $\Delta_{k+1,n}$ it necessarily meets the interior of the hypersimplex and hence the matroid must be loopless. □

Remark 5.2.18 Recall that the T-duality map is well-defined on positroid cells whose matroid is connected, and more generally, loopless. Proposition 5.2.17 implies that if we consider two cells $S_{\pi^{(1)}}$ and $S_{\pi^{(2)}}$ of $\mathrm{Gr}_{k+1,n}^{\geq 0}$ whose matroid is connected and whose moment map images (necessarily top-dimensional) intersect in a common facet, then that facet is the moment map image of a loopless cell $S_{\pi^{(3)}}$. Therefore we can apply the T-duality map to all three cells $S_{\pi^{(1)}}$, $S_{\pi^{(2)}}$, and $S_{\pi^{(3)}}$.

Conjecture 5.2.19 *Let $S_{\pi^{(1)}}$ and $S_{\pi^{(2)}}$ be two positroid cells in $\mathrm{Gr}_{k,n}^{\geq 0}$ corresponding to coloopless permutations $\pi^{(1)}$ and $\pi^{(2)}$. Let $\dim Z^{\circ}_{\pi^{(1)}} = \dim Z^{\circ}_{\pi^{(2)}} = 2k$ with $Z_{\pi^{(1)}} \cap Z_{\pi^{(2)}} = Z_{\pi^{(3)}}$, where $S_{\pi^{(3)}} \subset G^{+}_{k,n}$ is such that $\dim Z^{\circ}_{\pi^{(3)}} = 2k-1$. Then $\pi^{(3)}$ is a coloopless permutation.*

Remark 5.2.20 Conjecture 5.2.19 guarantees that if we consider two positroid cells with top-dimensional images in the amplituhedron $\mathcal{A}_{n,k,2}$, which have a facet in common, then the positroid cell corresponding to this facet is coloopless and therefore we can apply the T-duality map to it.

Finally we arrive at a conjecture connecting good dissections of hypersimplex and amplituhedron.

Conjecture 5.2.21 (Good Dissections of $\Delta_{k+1,n}$ and $\mathcal{A}_{n,k,2}$ are T-dual) *Let* $\{S_\pi\}$ *be a collection of cells of* $\mathrm{Gr}_{k+1,n}^{\geq 0}$. *Then this collection gives a good tiling (respectively, good dissection) of* $\Delta_{k+1,n}$ *if and only if the collection* $\{S_{\hat{\pi}}\}$ *of cells of* $\mathrm{Gr}_{k,n}^{\geq 0}$ *gives a good tiling (respectively, good dissection) of* $\mathcal{A}_{n,k,2}$.

In Sect. 5.2.2, we discussed the fact that arbitrary dissections of the hypersimplex and the amplituhedron can have rather unpleasant properties, with their maximal cells intersecting badly at their boundaries. We introduced the notion of *good dissections* for the hypersimplex and amplituhedron in Definitions 5.2.14 and 5.2.15. Our goal in this section is to introduce a large class of good dissections for the amplituhedron—these are the *regular positroid subdivisions*.

Regular positroid subdivisions of $\mathcal{A}_{n,k,2}$. Recall from Definition 3.2.15 that the *regular positroid subdivisions* of $\Delta_{k+1,n}$ are precisely the dissections of the form $\mathcal{D}_P$ (see Sect. 3.2.2) where $P = \{P_I\} \in \mathbb{R}^{\binom{[n]}{k+1}}$. We know from Theorem 5.1.9 that T-duality is a bijection between positroid tilings of $\Delta_{k+1,n}$ and the amplituhedron $\mathcal{A}_{n,k,2}$. Therefore, each maximal cone of the positive tropical Grassmannian $\mathrm{Trop}^+ \mathrm{Gr}_{k+1,n}$ induces a positroid tiling of the amplituhedron $\mathcal{A}_{n,k,2}(Z)$:

Corollary 5.2.22 (Tilings of $\mathcal{A}_{n,k,2}$ from $\mathrm{Trop}^+ \mathrm{Gr}_{k+1,n}$) *For* $P = \{P_I\}_{I \in \binom{[n]}{k+1}}$ *a positive tropical Plücker vector from a maximal cone of* $\mathrm{Trop}^+ \mathrm{Gr}_{k+1,n}$, *let* $\mathcal{D}_P = \{\Gamma_\pi\}$ *be the regular positroid tiling of* $\Delta_{k+1,n}$, *then* $\{Z_{\hat{\pi}}\}$ *is a positroid tiling of* $\mathcal{A}_{n,k,2}(Z)$.

Using Conjecture 5.1.14, we conjecture Corollary 5.2.22 extends to dissections:

Conjecture 5.2.23 (Dissections of $\mathcal{A}_{n,k,2}$ from $\mathrm{Trop}^+ \mathrm{Gr}_{k+1,n}$) *For* $P = \{P_I\}_{I \in \binom{[n]}{k+1}}$ *a positive tropical Plücker vector from* $\mathrm{Trop}^+ \mathrm{Gr}_{k+1,n}$, *let* $\mathcal{D}_P = \{\Gamma_\pi\}$ *be the regular positroid subdivision of* $\Delta_{k+1,n}$, *then* $\{Z_{\hat{\pi}}\}$ *is a positroid dissection of* $\mathcal{A}_{n,k,2}(Z)$.

Then we define:

Definition 5.2.24 We say that a positroid dissection $\{Z_{\hat{\pi}}\}$ of $\mathcal{A}_{n,k,2}$ is a *regular positroid subdivision* if $\{\Gamma_\pi\}$ is a regular positroid subdivision of $\Delta_{k+1,n}$.

In [9], the authors conjectured there are $\binom{n-2}{k}$ Grasstopes in a positroid tiling of $\mathcal{A}_{n,k,2}$. As noted in [3], this is also the number of positroid polytopes in a regular positroid tiling of $\Delta_{k+1,n}$ [10]. By Corollary 5.2.22, we have:

Corollary 5.2.25 *There are* $\binom{n-2}{k}$ *Grasstopes in any regular positroid tiling of* $\mathcal{A}_{n,k,2}(Z)$.

Remark 5.2.26 [3] showed that all *BCFW* tilings of $\mathcal{A}_{n,k,2}(Z)$ contain $\binom{n-2}{k}$ Grasstopes; there are BCFW tilings which are not regular and regular tilings which are not BCFW.

As every regular positroid subdivision of $\Delta_{k+1,n}$ is a polyhedral subdivision (and hence is good), Conjecture 5.2.21 implies the following.

Conjecture 5.2.27 *Every regular positroid subdivision of $\mathcal{A}_{n,k,2}$ is a good dissection.*

At the end of this section, we provide some computational evidence for Conjecture 5.2.27. For example, for $\mathcal{A}_{6,2,2}$ and $\mathcal{A}_{7,2,2}$, every regular positroid subdivision is good, and moreover, all good dissections are regular positroid subdivisions. (This appears to also be the case for $\mathcal{A}_{8,2,2}$; but we were only able to compute the number of *tilings* in this case.) One might hope to strengthen Conjecture 5.2.27 and conjecture that the regular positroid subdivisions are precisely the good dissections. However, the notion of regularity is rather subtle (as usual in polyhedral geometry), and starting from $\mathcal{A}_{9,2,2}$, there are some good dissections which are not regular.

The fan structure for regular positroid subdivisions. We now discuss the fan structure for regular positroid subdivisions of the hypersimplex and amplituhedron.

Definition 5.2.28 Given two subdivisions $\{\Gamma_\pi\}$ and $\{\Gamma_{\pi'}\}$ of $\Delta_{k+1,n}$, we say that $\{\Gamma_\pi\}$ refines $\{\Gamma_{\pi'}\}$ and write $\{\Gamma_\pi\} \preceq \{\Gamma_{\pi'}\}$ if every Γ_π is contained in some $\Gamma_{\pi'}$.

Similarly, given two subdivisions $\{Z_\pi\}$ and $\{Z_{\pi'}\}$ of $\mathcal{A}_{n,k,2}$, we say that $\{Z_\pi\}$ refines $\{Z_{\pi'}\}$ and write $\{Z_\pi\} \preceq \{Z_{\pi'}\}$ if every Z_π is contained in some $Z_{\pi'}$.

Recall from Sect. 3.2.3 that we have a fan structure on $\mathrm{Trop}^+ \mathrm{Gr}_{k+1,n}$ (the *secondary fan*, which coincides with the *Plücker fan*) which describes the regular positroid subdivisions of $\Delta_{k+1,n}$, ordered by refinement. We expect that this fan structure on $\mathrm{Trop}^+ \mathrm{Gr}_{k+1,n}$ also describes the regular positroid subdivisions of $\mathcal{A}_{n,k,2}$.

Conjecture 5.2.29 (Secondary Geometry of $\mathcal{A}_{n,k,2}$ from $\mathrm{Trop}^+ Gr_{k+1,n}$) *The regular positroid subdivisions of $\mathcal{A}_{n,k,2}$ are parametrized by the cones of* $\mathrm{Trop}^+ \mathrm{Gr}_{k+1,n}$*, with the natural partial order on the cones reflecting the refinement order on positroid subdivisions.*

Conjecture 5.2.29 is consistent with the following conjecture.

Conjecture 5.2.30 *Consider two regular positroid subdivisions $\{\Gamma_\pi\}$ and $\{\Gamma_{\pi'}\}$ of $\Delta_{k+1,n}$, and two corresponding positroid subdivisions $\{Z_{\hat{\pi}}\}$ and $\{Z_{\hat{\pi}'}\}$ of $\mathcal{A}_{n,k,2}$. Then we have that $\{\Gamma_\pi\} \preceq \{\Gamma_{\pi'}\}$ if and only if $\{Z_{\hat{\pi}}\} \preceq \{Z_{\hat{\pi}'}\}$*

Catalan Tilings. We here consider particular types of regular tilings in bijection with Catalan numbers C_{n-2}.

Definition 5.2.31 (*Colouring a fixed tree*) Let T be any planar trivalent tree with n leaves (which will necessarily have $n-2$ internal vertices), embedded in a disk with the leaves labelled from 1 to n in clockwise order. Let $\mathcal{T}_{n,k}(T)$ be the set of $\binom{n-2}{k}$ plabic graphs obtained from T by colouring precisely k of the internal vertices black, as in Fig. 5.8.

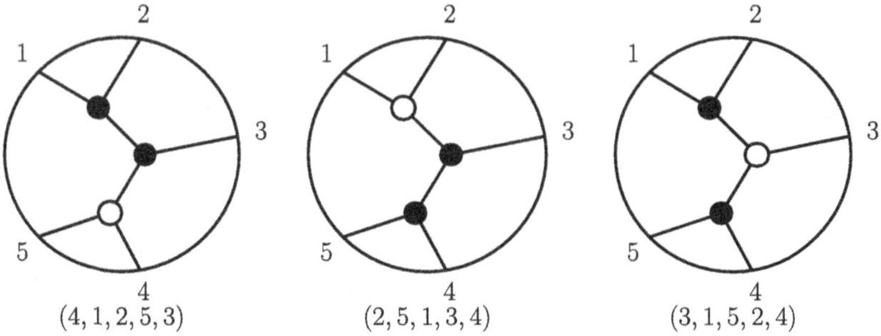

Fig. 5.8 A collection $\mathcal{T}_{5,2}$ of plabic graphs giving a regular subdivision of $\Delta_{3,5}$

Proposition 5.2.32 (Catalan Tilings of $\Delta_{k+1,n}$ and $\mathcal{A}_{n,k,2}$) *Let fix a tree T as in Definition 5.2.31. The positroid polytopes $\{\Gamma_G\}$ corresponding to the collection of plabic graphs $\{G\} = \mathcal{T}_{n,k}(T)$ is a regular tiling of $\Delta_{k+1,n}$. Therefore the collection of their T-dual Grasstopes $\{Z_{\hat{G}}\}$ is a regular tiling of $\mathcal{A}_{n,k,2}(Z)$.*

Proof We can use Theorem 5.2.3 (see Fig. 5.5) to inductively prove that the cells corresponding to $\mathcal{T}_{n,k}$ give a tiling of $\Delta_{k+1,n}$. The fact that the cells corresponding to the plabic graphs in $\mathcal{T}_{n,k}$ give a *regular* tiling of $\Delta_{k+1,n}$ follows from [11, Theorem 8.4]. Now using Theorem 5.1.9, it follows that the images of these cells under the T-duality give a tiling of $\mathcal{A}_{n,k,2}$. The fact that this tiling is regular now follows from Definition 5.2.24. □

Remark 5.2.33 The above construction gives us C_{n-2} regular tilings of $\mathcal{A}_{n,k,2}(Z)$, where $C_n = \frac{1}{n+1}\binom{2n}{n}$ is the Catalan number. We call these *Catalan tilings*.

At the end of this section, we will describe our computational data concerning the numbers of good dissections of the amplituhedron $\mathcal{A}_{n,k,2}$. We will notice the remarkable similarity with the data from the list of f-vectors in Sect. 3.2.3, giving evidence for Conjecture 5.2.29.

Remark 5.2.34 The coordinate ring of the Grassmannian has the structure of a *cluster algebra* [12]. In particular, $\mathrm{Gr}_{2,n}$, $\mathrm{Gr}_{3,6}$, $\mathrm{Gr}_{3,7}$, $\mathrm{Gr}_{3,8}$ have cluster structures of finite types A_n, D_4, E_6, and E_8, respectively. As discussed in [13], there is an intriguing connection between $\mathrm{Trop}^+ \mathrm{Gr}_{k,n}$ and the cluster structure. In particular, $F_{2,n}$ is the fan to the type A_n associahedron, while $F_{3,6}$ and $F_{3,7}$ are refinements of the fans associated to the D_4 and E_6 associahedra. Via our correspondence between $\mathrm{Trop}^+ \mathrm{Gr}_{k+1,n}$ and the amplituhedron $\mathcal{A}_{n,k,2}$, the Grassmannian cluster structure on $\mathrm{Gr}_{k+1,n}$ should be reflected in good subdivisions of $\mathcal{A}_{n,k,2}$. In particular the type A_n cluster structure should control $\mathcal{A}_{n,1,2}$ (this is apparent, since $\mathcal{A}_{n,1,2}$ is a projective polygon), while the type D_4, E_6, and E_8 cluster structures should be closely related to $\mathcal{A}_{6,2,2}$, $\mathcal{A}_{7,2,2}$, and $\mathcal{A}_{8,2,2}$.

Experimental Data. Checks for this section have been performed by using Mathematica. In particular, we used the packages `'positroid'` [6] and

Table 5.2 New results about the tilings of the amplituhedron $\mathcal{A}_{n,k,2}$ in relation to known results about the number of maximal cones of the positive tropical Grassmannian $\mathrm{Trop}^+ \mathrm{Gr}_{k+1,n}$

(k, n)	Tilings	Good tilings	$\mathrm{Trop}^+ \mathrm{Gr}_{k+1,n}$	Non-regular good tilings
$(1, n)$	C_{n-2}	C_{n-2}	C_{n-2}	0
$(2, 5)$	5	5	5	0
$(2, 6)$	120	48	48	0
$(2, 7)$	3073	693	693	0
$(2, 8)$	6 443 460	13 612	13 612	0
$(2, 9)$	?	346 806	346 710	96
$(3, 6)$	14	14	14	0
$(3, 7)$	3073	693	693	0
$(3, 8)$	?	91 496	90 608	888
$(3, 9)$	?	33 182 763	30 659 424	2 523 339

`'amplituhedronBoundaries'` [14]. This allowed us to find the complete poset of good dissections of $\mathcal{A}_{6,2,2}$ and $\mathcal{A}_{7,2,2}$, whose f-vectors read:

$$\mathcal{A}_{6,2,2} : (1, 48, 98, 66, 16, 1)$$
$$\mathcal{A}_{7,2,2} : (1, 693, 2163, 2583, 1463, 392, 42, 1)$$

which are exactly the f-vectors of the positive tropical Grassmannian $\mathrm{Trop}^+ \mathrm{Gr}_{3,6}$ and $\mathrm{Trop}^+ \mathrm{Gr}_{3,7}$, respectively. We summarize all our finding[3] about positroid tilings in Table 5.2. We observe that for $\mathcal{A}_{8,2,2}$ the number of good tilings agrees with the number of maximal cones in $\mathrm{Trop}^+ \mathrm{Gr}_{3,8}$. Starting from $n = 9$, the number of good tilings is larger than the number of maximal cones in positive tropical Grassmannian. It is indeed the first example where one can find good tilings which are *not regular*. In particular, out of 346806 good tilings, 96 are not regular. Similarly, for $k = 3$ and $n = 8$, 888 good tilings of $\mathcal{A}_{8,3,2}$ are not regular. We note that these correspond exactly to degenerate matrices found in [15].

5.2.3 *Tilings from Descents and Sign-Flips*

In Sect. 4.3 we used permutations and their cyclic descents to define both the w-simplices in $\Delta_{k+1,n}$ and the w-chambers in $\mathcal{A}_{n,k,2}(Z)$. In the same spirit, by refining the set of permutations based on the *positions* of the descents, we will obtain a distinguished positroid tiling of $\Delta_{k+1,n}$ and a distinguished positroid tiling of $\mathcal{A}_{n,k,2}(Z)$. These tilings are T-dual to each other.

[3] We also included there the results for $\mathrm{Gr}^{\geq 0}_{3,9}$ which, by using our conjectures, can be derived from [15].

Recall that

$$\Delta_{k+1,n} = \bigcup_{w \in D_{k+1,n}} \Delta_w.$$

Since 1 is always a cyclic descent of $w \in D_{k+1,n}$, we have that $D_{k+1,n}$ is the set of permutations $w \in S_n$ with k left descents and $w(n) = n$. The Eulerian numbers have a very natural refinement by descent set $\mathrm{Des}_{\mathrm{L}}(w)$. If $w \in S_n$ has $w(n) = n$, then neither 1 nor n is a left descent of w, so we have

$$E_{k,n-1} = \sum_{I \in \binom{[2,n-1]}{k}} \#\{w \in S_n : w(n) = n, \mathrm{Des}_{\mathrm{L}}(w) = I\}.$$

This inspires the following decomposition of $\Delta_{k+1,n}$. For $I \in \binom{[2,n-1]}{k}$, let

$$\Gamma_I := \bigcup_{\substack{w \in D_{k+1,n} \\ \mathrm{Des}_{\mathrm{L}}(w)=I}} \Delta_w.$$

Clearly, the collection of Γ_I cover the hypersimplex and their interiors are pairwise disjoint. There are also $\binom{n-2}{k}$ of them, which is exactly the number of full-dimensional positroid polytopes in a regular positroid tiling of $\Delta_{k+1,n}$ [10]. We will show that each Γ_I is in fact a positroid polytope, and that $\{\Gamma_I\}$ is a (regular) positroid tiling of $\Delta_{k+1,n}$. We will refer to it as the *descent tiling*.

On the other hand, given the sign-flip characterization of the amplituhedron from Theorem 2.4.1, it is natural to subdivide $\mathcal{A}_{n,k,2}(Z)$ into regions based on where the sequence $(\langle Y1a\rangle)_{a=1}^n$ has sign flips. That is, for each $I \in \binom{[2,n-1]}{k}$, we define[4]

$$Z_I^\circ := \{Y \in \mathcal{A}_{n,k,2}(Z) \mid \mathrm{Flip}(\langle Y11\rangle, \langle Y12\rangle, \langle Y13\rangle, \ldots, \langle Y1n\rangle) = I\}$$

and define Z_I to be the closure of Z_I°.

Arkani-Hamed et al. [16, Sect. 7] conjectured that $\{Z_I\}$ is a positroid tiling of $\mathcal{A}_{n,k,2}(Z)$. The authors referred to $\{Z_I\}$ as a *sign-flip (or kermit[5]) tiling*. In this section we prove this conjecture. Moreover we show that sign-flip tilings of $\mathcal{A}_{n,k,2}(Z)$ and descent tilings of the hypersimplex $\Delta_{k+1,n}$ are T-dual to each other and also regular.

Definition 5.2.35 (*Plabic tilings of kermit type*) Let $I = \{i_1, \ldots, i_k\} \in \binom{[2,n-1]}{k}$ and let $\mathcal{T}_I$ be the triangulated (k, n)-unpunctured plabic tiling whose triangles have vertices $\{1, i_\ell, i_\ell + 1\}$ for $\ell = 1, \ldots k$. We say $\mathcal{T}_I$ is *kermit type* and denote the plabic graph $\hat{G}(\mathcal{T})$ by K_I. We also denote the plabic graph $G(\mathcal{T})$ by C_I, and call it a *caterpillar* tree.

---

[4] Because of our conventions regarding sign flips, Z_I would be empty if $1, n \in I$.

[5] For $k = 2$, $\mathcal{A}_{n,2,2}(Z)$ provides the integrand for the 1-loop n-point scattering amplitude in $\mathcal{N} = 4$ SYM. The name 'kermit' comes from the resemblance of the pictorial expansion of such amplitude (e.g. see [17, p. 18]) with the Muppet character 'Kermit the Frog'.

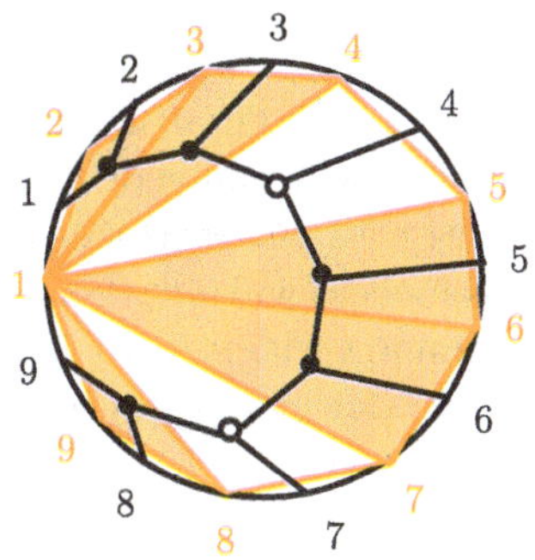

Fig. 5.9 In orange, the plabic tiling $\mathcal{T}_I$ of kermit-type for $I = \{2, 3, 5, 6, 8\}$. In black, the dual caterpillar tree C_I

For an example, see (Fig. 5.9).

Proposition 5.2.36 (Descent and sign-flips tilings are T-dual) *Let I run over $\binom{[2,n-1]}{k}$. The collections $\{\Gamma_I\}$ and $\{Z_I\}$ are T-dual regular positroid tilings of $\Delta_{k+1,n}$ and $\mathcal{A}_{n,k,2}(Z)$. Furthermore, $\Gamma_I = \Gamma_{C_I}$ and $Z_I = Z_{K_I}$ where C_I and K_I are as in Definition 5.2.35.* □

Proof By the sign description of $Z^{\circ}_{K_I}$ in Theorem 2.3.28, it is straightforward that $Z_I = Z_{K_I}$. Moreover, using Definition 4.3.8 and Corollary 4.3.20, we have

$$Z_{K_I} = \bigcup_{w:\mathrm{Des}_{\mathrm{L}}(w)=I} \hat{\Delta}_w(Z). \tag{5.2.37}$$

Using Proposition 4.2.6, it is not hard to check that the positroid polytope Γ_{C_I} satisfies

$$\Gamma_{C_I} = \bigcup_{w:\mathrm{Des}_{\mathrm{L}}(w)=I} \Delta_w. \tag{5.2.38}$$

But this is exactly Γ_I.

Finally, it is easy to check that $\{\Gamma_{C_I}\}_{I\in\binom{[2,n-1]}{k}}$ is a tiling of $\Delta_{k+1,n}$ of the sort appearing in [3, Proposition 10.7] ('Catalan type'), hence is a regular positroid tiling. It follows that $\{Z_{K_I}\}_{I\in\binom{[2,n-1]}{k}}$ is the T-dual regular positroid tiling.

Remark 5.2.39 Sign-flip tilings of $\mathcal{A}_{n,k,2}(Z)$ and descent tilings of $\Delta_{k+1,n}$ are of BFCW type (in particular, of 'Catalan type', see [3, Proposition 10.7]).

We end this section by describing each w-simplex (resp. w-chamber) as an intersection of cyclically shifted caterpillar positroid polytopes (resp. kermit Grasstopes).

Proposition 5.2.40 *Let Δ_w, $\hat{\Delta}_w(Z)$ be a w-simplex and a w-chamber in $\Delta_{k+1,n}$ and $\mathcal{A}_{n,k,2}(Z)$ respectively. Let $I_1, \ldots, I_n$ give the vertices of Δ_w, and let $J_a := (I_a \setminus \{a\}) - (a-1)$. Then:*

$$\Delta_w = \bigcap_{a\in[n]} \Gamma_{C^{(a)}_{J_a}} \quad \text{and} \quad \hat{\Delta}_w(Z) = \bigcap_{a\in[n]} Z_{K^{(a)}_{J_a}},$$

where $C_J^{(a)}$ (resp. $K_J^{(a)}$) denotes the cyclic shift of C_J (resp. K_J), such that $1 \mapsto a$.

Proof To see the statement about Δ_w, note that another way to phrase (5.2.38) is that Γ_{C_I} is the union of all w-simplices with 1st vertex given by $I \cup \{1\}$. Using the cyclic shift on the hypersimplex, it is not hard to see that $\Gamma_{C_{J_a}^{(a)}}$ is the union of all w-simplices with ath vertex given by I_a. So taking the intersection gives exactly the w-simplex with vertices $e_{I_1}, \ldots, e_{I_a}$.

The statement about $\hat{\Delta}_w(Z)$ follows from a similar argument, using (5.2.37) and the cyclic shift on $\mathrm{Gr}_{k,n}$. □

Example 5.2.41 Let us consider $w = 324156$. We have:

$$I_1 = \{1, 2, 3\}, \quad I_2 = \{2, 3, 5\}, \quad I_3 = \{1, 3, 4\}, \quad I_4 = \{1, 2, 4\}, \quad I_5 = \{1, 3, 5\}, \quad I_6 = \{2, 3, 6\};$$
$$J_1 = \{2, 3\}, \quad J_2 = \{2, 4\}, \quad J_3 = \{2, 5\}, \quad J_4 = \{4, 5\}, \quad J_5 = \{3, 5\}, \quad J_6 = \{3, 4\}.$$

Then Δ_w is the intersection of $\Gamma_{C_{J_1}^{(1)}}, \ldots, \Gamma_{C_{J_6}^{(6)}}$ and $\hat{\Delta}_w$ is the intersection of $Z_{K_{J_1}^{(1)}}, \ldots, Z_{K_{J_6}^{(6)}}$. The cyclically rotated kermit-type plabic tilings $\mathcal{T}_{J_1}^{(1)}, \ldots, \mathcal{T}_{J_6}^{(6)}$ are displayed below.

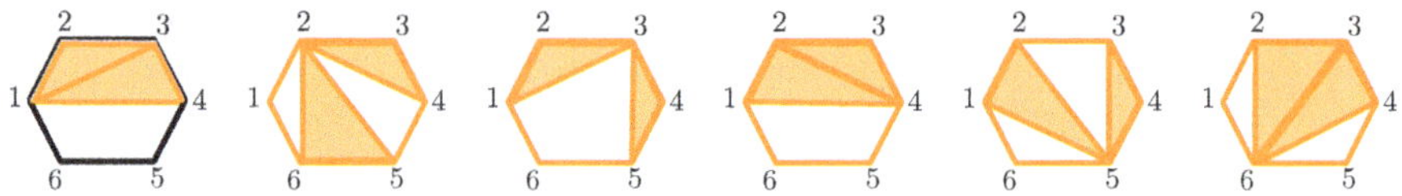

Notice that $\mathcal{T}_{J_1}^{(1)}$ is equivalent to $\mathcal{T}_{J_4}^{(4)}$. ◇

5.3 T-Duality, Cyclic Symmetry and Parity Duality

We end this chapter about T-duality by discussing its relation with *parity duality*—a duality between the amplituhedra $\mathcal{A}_{n,k,m}$ and $\mathcal{A}_{n,n-m-k,m}$. The definition of parity duality was originally inspired by the physical operation of parity conjugation in quantum field theory—more specifically, in the context of scattering amplitudes in $\mathcal{N} = 4$ Super-Yang-Mills, where amplitudes can be computed from the geometry of $\mathcal{A}_{n,k,4}$ [2]. Furthermore, the conjectural formula of Karp, Williams, and Zhang [9] for the number of cells in each tiling of the amplituhedron is invariant under the operation of swapping the parameters k and $n - m - k$ and hence is consistent with parity duality: this motivated further works [18, 19].

In Theorem 5.3.3, we will explain how parity duality for $m = 2$ amplituhedra is naturally induced by a composition of the usual duality for Grassmannians ($\mathrm{Gr}_{k,n} \simeq \mathrm{Gr}_{n-k,n}$) and the T-duality map (between loopless cells of $\mathrm{Gr}_{k+1,n}^{\geq 0}$ and coloopless cells of $\mathrm{Gr}_{k,n}^{\geq 0}$). The usual Grassmannian duality gives rise to a bijection between dissections of the hypersimplex $\Delta_{k+1,n}$ and dissections of the hypersimplex $\Delta_{n-k-1,n}$. By composing this Grassmannian duality with the T-duality map (on both sides), we obtain the parity duality between dissections of $\mathcal{A}_{n,k,2}$ and $\mathcal{A}_{n,n-k-2,2}$, respectively. This makes the notion of parity duality at $m = 2$ quite transparent!

Recall that our convention on dissections is that the images of all positroid cells are of full dimension $n - 1$. Therefore all positroids involved in a dissection must be connected, and the corresponding decorated permutations will be fixed-point-free.

Theorem 5.3.1 (Grassmannian duality for dissections of the hypersimplex) *Let* $\{\Gamma_\pi\}$ *be a collection of positroid polytopes which dissects the hypersimplex* $\Delta_{k+1,n}$. *Then the collection of positroid polytopes* $\{\Gamma_{\pi^{-1}}\}$ *dissects the hypersimplex* $\Delta_{n-k-1,n}$.

Proof If G is a plabic graph representing the positroid cell S_π, and if we swap the colors of the black and white vertices of G, we obtain a graph G' representing the positroid $S_{\pi^{-1}}$. It is not hard to see from [8] that G' and π^{-1} represent the *dual positroid* to G and π. But now the matroid polytopes Γ_π and $\Gamma_{\pi^{-1}}$ are isomorphic via the map dual : $\mathbb{R}^n \to \mathbb{R}^n$ sending $(x_1, \ldots, x_n) \mapsto (1 - x_1, \ldots, 1 - x_n)$. This maps relates the two dissections in the statement of the theorem. □

By composing the inverse map on decorated permutations $\pi \mapsto \pi^{-1}$ (which represents the Grassmannian duality of Theorem 5.3.1) with T-duality, we obtain the following map.

Definition 5.3.2 We define $\widetilde{U}_{k,n}$ to be the map between coloopless permutations on $[n]$ with k anti-excedances and coloopless permutations on $[n]$ with $n - k - 2$ anti-excedances such that $\widetilde{U}_{k,n}\hat{\pi} = \widehat{\pi^{-1}}$. Equivalently, we have $(\widetilde{U}_{k,n}\pi)(i) = \pi^{-1}(i-1) - 1$, where values of the permutation are considered modulo n, and any fixed points which are created are designated to be loops.

Theorem 5.3.3 (Parity duality from T-duality and Grassmannian duality) *Let* $\{Z_\pi\}$ *be a collection of Grasstopes which dissects the amplituhedron* $\mathcal{A}_{n,k,2}$. *Then the collection of Grasstopes* $\{Z_{\widetilde{U}_{k,n}\pi}\}$ *dissects the amplituhedron* $\mathcal{A}_{n,n-k-2,2}$.

We will prove Theorem 5.3.3 by using the cyclic symmetry of the positive Grassmannian and the amplituhedron, and showing (see Lemma 5.3.8) that up to a cyclic shift, our map $\widetilde{U}_{k,n}$ agrees with the parity duality map of [19].

The totally nonnegative Grassmannian exhibits a beautiful *cyclic symmetry* [20]. Let us represent an element of $\mathrm{Gr}_{k,n}^{\geq 0}$ by a $k \times n$ matrix, encoded by the sequence of n columns $\langle v_1, \ldots, v_n \rangle$. We define the *(left) cyclic shift map* σ to be the map which sends $\langle v_1, \ldots, v_n \rangle$ to the point $\langle v_2, \ldots, v_n, (-1)^{k-1} v_1 \rangle$, which one can easily verify lies in $\mathrm{Gr}_{k,n}^{\geq 0}$. Since the cyclic shift maps positroid cells to positroid cells, for π a decorated permutation, we define $\sigma\pi$ to be the decorated permutation such that $S_{\sigma\pi} = \sigma(S_\pi)$. It is easy to see that $\sigma\pi(i) = \pi(i+1) - 1$. (Note that under the cyclic shift, a fixed point of π at position $i + 1$ gets sent to a fixed point of $\sigma\pi$ at position i; we color fixed points accordingly.) Meanwhile the inverse operation, the *right cyclic shift* σ^{-1} satisfies $(\sigma^{-1}\pi)(i) = \pi(i-1) + 1$. We use σ^t (respectively, σ^{-t}) to denote the repeated application of σ (resp. σ^{-1}) t times, so that $(\sigma^t\pi)(i) := \pi(i+t) - t$ and $(\sigma^{-t}\pi)(i) := \pi(i-t) + t$.

The next result follows easily from the definitions.

Theorem 5.3.4 (Cyclic symmetry for dissections of the hypersimplex) *Let $\{\Gamma_\pi\}$ be a collection of positroid polytopes which dissects the hypersimplex $\Delta_{k+1,n}$. Then the collection of positroid polytopes $\{\Gamma_{\sigma\pi}\}$ dissects $\Delta_{k+1,n}$.*

Proof Let $\sigma_{\mathbb{R}} : \mathbb{R}^n \to \mathbb{R}^n$ be defined by $(x_1, \ldots, x_n) \mapsto (x_2, \ldots, x_n, x_1)$. Clearly $\sigma_{\mathbb{R}}$ is an isomorphism mapping the hypersimplex $\Delta_{k+1,n}$ back to itself. Moreover, applying the cyclic shift σ to a positroid has the effect of simply shifting all its bases, so the matroid polytope of $\sigma\pi$ satisfies $\Gamma_{\sigma\pi} = \sigma_{\mathbb{R}}(\Gamma_\pi)$. The result now follows. □

The above cyclic symmetry for dissections of the hypersimplex also has an analogue for the amplituhedron.

Theorem 5.3.5 (Cyclic symmetry for dissections of the amplituhedron)*[7, Corollary 3.2] Let $\{Z_\pi\}$ be a collection of Grasstopes which dissects the amplituhedron $\mathcal{A}_{n,k,m}$, with m even. Then the collection of Grasstopes $\{Z_{\sigma\pi}\}$ also dissects $\mathcal{A}_{n,k,m}$.*

In order to make contact with [19], we introduce a map $U_{k,n}$ on (coloopless) decorated permutations as follows.

Definition 5.3.6 We define $U_{k,n}$ to be the map from coloopless permutations on $[n]$ with k anti-excedances to coloopless permutations on $[n]$ with $n-k-2$ anti-excedances such that $(U_{k,n}\pi)(i) = \pi^{-1}(i+k) + (n-k-2)$, where values of the permutation are considered modulo n, and any fixed points which are created are designated to be loops.

It is not hard to see that this map is equivalent to the parity duality from [19] for $m = 2$. In particular we have the following theorem:

Theorem 5.3.7 [19, Theorem 7.2] *Let $\{Z_\pi\}$ be a collection of Grasstopes which dissects the amplituhedron $\mathcal{A}_{n,k,2}$. Then the collection of Grasstopes $\{Z_{U_{k,n}\pi}\}$ dissects the amplituhedron $\mathcal{A}_{n,n-k-2,2}$.*

Lemma 5.3.8 *For fixed n and k, the maps $\widetilde{U}_{k,n}$ and $U_{k,n}$ are related by the cyclic shift map*

$$\widetilde{U}_{k,n} = \sigma^{-(k+1)} \circ U_{k,n}. \tag{5.3.9}$$

Proof Since $(U_{k,n}\pi)(i) = \pi^{-1}(i+k) + (n-k-2)$, we have that $(\sigma^{-(k+1)} \circ U_{k,n}\pi)(i) = \pi^{-1}((i+k)-(k+1)) + (n-k-2) + (k+1) = \pi^{-1}(i-1) + n - 1$, which is exactly $\widetilde{U}_{k,n}$ (mod n). □

We now prove Theorem 5.3.3.

Proof This result follows immediately from Theorems 5.3.5, 5.3.7, and Lemma 5.3.8. □

Remark 5.3.10 From Theorems 5.3.4 and 5.3.5 it is clear that if we redefine the T-duality map in Definition 4.1.1 by composing it with any cyclic shift σ^a (for a an integer), the main properties of the map will be preserved. In particular, any statement about dissections of the hypersimplex versus the corresponding ones of the amplituhedron will continue to hold, along with the parity duality.

Remark 5.3.11 Parity duality has a nice graphical interpretation when we represent a positroid tile $Z_{\mathcal{T}}$ of $\mathcal{A}_{n,k,2}$ in terms of a (k, n)-unpunctured plabic tiling $\mathcal{T}$. The Grassmannian duality of $\mathrm{Gr}_{k+1,n}^{\geq 0}$ amounts to swapping black and white vertices in the plabic graphs. Therefore, when we compose it with the T-duality map, the parity dual positroid tile of $Z_{\mathcal{T}}$ is $Z_{\overline{\mathcal{T}}}$, where $\overline{\mathcal{T}}$ is the $(n-k-2, n)$-unpunctured plabic tiling obtained from $\mathcal{T}$ by swapping black and white polygons.

It is also natural to think about parity duality between $\mathcal{A}_{n,k,m}$ and $\mathcal{A}_{n,n-k-m,m}$ in terms of the composition of Grassmannian duality and T-duality. Imitating Definition 5.3.2, let us define $\widetilde{U}_{k,n,m}(\hat{\pi}) := \widehat{\pi^{-1}}$.

Remark 5.3.12 In the context of amplituhedra for general (even) m, when we refer to *T-duality* for $\mathcal{A}_{n,k,m}$, we always mean the composition of T-duality in Sect. 4.1 $m/2$ times. With abuse of notation, the 'hat' notion will therefore reflect this convention.

Then we have the following theorem:

Theorem 5.3.13 (Parity duality from T-duality and Grassmannian duality) *Let* $\{S_\pi\}$ *be a collection of cells in* $\mathrm{Gr}_{k,n}^{\geq 0}$ *which gives a dissection of the amplituhedron* $\mathcal{A}_{n,k,m}$. *Then the collection of cells* $\{S_{\widetilde{U}_{k,n,m}\pi}\}$ *in* $\mathrm{Gr}_{n-k-m,n}^{\geq 0}$ *gives a dissection of the amplituhedron* $\mathcal{A}_{n,n-k-m,m}$.

Proof The parity duality $U_{k,n,m}$ in [19] was defined for any (even) m as: $U_{k,n,m}(\pi) := (\pi - k)^{-1} + (n - k - m)$. Then it easy to show that $U_{k,n,m} = \sigma^{k+\frac{m}{2}} \circ \widetilde{U}_{k,n,m}$. Using Theorem 5.3.5, the prove follows immediately. □

References

1. E. Tsukerman, L. Williams, Bruhat interval polytopes. Adv. Math. **285**, 766–810 (2015)
2. N. Arkani-Hamed, J. Trnka, The amplituhedron. J. High Energy Phys. (10), 33 (2014)
3. T. Łukowski, M. Parisi, L.K. Williams, The positive tropical Grassmannian, the hypersimplex, and the $m = 2$ amplituhedron. Int. Math. Res. Not. 03 2023. rnad010
4. M. Parisi, M. Sherman-Bennett, L. Williams, The $m = 2$ amplituhedron and the hypersimplex: signs, clusters, triangulations, Eulerian numbers (2023). Accepted and to be published in *Communications of the American Mathematical Society*
5. T. Lam, A. Postnikov, Alcoved polytopes. I. Discrete Comput. Geom. **38**(3), 453–478 (2007)
6. J.L. Bourjaily, Positroids, Plabic Graphs, and Scattering Amplitudes in Mathematica, 12 2012
7. H. Bao, X. He, The $m = 2$ amplituhedron (2019)
8. F. Ardila, F. Rincón, L. Williams, Positroids and non-crossing partitions. Trans. Am. Math. Soc. **368**(1), 337–363 (2016)
9. S.N. Karp, L.K. Williams, Y.X. Zhang, Decompositions of amplituhedra (2017)
10. D. Speyer, L. Williams, The positive Dressian equals the positive tropical Grassmannian, To appear in Transactions of the AMS
11. D.E. Speyer, Variations on a theme of Kasteleyn, with application to the totally nonnegative Grassmannian. Electron. J. Combin. **23**(2), pp. Paper 2.24, 7, 2016
12. J.S. Scott, Grassmannians and cluster algebras, in *Proceedings of the London Mathematical Society*, vol. 92, pp. 345–380, 03 2006

13. D. Speyer, L. Williams, The tropical totally positive Grassmannian. J. Algebraic Combin. **22**(2), 189–210 (2005)
14. T. Łukowski, R. Moerman, Boundaries of with amplituhedron Boundaries. Comput. Phys. Commun. **259**, 107653 (2021)
15. F. Cachazo, A. Guevara, B. Umbert, Y. Zhang, Planar Matrices and Arrays of Feynman Diagrams (2019)
16. N. Arkani-Hamed, H. Thomas, J. Trnka, Unwinding in binary. Preprint, http://arxiv.org/abs/1704.05069
17. N. Arkani-Hamed, J.L. Bourjaily, F. Cachazo, S. Caron-Huot, J. Trnka, The all-loop integrand for scattering amplitudes in planar $\mathrm{N} = 4$ SYM. JHEP **1101**, 041 (2011)
18. L. Ferro, T. Lukowski, M. Parisi, Amplituhedron meets Jeffrey-Kirwan residue. J. Phys. A **52**(4), 045201 (2018)
19. P. Galashin, T. Lam, Parity duality for the amplituhedron (2018)
20. A. Postnikov, Total positivity, Grassmannians, and networks. Preprint, http://math.mit.edu/~apost/papers/tpgrass.pdf

Chapter 6
The Momentum Amplituhedron

In previous chapters we have explored the remarkable connection between the hypersimplex and the $m = 2$ amplituhedron. This was established via the T-duality map which allowed to relate positroid tiles, tilings (dissections), and finer decompositions of both objects. It is natural to wonder whether the story generalizes for any (even) m. In particular, for $m = 4$, this is a deep question ultimately rooted in *T-duality* from String Theory.

Scattering Amplitudes in planar $\mathcal{N} = 4$ SYM have a dual formulation: on one side we can express them in momentum space (e.g., via spinor-helicity variables and twistors). On the other side, we can use the dual space (e.g., via dual variables and momentum twistors). These were reviewed in Sect. 1.4.1. Surprisingly, scattering amplitudes in planar $\mathcal{N} = 4$ SYM enjoy a version of superconformal symmetry in both formulations, which combined together form an infinite dimensional symmetry—the *Yangian* (see Sect. 1.3.1). At the core of this there is the fact that a certain *T-duality* from String Theory takes planar $\mathcal{N} = 4$ SYM to itself, interchanging the dual coordinates and space-time coordinates.[1] In particular, this is responsible for the *Amplitude/Wilson Loop duality* [2, 3] which converts a scattering process with momenta $\{p_i\}$ to the expectation value of a polygonal Wilson loop with vertices at the points $\{x_i = p_i - p_{i-1}\}$. The 'hidden' symmetries of the scattering amplitudes have now become ordinary superconformal transformations of this Wilson loop.

In this work we show that this duality emerges within the paradigm of positive geometries, see Fig. 6.1. On one side, tree-level scattering amplitudes in momentum twistor space were shown to be encoded in the geometry of the $m = 4$ amplituhedron $\mathcal{A}_{n,k,m=4}$ [4]. On the other side, we introduce a new object—the *momentum amplituhedron* $\mathcal{M}_{n,k',m}$—whose geometry and combinatorics encode scattering amplitudes in momentum space for $k' = k + 2$ and $m = 4$. As the amplituhedron $\mathcal{A}_{n,k,4}$ is the

[1] For more details, see [1]. The authors showed that certain combination of *bosonic* and *fermionic* *T-dualities* maps the full superstring theory on $AdS_5 \times S_5$ back to itself.

M. Parisi, *Combinatorial Aspects of Scattering Amplitudes*, Springer Theses,
https://doi.org/10.1007/978-3-031-41069-7_6

image of a positive Grassmannian $\mathrm{Gr}_{k,n}^{\geq 0}$ under the $\tilde{Z}$ map, also $\mathcal{M}_{n,k',m}$ is the image of a positive Grassmannian $\mathrm{Gr}_{k',n}^{\geq 0}$ under a related map $\Phi_{\Lambda,\tilde{\Lambda}}$. Both $\tilde{Z}$ and Φ maps are induced by some totally positive matrices.

Before the discovery of amplituhedra, the duality was observed in connection with (positive) Grassmannians [5–7]: scattering amplitudes at tree level are computed by performing a contour integral around specific cycles inside the Grassmannian—referred to as a 'BCFW contour'. In momentum space, one has to integrate over cycles corresponding to collections of $(2n-4)$-dimensional positroid cells of $\mathrm{Gr}_{k+2,n}^{\geq 0}$. Whereas, if we are in momentum twistor space, the integral is over collections of $4k$-dimensional positroid cells of $\mathrm{Gr}_{k,n}^{\geq 0}$. In particular, this implied the existence of a map between certain $(2n-4)$-dimensional positroid cells of $\mathrm{Gr}_{k+2,n}^{\geq 0}$ and certain $4k$-dimensional positroid cells of $\mathrm{Gr}_{k,n}^{\geq 0}$, which was defined in [7, Formula (8.25)]. This map is, up to a cyclic shift, the T-duality map for the $m=4$ case, i.e., the double iteration of the T-duality map for the $m=2$ case (Definition 4.1.1).

On one hand, collections of $4k$-dimensional 'BCFW' positroid cells of $\mathrm{Gr}_{k,n}^{\geq 0}$ defined from physics were conjectured to provide a positroid tiling of $\mathcal{A}_{n,k,4}$ [4]. This was recently proved by [8] for a specific collection of BCFW cells and then by [9] for general collections of BCFW cells coming from BCFW recursions. On the other hand, there was no known mathematical object which could be tiled by the collection of T-dual $(2n-4)$-dimensional 'BCFW' positroid cells of $\mathrm{Gr}_{k+2,n}^{\geq 0}$. In this chapter we introduce such an object—the *momentum amplituhedron* $\mathcal{M}_{n,k',m}$—which we conjecture to be related by T-duality to the amplituhedron $\mathcal{A}_{n,k,m}$ for $k'=k+m/2$ and even m. In particular, for $m=4$ we conjecture that their positroid tilings are T-dual. Note that tilings of BCFW type are only a subset of the total tilings. Moreover, T-duality goes beyond being purely a combinatorial bijection between tilings. It can shed light on how to relate geometric properties of the two objects—such as their positroid tiles, facets, chamber decompositions, and their connection to cluster algebras.

Summary of the Chapter. This chapter is based on the following works by the author: [11, Sect. 12] and [12]. Note that: Sect. 6.1.2 is a new addition, in the light of the author's work [13]; Sect. 6.1.3 has been rephrased in a more mathematical setting compared to [12].

In Sect. 6.1 we introduce and discuss the main properties of the momentum amplituhedron $\mathcal{M}_{n,k',m}$. In Sect. 6.1.1 we give the definition of $\mathcal{M}_{n,k',m}$ as the image of the non-negative Grassmannian $\mathrm{Gr}_{k',n}^{\geq 0}$ under the momentum amplituhedron map $\Phi_{\Lambda,\tilde{\Lambda}}$. Then, as for the amplituhedron, in Sect. 6.1.2 we consider twistor coordinates and their signs to define a sign stratification of $\mathcal{M}_{n,k',m}$ and give a sign-flip description. In Sect. 6.1.3, we define the momentum amplituhedron-analogue of *Mandelstam variables* and, for the $m=4$ momentum amplituhedron $\mathcal{M}_{n,k+2,4}$, we discuss its facet structure in relation to the factorization channels of $\mathrm{N}^k\mathrm{MHV}_n$ scattering amplitudes in $\mathcal{N}=4$ SYM. In Sect. 6.1.4 we explain how T-duality conjecturally relates the momentum amplituhedron $\mathcal{M}_{n,k+m/2,m}$ and the amplituhedron $\mathcal{A}_{n,k,m}$—which geometrize and generalize the physical *Amplitudes/Wilson loops* duality. Finally, in

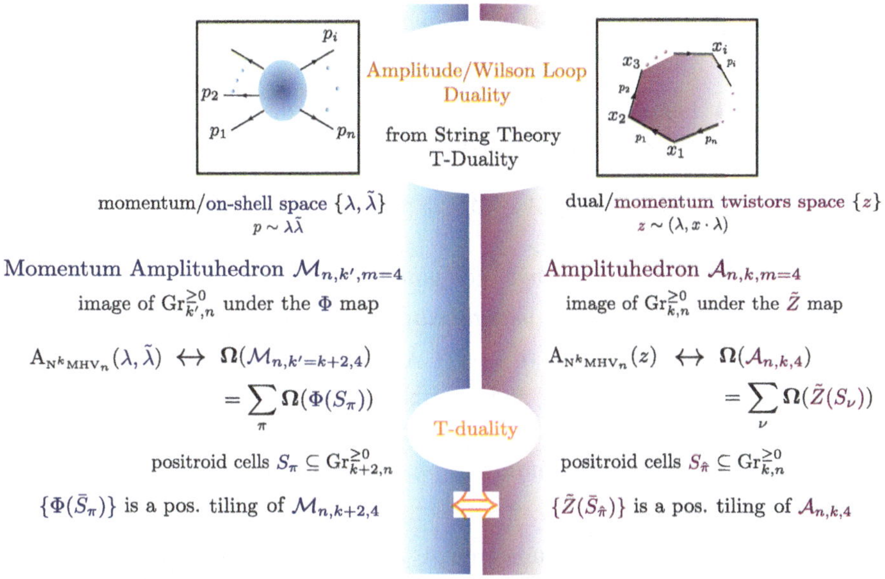

Fig. 6.1 Physical and Combinatorial T-duality for $m = 4$

Sect. 6.2 we explain how to extract $\mathrm{N}^k\mathrm{MHV}_n$ scattering amplitudes in spinor helicity space from $\mathcal{M}_{n,k+2,4}$.

6.1 The Momentum Amplituhedron

6.1.1 The Definition of $\mathcal{M}_{n,k',m}$

In the following we fix[2] positive integers n, k', n such that even m and $k' + \frac{m}{2} \leq n$.

Definition 6.1.1 For a, b such that $a \leq b$, define the *twisted* totally non-negative part of $\mathrm{Gr}_{a,b}$ as:

$$\mathrm{Gr}^{\geq 0,\tau}_{a,b} := \{X \in \mathrm{Gr}_{a,b} : \epsilon_{[b]\setminus I, I}\, p_I(X) \geq 0\}. \tag{6.1.2}$$

The lemma below can be found in [14, Lemma 1.11], which sketched a proof and attributed it to Hochster and Hilbert.

[2] In this chapter, we will use k' in order not confuse the notation with 'k' used in the previous chapters.

Lemma 6.1.3 *An element V is in* $\mathrm{Gr}_{a,b}^{\geq 0}$ *if and only if the kernel* $V^{\perp} \in \mathrm{Gr}_{b-a,b}$ *of V lies in* $\mathrm{Gr}_{b-a,b}^{\geq 0,\tau}$.

Definition 6.1.4 For a, b such that $a \leq b$, define $\mathrm{Mat}_{a,b}^{>0}$ the set of real $a \times b$ matrices whose $a \times a$ minors are all positive and its *twisted positive part* as

$$\mathrm{Mat}_{a,b}^{>0,\tau} := \{A \in \mathrm{Mat}_{a,b} : \epsilon_{[b]\setminus I,I}\, p_I(A) > 0\} \tag{6.1.5}$$

Definition 6.1.6 (*The momentum amplituhedron*) Let $\tilde{\Lambda} \in \mathrm{Mat}_{n,k'+\frac{m}{2}}^{>0}$, $\Lambda \in \mathrm{Mat}_{n,n-k'+\frac{m}{2}}^{>0,\tau}$. The *momentum amplituhedron map* $\Phi_{\tilde{\Lambda},\Lambda} : \mathrm{Gr}_{k',n}^{\geq 0} \to \mathrm{Gr}_{k',k'+\frac{m}{2}} \times \mathrm{Gr}_{n-k',n-k'+\frac{m}{2}}$ is defined by $\Phi_{\tilde{\Lambda},\Lambda}(C) := (C\tilde{\Lambda}, C^{\perp}\Lambda)$, where C and $C^{\perp}$ are matrices representing an element of $\mathrm{Gr}_{k',n}^{\geq 0}$ and its orthogonal in $\mathrm{Gr}_{k',n}^{\geq 0,\tau}$ respectively, and $C\tilde{\Lambda}$ and $C^{\perp}\Lambda$ matrices representing an element of $\mathrm{Gr}_{k',k'+\frac{m}{2}}$ and $\mathrm{Gr}_{n-k',n-k'+\frac{m}{2}}$ respectively. The *momentum amplituhedron* $\mathcal{M}_{n,k',m}(\Lambda, \tilde{\Lambda}) \subseteq \mathrm{Gr}_{k',k'+\frac{m}{2}} \times \mathrm{Gr}_{n-k',n-k'+\frac{m}{2}}$ is the image $\Phi_{\tilde{\Lambda},\Lambda}(\mathrm{Gr}_{k',n}^{\geq 0})$.

Remark 6.1.7 As for the amplituhedron $\mathcal{A}_{n,k,m}(Z)$, some combinatorial properties of $\mathcal{M}_{n,k',m}(\Lambda, \tilde{\Lambda})$ are expected to be independent from the choice of the matrices Λ and $\tilde{\Lambda}$. We will write $\mathcal{M}_{n,k',m}$ in such cases, and $\mathcal{M}_{n,k',m}(\Lambda, \tilde{\Lambda})$ otherwise.

Definition 6.1.8 (*Momentum conservation*) Let $(\tilde{Y}, Y)$ be a point in $\mathrm{Gr}_{k',k'+\frac{m}{2}} \times \mathrm{Gr}_{n-k',n-k'+\frac{m}{2}}$ and let $\tilde{Y}^{\perp}$ and $Y^{\perp}$ be matrices representing the orthogonal complements of Y and $\tilde{Y}$, respectively. We define the hypersurface $P(Y, \tilde{Y}) = 0$ of *momentum conservation* as

$$P(Y, \tilde{Y}) := (Y^{\perp}\Lambda^T) \cdot (\tilde{Y}^{\perp}\tilde{\Lambda}^T)^T = 0 \tag{6.1.9}$$

Proposition 6.1.10 *The momentum amplituhedron $\mathcal{M}_{n,k',m}$ lies on the hypersurface of momentum conservation $P(Y, \tilde{Y}) = 0$.*

Proof From the identity

$$0 = Y^{\perp}Y^T = Y^{\perp}\Lambda^T(C^{\perp})^T \tag{6.1.11}$$

we deduce that the row-span of $Y^{\perp}\Lambda^T$ is included in the row-span of the orthogonal of $C^{\perp}$, i.e., C. Analogously, from

$$0 = \tilde{Y}^{\perp}\tilde{Y}^T = \tilde{Y}^{\perp}\tilde{\Lambda}^T C \tag{6.1.12}$$

we deduce that the row-span of $\tilde{Y}^{\perp}\tilde{\Lambda}^T$ is included in the row-span of the $C^{\perp}$. Therefore $Y^{\perp}\Lambda^T$ and $\tilde{Y}^{\perp}\tilde{\Lambda}^T$ belong to orthogonal subspaces and satisfy

$$(Y^{\perp}\Lambda^T) \cdot (\tilde{Y}^{\perp}\tilde{\Lambda}^T)^T = 0. \tag{6.1.13}$$

□

Remark 6.1.14 We observe that:

$$\dim\left(\mathrm{Gr}_{k',k'+\frac{m}{2}}\times\mathrm{Gr}_{n-k',n-k'+\frac{m}{2}}\right)=\frac{m}{2}k'+\frac{m}{2}(n-k')=\frac{m}{2}n. \tag{6.1.15}$$

Therefore, Proposition 6.1.10 implies that the momentum amplituhedron $\mathcal{M}_{n,k',m}$ is included in a codimension $(\frac{m}{2})^2$ sub-variety of $\mathrm{Gr}_{k',k'+\frac{m}{2}}\times\mathrm{Gr}_{n-k',n-k'+\frac{m}{2}}$. The dimension of $\mathcal{M}_{n,k,m}$ is conjectured to be:

$$\frac{m}{2}n-\left(\frac{m}{2}\right)^2=\frac{m}{2}\left(n-\frac{m}{2}\right). \tag{6.1.16}$$

We observe that, for $m=2$, this dimension is exactly $n-1$, which is the dimension of the hypersimplex $\Delta_{k+1,n}$; whereas, for $m=4$, the dimension is $2n-4$, which is the one of BCFW cells in momentum space.

Remark 6.1.17 ($m=2$ case) For $m=2$, Definition 6.1.6 reads:

$$\Phi_{\tilde{\Lambda},\Lambda}:\mathrm{Gr}^{+}_{k+1,n}\to\mathrm{Gr}_{k+1,k+2}\times\mathrm{Gr}_{n-k-1,n-k}\cong\mathbb{P}^{k+1}\times\mathbb{P}^{n-k-1}. \tag{6.1.18}$$

Moreover, the conditions in Proposition 6.1.10 are equivalent to:

$$\lambda\cdot\tilde{\lambda}=0 \tag{6.1.19}$$

where we used the dot product in $\mathbb{R}^n$ of the vectors $\lambda:=\Lambda(Y^{\perp})^T$ and $\tilde{\lambda}:=\tilde{\Lambda}(\tilde{Y}^{\perp})^T$.

Note that the $m=2$ momentum amplituhedron is not equal to the hypersimplex, but these objects seem to be closely related.

Remark 6.1.20 ($m=4$ case) For $m=4$ and $k'=k+2$, we conjecture $\mathcal{M}_{n,k',4}$ encodes tree-level $\mathrm{N}^k\mathrm{MHV}_n$ amplitudes of $\mathcal{N}=4$ SYM in spinor helicity space, see Sect. 6.2.

6.1.2 Twistor Coordinates and Sign Variation

In analogy with the amplituhedron, we can define twistor coordinates for the momentum amplituhedron $\mathcal{M}_{n,k',m}(\Lambda,\tilde{\Lambda})$. In the following, we will denote the rows of Λ as $\Lambda_1,\ldots,\Lambda_n$ and the rows of $\tilde{\Lambda}$ as $\tilde{\Lambda}_1,\ldots,\tilde{\Lambda}_n$.

Definition 6.1.21 (*Twistor Coordinates*) Let the matrix Y with rows $y_1,\ldots,y_{n-k'}$ representing an element of $\mathrm{Gr}_{n-k',n-k'+m/2}$, and $i_1,\ldots,i_{m/2}$ a sequence of elements of $[n]$. Then we define $\langle Yi_1i_2\ldots i_{m/2}\rangle$ to be the determinant of the $(n-k'+m/2)\times(n-k'+m/2)$ matrix whose rows are $y_1,\ldots,y_{n-k'},\Lambda_{i_1},\ldots,\Lambda_{i_{m/2}}$. Analogously, let the matrix $\tilde{Y}$ with rows $\tilde{y}_1,\ldots,\tilde{y}_{k'}$ representing an element of $\mathrm{Gr}_{k',k'+m/2}$. Then we define $[\tilde{Y}i_1i_2\ldots i_{m/2}]$ to be the determinant of the $(k'+m/2)\times(k'+m/2)$ matrix whose rows are $\tilde{y}_1,\ldots,\tilde{y}_{k'},\tilde{\Lambda}_{i_1},\ldots,\tilde{\Lambda}_{i_{m/2}}$.

Using Theorem 2.4.1 for the $m = 2$ amplituhedron, we can give the following description of $\mathcal{M}_{n,k',4}$ in terms of sign variations.

Theorem 6.1.22 (Sign Variation for $\mathcal{M}_{n,k',4}$) *Fix positive k', n such that $k' + 2 \leq n$ and let $\tilde{\Lambda} \in \text{Mat}^{>0}_{n,k'+2}$, $\Lambda \in \text{Mat}^{>0,\tau}_{n,n-k'+2}$. Let $\mathcal{H}^{\circ}_{n,k',4}(\Lambda, \tilde{\Lambda})$ be the set of points of $(Y, \tilde{Y})$ in $\text{Gr}_{k',k'+2} \times \text{Gr}_{n-k',n-k'+2}$ such that*

$$\text{(1)} \quad \langle Y i, i+1 \rangle > 0 \text{ for } 1 \leq i \leq n-1, \text{ and } \langle Y n \hat{1} \rangle > 0, \tag{6.1.23}$$

$$\text{and } \operatorname{var}(\langle Y 12 \rangle, \langle Y 13 \rangle, \ldots \langle Y 1n \rangle) = k' - 2. \tag{6.1.24}$$

$$\text{(2)} \quad [\tilde{Y} i, i+1] > 0 \text{ for } 1 \leq i \leq n-1, \text{ and } [\tilde{Y} n \hat{1}] > 0, \tag{6.1.25}$$

$$\text{and } \operatorname{var}([\tilde{Y} 12], [\tilde{Y} 13], \ldots [\tilde{Y} 1n]) = k'. \tag{6.1.26}$$

Then $\mathcal{M}_{n,k',4}(\Lambda, \tilde{\Lambda}) \subset \overline{\mathcal{H}^{\circ}_{n,k',4}(\Lambda, \tilde{\Lambda})}$.

Proof By definition, if $(\tilde{Y}, Y)$ represents a point in $\mathcal{M}^{\circ}_{n,k',2}(\tilde{\Lambda}, \Lambda) := \Phi_{\Lambda,\tilde{\Lambda}}(\text{Gr}^{>0}_{k',n})$, then $\tilde{Y}$ lies in $\mathcal{A}^{\circ}_{n,k',2}(\tilde{\Lambda})$. We recall that $\mathcal{A}^{\circ}_{n,k,m}(Z) := \tilde{Z}(\text{Gr}^{>0}_{k,n})$. By Theorem 2.4.1 (2) immediately follows.

Let $\Lambda^{\perp}$ be a matrix representing the orthogonal complement of the column-span of Λ in $\text{Gr}^{\geq 0}_{\tilde{k}'-2,n}$ and let $Y = C^{\perp}\Lambda$, with C a matrix representing an element of $\text{Gr}^{>0}_{\tilde{k}',n}$. We consider the matrix $\Lambda^{\perp}C^{T}$ representing an element of $\text{Gr}_{k'-2,k'}$ and we denote it as X. Then we note that X represents an element of the amplituhedron $\mathcal{A}^{\circ}_{n,k'-2,2}(C^{T})$, therefore by Theorem 2.4.1 it lies in the region where

$$\langle X c_i, c_{i+1} \rangle > 0 \text{ for } 1 \leq i \leq n-1, \text{ and } \langle X c_n \hat{c}_1 \rangle > 0,$$
$$\text{and } \operatorname{var}(\langle X c_1 c_2 \rangle, \langle X c_1 c_3 \rangle \ldots, \langle X c_1 c_n \rangle) = k' - 2.$$

We now show that $\langle Y i j \rangle = \langle X c_i c_j \rangle$.

$$\langle X c_i c_j \rangle = \sum_{\{j_1 < \cdots < j_{k'-2}\} \in \binom{[n]}{k'-2}} p_J(\Lambda^{\perp}) \langle c_{j_1}, \ldots, c_{j_{k'-2}}, c_i, c_j \rangle. \tag{6.1.27}$$

Let us denote $I = \{i_1, \ldots, i_{n-k'}\} := [n] \setminus (J \cup \{i, j\})$. Then

$$p_J(\Lambda^{\perp}) = \epsilon_{J,[n]\setminus J}\, p_{[n]\setminus J}(\Lambda) = \epsilon_{J,\{i_1,\ldots,i_{n-k'},i,j\}}\, \langle \Lambda_{i_1}, \cdots, \Lambda_{i_{n-k'}}, \Lambda_i, \Lambda_j \rangle$$

and

$$\langle c_{j_1}, \ldots, c_{j_k}, c_i, c_j \rangle = \epsilon_{\{j_1,\ldots,j_{k'-2},i,j\},\{i_1,\ldots,i_{n-k'}\}}\, p_{\{i_1,\ldots,i_{n-k'}\}}(C^{\perp}).$$

Therefore, we obtain:

$$\langle X c_i c_j \rangle = \sum_{\{i_1 < \cdots < i_{n-k'}\} \in \binom{[n]}{n-k'}} p_I(C^{\perp}) \langle \Lambda_{i_1}, \cdots, \Lambda_{i_{n-k'}}, \Lambda_i, \Lambda_j \rangle = \langle Y i j \rangle, \tag{6.1.28}$$

where we used $Y = C^{\perp}\Lambda$ and $\epsilon_{\{j_1,\ldots,j_{k'-2},i,j\},\{i_1,\ldots,i_{n-k'}\}} \cdot \epsilon_{J,\{i_1,\ldots,i_{n-k'},i,j\}} = 1$. As

$$\langle Xc_n\hat{c}_1\rangle = (-1)^{k'-2-1}\langle Xc_nc_1\rangle = (-1)^{k'-1}\langle Yn1\rangle = \langle Yn\hat{1}\rangle,$$

this completes the proof for (1). Finally, as $\Phi_{\Lambda,\tilde{\Lambda}}$ is continuous, $\mathcal{M}_{n,k,4} \subseteq \overline{\mathcal{M}^{\circ}_{n,k,4}} \subseteq \overline{\mathcal{H}^{\circ}_{n,k',4}}$. □

Theorem 6.1.22 suggests the following conjecture for general m:

Conjecture 6.1.29 (Sign Variation for $\mathcal{M}_{n,k',m}$) Fix positive k', n and even m such that $k' + \frac{m}{2} \leq n$ and let $\tilde{\Lambda} \in \mathrm{Mat}^{>0}_{n,k'+\frac{m}{2}}$, $\Lambda \in \mathrm{Mat}^{>0,\tau}_{n,n-k'+\frac{m}{2}}$. Let $\mathcal{H}^{\circ}_{n,k',m}(\Lambda,\tilde{\Lambda})$ be the set of points of $(Y,\tilde{Y})$ in $\mathrm{Gr}_{k',k'+\frac{m}{2}} \times \mathrm{Gr}_{n-k',n-k'+\frac{m}{2}}$ such that

(1) $\langle Yi_1, i_1+1, \ldots i_{\frac{m}{2}}, i_{\frac{m}{2}}+1\rangle > 0$ (6.1.30)

and $\mathrm{var}(\langle Y1,\ldots m-1,m\rangle, \langle Y1\ldots m-1,m+1\rangle,\ldots \langle Y1\ldots m-1,n\rangle) = k' - \frac{m}{2}$. (6.1.31)

(2) $[\tilde{Y}i_1, i_1+1, \ldots i_{\frac{m}{2}}, i_{\frac{m}{2}}+1] > 0$ (6.1.32)

and $\mathrm{var}([\tilde{Y}1\ldots m-1,m], [\tilde{Y}1\ldots m-1,m+1],\ldots [\tilde{Y}1\ldots m-1,n]) = k'$. (6.1.33)

Then $\mathcal{M}_{n,k',m}(\Lambda,\tilde{\Lambda}) \subset \overline{\mathcal{H}^{\circ}_{n,k',m}(\Lambda,\tilde{\Lambda})}$.

Question 6.1.34 *Is there a characterization of the momentum amplituhedron $\mathcal{M}_{n,k',m}$ in terms of signs of (polynomials of) twistor coordinates?*

We end this section by defining the *sign stratification* of $\mathcal{M}_{n,k',m}(\Lambda,\tilde{\Lambda})$. This is in analogy with the sign stratification of the amplituhedron $\mathcal{A}_{n,k,m}(Z)$ in Sect. 2.2.2. *Sign strata* are regions where *all* twistor coordinates have a fixed sign prescribed by certain sign vectors:

Definition 6.1.35 (*Sign stratification of $\mathcal{M}_{n,k',m}$*) Fix positive $k' < n$ and m such that $k' + \frac{m}{2} \leq n$. Let $\sigma = (\sigma_{i_1,\ldots,i_{\frac{m}{2}}})$ and $\tilde{\sigma} = (\tilde{\sigma}_{i_1,\ldots,i_{\frac{m}{2}}})$ in $\{0,+,-\}^{\binom{n}{m/2}}$ be nonzero sign vectors, considered[3] *modulo multiplication by* ± 1. Let $\mathcal{M}^{\sigma,\tilde{\sigma}}_{n,k',m}(\Lambda,\tilde{\Lambda})$ be the set

$$\{(Y,\tilde{Y}) \in \mathcal{M}_{n,k',m}(\Lambda,\tilde{\Lambda}) \mid \mathrm{sgn}\langle Yi_1\ldots i_{m/2}\rangle = \sigma_{i_1,\ldots,i_{m/2}},\ \mathrm{sgn}[\tilde{Y}i_1\ldots i_{m/2}] = \tilde{\sigma}_{i_1,\ldots,i_{m/2}}\}.$$

We call $\mathcal{M}^{\sigma,\tilde{\sigma}}_{n,k',m}(\Lambda,\tilde{\Lambda})$ a *(momentum amplituhedron) sign stratum*. Clearly

$$\mathcal{M}_{n,k',m}(\Lambda,\tilde{\Lambda}) = \bigsqcup_{\sigma,\tilde{\sigma}} \mathcal{M}^{\sigma,\tilde{\sigma}}_{n,k',m}(\Lambda,\tilde{\Lambda}).$$

[3] This is because Plücker and twistor coordinates are only defined up to multiplication by a common scalar.

If $\sigma, \tilde{\sigma} \in \{+,-\}^{\binom{n}{m/2}}$, we call $\mathcal{M}^{\sigma,\tilde{\sigma}}_{n,k',m}(\Lambda, \tilde{\Lambda})$ an open *(momentum amplituhedron) chamber*.

In analogy with Definition 2.2.10 for $\mathcal{A}_{n,k,m}(Z)$, we introduce the following definition:

Definition 6.1.36 (*Realizable sign strata of* $\mathcal{M}_{n,k',m}$) A sign vector $(\sigma, \tilde{\sigma})$ (or sign stratum $\mathcal{M}^{\sigma,\tilde{\sigma}}_{n,k',m}$) is *realizable* for $\mathcal{M}_{n,k',m}$ if $\mathcal{M}^{\sigma,\tilde{\sigma}}_{n,k',m}(\Lambda, \tilde{\Lambda})$ is nonempty for some $\Lambda, \tilde{\Lambda}$.

In Corollary 4.4.13 we showed that for the $m = 2$ amplituhedron $\mathcal{A}_{n,k,m=2}(Z)$ the realizable sign chambers are exactly the w-chambers labelled by the Eulerian numbers. Moreover, those are in bijection with w-simplices in which the hypersimplex $\Delta_{k+1,n}$ decomposes.

Question 6.1.37 *Can we use the w-chambers of the $m = 2$ amplituhedron to understand the realizable sign chambers of the $m = 4$ momentum amplituhedron? Is there a relation between them and the realizable sign chambers of the $m = 4$ amplituhedron?*

6.1.3 Mandelstam Variables and Facets

Definition 6.1.38 (*Mandelstam variables*) Let $(Y, \tilde{Y})$ be a point in $\mathrm{Gr}_{k',k'+\frac{m}{2}} \times \mathrm{Gr}_{n-k',n-k'+\frac{m}{2}}$, and let $I \subseteq [n]$. Then we define the *Mandelstam variable* $S_I(Y, \tilde{Y})$ as the following quadratic polynomial in twistor coordinates:

$$S_I(Y, \tilde{Y}) := \sum_{J \in \binom{I}{\frac{m}{2}}} \langle Y J \rangle [\tilde{Y} J] \tag{6.1.39}$$

Remark 6.1.40 Mandelstam variables $S_I(Y, \tilde{Y})$ can be expressed in terms of Plückers of $C, \Lambda, \tilde{\Lambda}$ using Eq. (6.1.27) as:

$$S_I(Y, \tilde{Y}) = \sum_{L \in \binom{[n]}{k'-m/2}} \sum_{R \in \binom{[n]}{k'}} \sum_{J \in \binom{I}{\frac{m}{2}}} \epsilon_{L,J}\, \epsilon_{R,J}\, p_{L \cup J}(C)\, p_R(C)\, p_L(\Lambda^{\perp})\, p_{R \cup J}(\tilde{\Lambda}). \tag{6.1.41}$$

Let us denote $[i, j]$ as the cyclic interval $\{i, i+1, \dots, j\}$, with $i, j \in [n]$.

Proposition 6.1.42 *Let $(Y, \tilde{Y})$ represent a point in $\mathcal{M}_{n,k',4}(\Lambda, \tilde{\Lambda})$ and $i, j \in [n]$. Then:*

(1) $\langle Y i, i+1 \rangle \geq 0, \quad [\tilde{Y} i, i+1] \geq 0.$ (6.1.43)

(2) $S_{[i,j]}(Y, \tilde{Y}) \geq 0$, *for $k' = 2$ (MHV case).* (6.1.44)

(3) *For $k' \neq 2$, the sign definiteness of $S_{[i,j]}$ depends on the choice of $\Lambda, \tilde{\Lambda}$.* (6.1.45)

Proof **Part (1).** $[\tilde{Y}i, i+1] \geq 0$ is straightforward.

$$\langle Yi, i+1\rangle = \sum_{I\in\binom{[n]}{n-k'}} p_I(C^\perp)\langle I, i, i+1\rangle = \sum_{I\in\binom{[n]}{n-k'}} \epsilon_{I,[n]\setminus I}\, p_{[n]\setminus I}(C)\langle I, i, i+1\rangle, \tag{6.1.46}$$

changing summation indices, we have

$$\sum_{J\in\binom{[n]}{k'}} \epsilon_{[n]\setminus J, J}\, p_J(C)\langle [n]\setminus J, i, i+1\rangle = \sum_{J\in\binom{[n]}{k'}} p_J(C) p_{J\setminus\{i,i+1\}}(\Lambda^\perp) \geq 0. \tag{6.1.47}$$

Part (2). For $k' = 2$, we have:

$$\langle Yij\rangle = p_I(C^\perp)\langle \Lambda_{i_1}, \cdots, \Lambda_{i_{n-2}}, \Lambda_i, \Lambda_j\rangle. \tag{6.1.48}$$

where we denoted $I = \{i_1, \ldots, i_{n-k'}\} := [n]\setminus\{i, j\}$. Using $p_I(C^\perp) = \epsilon_{I,\{i,j\}}\, p_{ij}(C)$ and $\epsilon_{I,\{i,j\}}\langle \Lambda_{i_1}, \cdots, \Lambda_{i_{n-2}}, \Lambda_i, \Lambda_j\rangle = p_{[n]}(\Lambda) > 0$, we obtain:

$$\langle Yij\rangle = p_{[n]}(\Lambda)\, p_{ij}(C). \tag{6.1.49}$$

Then we can rewrite:

$$S_{[i,j]}(Y, \tilde{Y}) = p_{[n]}(\Lambda) \sum_{j_1<j_2\in[i,j], a<b\in[n]} p_{j_1 j_2}(C) p_{ab}(C)\langle \Lambda_a, \Lambda_b, \Lambda_{j_1}, \Lambda_{j_2}\rangle\,. \tag{6.1.50}$$

There are only two cases for which the bracket $\langle \Lambda_a, \Lambda_b, \Lambda_{j_1}, \Lambda_{j_2}\rangle$ is negative: $a < j_1 < b < j_2$ or $j_1 < a < j_2 < b$. In particular, we observe that if $a, b \notin I$ then $\langle \Lambda_a, \Lambda_b, \Lambda_{j_1}, \Lambda_{j_2}\rangle > 0$. For $b \in I$ and $a < j_1 < b < j_2$, the terms proportional to $\langle \Lambda_a, \Lambda_{j_1}, \Lambda_b, \Lambda_{j_2}\rangle$ are:

$$\left(p_{aj_1}(C)\, p_{bj_2}(C) - p_{j_1 j_2}(C)\, p_{ab}(C) + p_{aj_2}(C)\, p_{j_1 b}(C)\right)\langle \Lambda_a, \Lambda_{j_1}, \Lambda_b, \Lambda_{j_2}\rangle, \tag{6.1.51}$$

which is zero by the three-term Plücker relation of the C matrix. Analogously, for the case $j_1 < a < j_2 < b$. We therefore conclude that all non-positive terms are cancelled and $S_{[i,j]}(Y, \tilde{Y})$ is non-negative.

Part (3). Let us consider $\mathcal{M}_{6,3,4}$, i.e., $n = 6$ and $k' = 3$ (NMHV case). If we expand S_{123} using (6.1.40), we obtain many non-negative terms plus the following term proportional to $p_{123}(C)\, p_{456}(C)$:

$$\sum_{i=1}^{6} (-1)^{i+1} p_i(\Lambda^\perp)\, p_{[6]\setminus\{i\}}(\tilde{\Lambda}). \tag{6.1.52}$$

Therefore, positivity of this expression is a sufficient condition for S_{123} to be non-negative. Moreover, it is possible to choose Λ and $\tilde{\Lambda}$ such that this expression is negative and such that spanning over all C in $\mathrm{Gr}_{k',n}^{\geq 0}$ the resulting S_{123} is negative. □

Despite Part (3) of Proposition 6.1.42, we still make the following conjecture.

Conjecture 6.1.53 Let $k'+2 \leq n$. Then there is a choice of $\tilde{\Lambda} \in \mathrm{Mat}^{>0}_{n,k'+2}$ and $\Lambda \in \mathrm{Mat}^{>0,\tau}_{n,n-k'+2}$ such that all points $(Y,\tilde{Y})$ in $\mathcal{M}_{n,k',4}(\Lambda,\tilde{\Lambda})$ satisfy $S_{[i,j]}(Y,\tilde{Y}) \geq 0$, for all $i, j \in [n]$.

Remark 6.1.54 We have checked using `Mathematica` that the following choice of Λ and $\tilde{\Lambda}$ satifies the condition of Conjecture 6.1.53 for all $\mathcal{M}_{n,k',4}(\Lambda,\tilde{\Lambda})$ with $n \leq 10$:

$$(\Lambda^{\perp})_{i,a} = i^{a-1}, \quad (\tilde{\Lambda})_{i,\tilde{a}} = i^{\tilde{a}-1}, \qquad i \in [n], a \in [k'-2], \tilde{a} \in [k+2]. \quad (6.1.55)$$

Note that the rows of Λ and $\tilde{\Lambda}$ are vectors lying on moment curves.

Remark 6.1.56 We also conjecture a set of sufficient conditions on Λ and $\tilde{\Lambda}$ such that Conjecture 6.1.53 is satified for $k'=3$ (NHMV case). Let $I = \{i_1 < i_2 < i_3\}$, $J = \{j_1 < j_2 < j_3\}$ be subsets of $[n]$ and disjoint. Let us define:

$$Q_{I,J}(\Lambda,\tilde{\Lambda}) := \sum_{s=1}^{3}(-1)^{s+1}\left(p_{i_s}(\Lambda^{\perp})p_{I\cup J\setminus\{i_s\}}(\tilde{\Lambda}) + p_{j_s}(\Lambda^{\perp})p_{I\cup J\setminus\{j_s\}}(\tilde{\Lambda})\right). \quad (6.1.57)$$

Let $i, j \in [n]$, we conjecture that if for all $I \subseteq [i,j]$ and $J \in [n] \setminus [i,j]$

$$Q_{I,J}(\Lambda,\tilde{\Lambda}) > 0, \quad (6.1.58)$$

then $S_{[i,j]}(Y,\tilde{Y}) \geq 0$ inside the momentum amplituhedron $\mathcal{M}_{n,3,4}(\Lambda,\tilde{\Lambda})$. For example, for $n=6$ and $S_{[1,3]}$ the above condition reads exactly as in Eq. (6.1.52).

Remark 6.1.59 The properties in Eq. (6.1.44) and in Conjecture 6.1.53 are the momentum amplituhedron analogue of the conjecture in [15, Sect. 5]. Let us consider the space of spinor-helicity variables (cf. Sect. 1.4.1) $\{\lambda, \tilde{\lambda}\}$, and consider the brackets $\langle ij\rangle_\lambda := \langle \lambda_{ij}\rangle$ and $[ij]_{\tilde{\lambda}} := \langle \tilde{\lambda}_{ij}\rangle$. Let us also define *Mandelstam variables*

$$s_I := \sum_{J \in \binom{I}{2}} \langle J\rangle_\lambda [J]_{\tilde{\lambda}}\,. \quad (6.1.60)$$

Then the conjecture of [15] states that the positive region would be defined by the following conditions:

$$\begin{aligned}
&(1)\langle i,i+1\rangle_\lambda > 0 \text{ for } 1 \leq i \leq n-1, \text{ and } \langle n\hat{1}\rangle_\lambda > 0,\\
&\quad \text{and } \mathrm{var}(\langle 12\rangle_\lambda, \langle 13\rangle_\lambda, \ldots \langle 1n\rangle_\lambda) = k'-2.\\
&(2)[i,i+1]_{\tilde{\lambda}} > 0 \text{ for } 1 \leq i \leq n-1, \text{ and } [n\hat{1}]_{\tilde{\lambda}} > 0,\\
&\quad \text{and } \mathrm{var}([12]_{\tilde{\lambda}}, [13]_{\tilde{\lambda}}, \ldots [1n]_{\tilde{\lambda}}) = k'.\\
&(3)s_{[i,j]} > 0, \quad i,j \in [n].
\end{aligned}$$

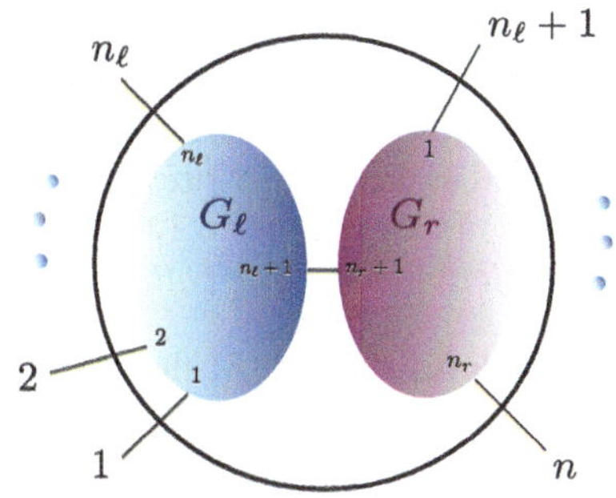

Fig. 6.2 Plabic graph $G_\ell(1,\dots,n_\ell) \otimes G_r(n_\ell + 1,\dots,n)$

A conjectural description of all faces of the momentum amplituhedron $\mathcal{M}_{n,k',4}$ is presented in [16].

Definition 6.1.61 Let fix a partition of $[n]$ into the intervals $I_\ell = [1, n_\ell]$ and $I_r = [n_\ell + 1, n]$ of size n_ℓ and n_r, respectively, with $n_\ell + n_r = n$. Let G_ℓ and G_r be reduced plabic graphs of the top cells of $\mathrm{Gr}^{\geq 0}_{k_\ell,n_\ell+1}$ and $\mathrm{Gr}^{\geq 0}_{k_r,n_r+1}$, respectively. We define the plabic graph $G = G_\ell(I_\ell) \otimes G_r(I_r)$ by:

1. relabel the boundary vertices $\{1,\dots,n_r+1\}$ of G_r as $\{n_\ell+1, n_\ell+2,\dots,n\}$;
2. connect the boundary vertex $n_\ell+1$ of G_ℓ with the boundary vertex $n_\ell+1$ of G_r.

See Fig. 6.2. By cyclic rotation, we analogously define $G_\ell(I_\ell) \otimes G_r(I_r)$ for arbitrary partition of $[n]$ into two intervals I_ℓ and I_r.

Remark 6.1.62 It is immediate to see that G as in Definition 6.1.61 labels a cell in $\mathrm{Gr}^{\geq 0}_{k',n}$, with $k' = k_\ell + k_r - 1$, $n = n_\ell + n_r$, and has dimension $k_\ell(n_\ell + 1 - k_\ell) + k_r(n_r + 1 - k_r) - 1$.

Remark 6.1.63 The construction in Definition 6.1.61 was referred to as *amalgamation of on-shell diagrams* inside the positive Grassmannian in [17].

Conjecture 6.1.64 (Facets of $\mathcal{M}_{n,k,4}$) Let fix $k < n$ and let $\{I_\ell, I_r\}$ a partition of $[n]$ into two intervals of sizes $n_\ell, n_r > 1$. Let $k_\ell < n_\ell$, $k_r < n_r$ such that $k_\ell + k_r - 1 = k$. Then $\mathcal{M}_{n,k',4}$ has a facet which is the image under the momentum amplituhedron map of the positroid cell labelled by $G_\ell(I_\ell) \otimes G_r(I_r)$. Moreover, the facet lies on the hypersurface:

1. $\langle Y I_\ell\rangle = 0$, if $k_\ell = 1$ and $n_\ell = 2$ (or $\langle Y I_r\rangle = 0$, if $k_r = 1$ and $n_r = 2$);
2. $[\tilde{Y} I_\ell] = 0$, if $k_\ell = 2$ and $n_\ell = 2$ (or $[\tilde{Y} I_r] = 0$, if $k_r = 2$ and $n_r = 2$);
3. $S_{I_\ell}(Y, \tilde{Y}) = 0$, otherwise (or, by momentum conservation, $S_{I_r}(Y, \tilde{Y}) = 0$).

Finally, all facets of $\mathcal{M}_{n,k',4}$ arise in this way.

Facets of $\mathcal{M}_{n,k',4}$ are in bijection with factorization channels of scattering amplitudes of $\mathcal{N} = 4$ SYM in spinor helicity space. Given a facet of $\mathcal{M}_{n,k',4}$, by Conjecture

6.1.64 it is the image of the positroid cell labelled by $G_\ell(I_\ell) \otimes G_r(I_r)$. This corresponds to the factorization $A_{n,k} = A_{n_\ell,k_\ell}(I_\ell) \otimes A_{n_r,k_\ell}(I_r)$. In particular, case (1) in Conjecture 6.1.64 is associated to $A_{n,k} = A_{\overline{MHV}_3}(I_\ell) \otimes A_{n-2,k}(I_r)$, and can be regarded as the *helicity-preserving collinear limit* of the two particles in I_ℓ. Whereas, case (2) in Conjecture 6.1.64 is associated to $A_{n,k} = A_{MHV_3}(I_\ell) \otimes A_{n-2,k-1}(I_r)$ is the *helicity-decreasing collinear limit* of the two particles in I_ℓ.

Remark 6.1.65 In [18], the author shows that all facets of the momentum amplituhedron in Conjecture 6.1.64 arise from the factorizations of twistor strings amplitudes on the worldsheet via the *polarized scattering equations*.

Example 6.1.66 (NMHV Amplitudes) Let us consider $n = 6$ and $k = 3$. There are 3 types of facets.

Type 1. We consider $k_r = 1$ and $I_r = \{5, 6\}$. Then by Part (1) of Conjecture 6.1.64, we get a facet on $\langle Y56\rangle = 0$. The corresponding positroid cell is swiped by matrices of the type:

$$C_1(\alpha) = \left(\begin{array}{ccccc|c} 1 & \alpha_5+\alpha_7 & \alpha_5\alpha_6 & 0 & 0 & 0 \\ 0 & 1 & \alpha_3+\alpha_6 & \alpha_3\alpha_4 & 0 & 0 \\ \cline{4-6} 0 & 0 & 1 & \alpha_4 & \alpha_2 & \alpha_1 \end{array}\right), \tag{6.1.67}$$

where we see that the submatrix obtained from C_1 by takining its rows $\{1, 2, 3\}$ and columns $\{1, 2, 3, 4, 5\}$ represents a point in $\mathrm{Gr}_{3,5}^{\geq 0}$ and, by varing over α, it spans its top cell. Analogously, the submatrix obtained from row $\{3\}$ and columns $\{4, 5, 6\}$ represents a point in $\mathrm{Gr}_{1,3}^{\geq 0}$. This facet corresponds to the factorization channel $A_{6,3} = A_{NMHV_5}(1, 2, 3, 4, *) \otimes A_{\overline{MHV}_3}(*, 5, 6)$.

Type 2. We now consider $k_r = 2$ and $I_r = \{5, 6\}$. Then by Part (2) of Conjecture 6.1.64, we get a facet on $[\tilde{Y}56] = 0$. The corresponding positroid cell is swiped by matrices of the type:

$$C_2(\alpha) = \left(\begin{array}{ccc|ccc} 1 & \alpha_3+\alpha_5+\alpha_7 & (\alpha_3+\alpha_5)\alpha_6 & \alpha_3\alpha_4 & 0 & 0 \\ 0 & 1 & \alpha_6 & \alpha_4 & \alpha_2 & 0 \\ \hline 0 & 0 & 0 & 0 & 1 & \alpha_1 \end{array}\right), \tag{6.1.68}$$

where we see that the submatrix obtained from C_2 by takining its rows $\{1, 2\}$ and columns $\{1, 2, 3, 4, 5\}$ represents a point in $\mathrm{Gr}_{2,5}^{\geq 0}$ and, by varing over α, it spans its top cell. Analogously, the submatrix obtained from row $\{2, 3\}$ and columns $\{4, 5, 6\}$ represents a point in $\mathrm{Gr}_{2,3}^{\geq 0}$. This facet corresponds to the factorization channel $A_{6,3} = A_{MHV_5}(1, 2, 3, 4, *) \otimes A_{MHV_3}(*, 5, 6)$.

Type 3. Finally, we consider $I_\ell = \{1, 2, 3\}$ and $k_\ell = 2$. Then by Part (2) of Conjecture 6.1.64, we get a facet on $S_{123}(Y, \tilde{Y}) = 0$. The associated plabic graph $G_\ell(I_\ell) \otimes G_r(I_r)$ is depicted in Fig. 6.3. The corresponding positroid cell has vanishing plückers p_{123}, p_{456}. Points in this cell can be represented in terms of the following matrix

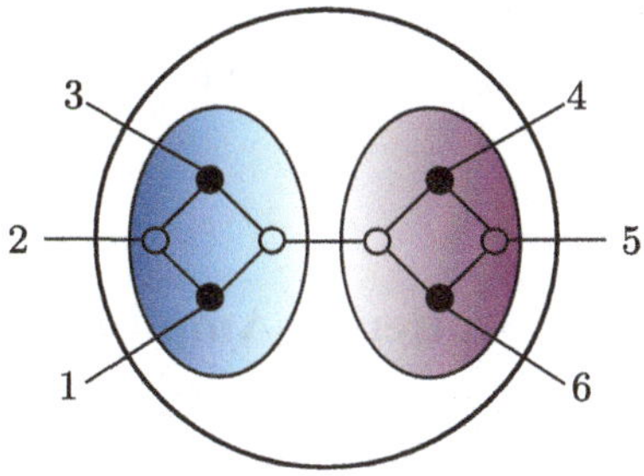

Fig. 6.3 Plabic graph $G_\ell(1,2,3) \otimes G_r(4,5,6)$

$$C_3(\alpha) = \left(\begin{array}{cc|cc|cc} 1 & \alpha_5+\alpha_7 & \alpha_5\alpha_6 & 0 & 0 & 0 \\ 0 & 1 & \alpha_6 & \alpha_2+\alpha_4 & \alpha_2\alpha_3 & 0 \\ 0 & 0 & 0 & 1 & \alpha_3 & \alpha_1 \end{array}\right), \tag{6.1.69}$$

where we see that the submatrix obtained from C_3 by takining its rows $\{1, 2\}$ and columns $\{1, 2, 3, 4\}$ represents a point in $\text{Gr}_{2,4}^{\geq 0}$ and, by varing over α, it spans its top cell. Analogously, for the submatrix obtained from rows $\{2, 3\}$ and columns $\{3, 4, 5, 6\}$. This facet corresponds to the factorization channel $A_{6,3} = A_{MHV_4}(1, 2, 3, *) \otimes A_{MHV_4}(*, 4, 5, 6)$. ◇

6.1.4 The Momentum Amplituhedron and T-Duality

The momentum amplituhedron $\mathcal{M}_{n,k',m}$ is the image of the positive Grassmannian $\text{Gr}_{k',n}^{\geq 0}$ under the momentum amplituhedron map Φ. As for the amplituhedron or the hypersimplex, it is then natural to subdivide $\mathcal{M}_{n,k',m}$ using (the closure of) images of positroid cells under the Φ map. Following the definitions in Chap. 5, we can call a *positroid dissection* of the momentum ampltituhedron $\mathcal{M}_{n,k',m}$ a collection $\{\Phi_\pi\}$ of (the closure of) full-dimensional images $\Phi_\pi := \overline{\Phi(S_\pi)}$ of positroid cells $S_\pi \subseteq \text{Gr}_{k',n}^{\geq 0}$ under the Φ map with disjoint interiors which cover the whole space. If furthermore the Φ map is injective on each cell S_π of the collection, then $\{\Phi_\pi\}$ is called *positroid tiling*, and each Φ_π is referred to as a *positroid tile* of $\mathcal{M}_{n,k',m}$.

Then the following conjecture is the combinatorial and geometric counterpart of the Amplitude/Wilson Loop duality of $\mathcal{N} = 4$ SYM:

Conjecture 6.1.70 (Tilings of $\mathcal{M}_{n,k+2,4}$ and $\mathcal{A}_{n,k,4}$ are T-dual) The collection $\mathcal{C} = \{\Phi_\pi\}$ is a positroid tiling of $\mathcal{M}_{n,k+2,4}$ if and only if the collection of T-dual Grasstopes $\hat{\mathcal{C}} = \{Z_{\hat{\pi}}\}$ is a positroid tiling of $\mathcal{A}_{n,k,4}$.

We expect the same conjecture to hold true if we replace the word 'tiling' with 'dissection'. We recall the notation for T-duality as in Remark 5.3.12. Moreover, we speculate that a similar conjecture holds for general even $m \geq 4$, i.e., the T-duality map provides a bijection between positroid tilings (dissections) of $\mathcal{M}_{n,k',m}$ and $\mathcal{A}_{n,k,m}$, when $k' = k + m/2$.

Remark 6.1.71 For $m = 2$ we showed that $\mathcal{A}_{n,k,m=2}$ is T-dual to the hypersimplex $\Delta_{k+1,n}$, which is *not* $\mathcal{M}_{n,k+1,m=2}$—even if they are somehow related. See [19] for more details.

6.2 Scattering Amplitudes from $\mathcal{M}_{n,k+2,4}$

The $m = 4$ momentum amplituhedron $\mathcal{M}_{n,k',4}$ encodes the N$^{k'-2}$MHV n-particle tree-level amplitudes in $\mathcal{N} = 4$ SYM. The starting point is the *non-chiral superspace* which is parametrized by spinor helicity variables, $\{\lambda^a, \tilde{\lambda}^{\dot{a}}\}$, $a, \dot{a} = 1, 2$, together with the Grassmann odd parameters[4] $\{\eta^r, \tilde{\eta}^{\dot{r}}\}$, $r, \dot{r} = 1, 2$. Then the amplitude $\mathcal{A}_{\text{N}^k\text{MHV}_n}$ is a function on n copies of this superspace with coordinates $\{\lambda_i, \eta_i | \tilde{\lambda}_i, \tilde{\eta}_i\}$. Let us remark that in this space the supercharges take the form:

$$\tilde{q}^{\dot{a}r} = \sum_{i=1}^{n} \tilde{\lambda}_i^{\dot{a}} \eta_i^r \quad , \quad q^{ar} = \sum_{i=1}^{n} \lambda_i^a \tilde{\eta}_i^{\dot{r}} \,. \tag{6.2.1}$$

and the amplitude is a function of degree $(2(n-k'), 2k')$ in $(\eta, \tilde{\eta})$. This is a similar situation to the one which we have encountered for the momentum twistors, where the amplitude is a function of degree $4k = 4(k'-2)$ in the fermionic part the momentum twistor superspace. Equivalently, it was possible to introduce a bosonized momentum twistor space by introducing auxiliary Grassmann-odd parameters. We will now repeat this construction for the spinor helicity variables.

Let us introduce $2(n-k')$ auxiliary Grassmann-odd parameters ϕ_a^α, $\alpha = 1, \ldots, n-k'$, $a \in [2]$ and $2k'$ auxiliary Grassmann-odd parameters $\tilde{\phi}_{\dot{a}}^{\dot{\alpha}}$, $\dot{\alpha} = 1, \ldots, k'$, $\dot{\alpha} \in [2]$. We define bosonized spinor helicity variables $\Lambda \in \text{Mat}_{n,n-k'+2}$, $\tilde{\Lambda} \in \text{Mat}_{n,k'+2}$ as

$$(\Lambda)_{iA} = \begin{pmatrix} \lambda_i^a \\ \phi^\alpha \cdot \eta_i \end{pmatrix}, \qquad A = (a|\alpha) \in [n-k'+2]\,, \tag{6.2.2}$$

$$(\tilde{\Lambda})_{i\dot{A}} = \begin{pmatrix} \tilde{\lambda}_i^{\dot{a}} \\ \tilde{\phi}^{\dot{\alpha}} \cdot \tilde{\eta}_i \end{pmatrix}, \qquad \dot{A} = (\dot{a}|\dot{\alpha}) \in [k'+2],\ i \in [n]. \tag{6.2.3}$$

In order to make connection with scattering amplitudes we identify the matrices $\tilde{\Lambda} \in \text{Mat}_{n,k'+2}^{>0}$, $\Lambda \in \text{Mat}_{n,n-k'+2}^{>0,\tau}$ in Definition 6.1.6 with the above matrices. For this reason, we refer to the pair $(\Lambda, \tilde{\Lambda})$ as the *kinematic data*.

The next step is to find the canonical form $\mathbf{\Omega}(\mathcal{M}_{n,k',4})$ of $\mathcal{M}_{n,k',4}$, *i.e.*, the differential form with logarithmic singularities on all its boundaries. One way to do this is to find a positroid tiling of $\mathcal{M}_{n,k',4}$ with each positroid tile being the image of a $(2n-4)$-dimensional cell of $\text{Gr}_{k',n}^{\geq 0}$ on which the momentum amplituhedron

[4] $\{\tilde{\eta}^{\dot{r}}\}_{\dot{r}=1,2}$ is obtained from $\{\eta^{\mathsf{A}}\}_{\mathsf{A}=3,4}$ of the chiral super-space (cf. Sect. 1.3.3) by a (Grassmann) Fourier tranform. See also [15, 20].

map $\Phi_{(\Lambda,\tilde{\Lambda})}$ is injective. Conjectural positroid tilings of $\mathcal{M}_{n,k',4}$ are provided by the so-called BCFW tilings.[5] The canonical form of $\mathcal{M}_{n,k',4}$ is the sum of push-forwards of the canonical forms of cells in such tilings. As for the amplituhedron $\mathcal{A}_{n,k,4}$, the explicit answer is a sum of rational forms where the denominators can contain spurious singularities, corresponding to spurious boundaries in a given tiling. These singularities disappear in the complete sum and the only poles of $\mathbf{\Omega}(\mathcal{M}_{n,k',4})$ correspond to the facets of $\mathcal{M}_{n,k',4}$ (cf. Conjecture 6.1.64).

Let us now describe how to obtain the amplitude $\mathcal{A}_{\text{N}^k\text{MHV}_n}(\lambda_i, \eta_i, \tilde{\lambda}_i, \tilde{\eta}_i)$ from the canonical form $\mathbf{\Omega}(\mathcal{M}_{n,k',4})$. Let us recall that the momentum amplituhedron $\mathcal{M}_{n,k',4}$ is $(2n-4)$-dimensional and therefore the degree of $\mathbf{\Omega}(\mathcal{M}_{n,k',4})$ is $(2n-4)$. Let $\mu_{\mathcal{M}_{n,k',4}}$ and ω forms such that:

$$\mu_{\text{Gr}_{n-k',n-k'+2}}(Y) \wedge \mu_{\text{Gr}_{k',k'+2}}(\tilde{Y}) = \mu_{\mathcal{M}_{n,k',4}}(Y, \tilde{Y}) \wedge \text{d}^4 P + \omega \tag{6.2.4}$$

and ω vanishes when restricted to $P(Y, \tilde{Y}) = 0$. Then we can write

$$\mathbf{\Omega}(\mathcal{M}_{n,k',4}) = \mu_{\mathcal{M}_{n,k',4}}|_{P=0}(Y, \tilde{Y})\, \Omega(\mathcal{M}_{n,k',4}), \tag{6.2.5}$$

where we refer to $\Omega(\mathcal{M}_{n,k',4})$ as the *canonical function* of $\mathcal{M}_{n,k',4}$. This is well-defined only on the hypersurface $P = 0$. We can also think it as a distribution on $\text{Gr}_{k',k'+2} \times \text{Gr}_{n-k',n-k'+2}$ supported on $P = 0$, i.e., as $\delta^4(P)\,\Omega(\mathcal{M}_{n,k',4})$. Then, the procedure to extract the amplitude from the canonical form $\mathbf{\Omega}(\mathcal{M}_{n,k',4})$ is similar to the ordinary amplituhedron, *i.e.*, we localize the Y and $\tilde{Y}$ on reference subspaces[6]

$$Y^* = \begin{pmatrix} 0_{2\times(n-k)} \\ \hline 1_{(n-k)\times(n-k)} \end{pmatrix}, \qquad \tilde{Y}^* = \begin{pmatrix} 0_{2\times k} \\ \hline 1_{k\times k} \end{pmatrix}, \tag{6.2.6}$$

obtaining

$$\mathcal{A}_{\text{N}^k\text{MHV}_n}(\lambda_i, \eta_i, \tilde{\lambda}_i, \tilde{\eta}_i) = \delta^4(p) \int d\phi_a^1 \dots d\phi_a^{n-k} \int d\tilde{\phi}_{\dot{a}}^1 \dots d\tilde{\phi}_{\dot{a}}^k\, \Omega_{n,k}(\mathcal{M}_{n,k',4})|_{(Y,\tilde{Y})=(Y^*,\tilde{Y}^*)}, \tag{6.2.7}$$

where $\delta^4(p)$ comes from the localization of $\delta^4(P)$ on Y^*, $\tilde{Y}^*$. In the following section we will show how extracting the amplitude works in practice in a few examples.

Finally, in analogy with the ordinary amplituhedron (see Eq. (2.5.12)), we can introduce an integral representation of the canonical function $\delta^4(P)\,\Omega(\mathcal{M}_{n,k',4})$ as a contour integral. Let us fix a point $(Y, \tilde{Y})$ in $\mathcal{M}_{n,k',4}$ and let $\omega(\mathcal{M}_{n,k',4})$ and α be differential forms on $\text{Gr}_{k',n}^{\geq 0}$ such that:

$$\mathbf{\Omega}(\text{Gr}_{k',n}^{\geq 0}) = \omega(\mathcal{M}_{n,k',4}) \wedge \mu^*_{\mathcal{M}_{n,k',4}}|_{P=0} + \alpha \tag{6.2.8}$$

---

[5] The package `positroid` Mathematica™ package [21] can provide the collection of BCFW cells which give a tiling via the function `treeContour[n,k]`.

[6] This choice of Y^*, $\tilde{Y}^*$ is compatible with the embedding of λ, $\tilde{\lambda}$ in Λ, $\tilde{\Lambda}$ as in (6.2.2), (6.2.3).

and the restriction of α vanishes on the momentum amplituhedron fiber $\Phi^{-1}_{\Lambda,\tilde{\Lambda}}(Y,\tilde{Y})$. Here $\mu^*_{\mathcal{M}_{n,k',4}}|_{P=0}$ is the pull-back of $\mu_{\mathcal{M}_{n,k',4}}|_{P=0}(Y,\tilde{Y})$ under $\Phi_{\Lambda,\tilde{\Lambda}}$. We then consider the *fiber volume form* of $\mathcal{M}_{n,k',4}$ of degree $(k'-2)(n-k'-2) = k(n-k-4)$, by restricting $\omega(\mathcal{M}_{n,k',4})$ on the fiber $\Phi^{-1}_{\Lambda,\tilde{\Lambda}}(Y,\tilde{Y})$ (see Definition B.0.13). Then one obtains the canonical function as a contour integral of the fiber volume form:

$$\Omega(\mathcal{M}_{n,k',4})(Y,\tilde{Y}) = \int_\gamma \omega(\mathcal{M}_{n,k',4})|_{\Phi^{-1}_{\Lambda,\tilde{\Lambda}}(Y,\tilde{Y})}. \tag{6.2.9}$$

Remark 6.2.10 This procedure is equivalent to the formulas in [12], where $\delta^4(P)\,\Omega(\mathcal{M}_{n,k',4})$ is obtained from the following contour integral:

$$\int_\gamma \frac{\mathrm{d}^{(n-k')\cdot(n-k')}g}{(\det g)^{n-k'}} \wedge \boldsymbol{\Omega}(\mathrm{Mat}^{\geq 0}_{k,n})\,\delta^{(n-k')(n-k'+2)}(Y - g\,C^{\perp}\Lambda)\,\delta^{k'(k'+2)}(Y - C\,\tilde{\Lambda})\,, \tag{6.2.11}$$

where

$$\boldsymbol{\Omega}(\mathrm{Mat}^{\geq 0}_{k',n}) := \frac{d^{k'\cdot n}C}{p_{[1,k']}(C)\,p_{[2,k'+1]}(C)\ldots p_{[n,k'-1]}(C)}\,. \tag{6.2.12}$$

Example 6.2.13 (MHV and $\overline{\text{MHV}}$ Amplitudes) We now move to study examples of momentum amplituhedra, starting with MHV and $\overline{\text{MHV}}$ amplitudes. Already in this case the canonical function takes a new and interesting form. The dimension of the momentum amplituhedron $\mathcal{M}_{n,2,4}$ is the same as the dimension of the positive Grassmannian $\mathrm{Gr}_{2,n}$ and therefore there is no need to tile the momentum amplituhedron, it is enough to take the image of the top-dimensional positroid cell of $\mathrm{Gr}^{\geq 0}_{2,n}$, i.e., $\mathrm{Gr}^{>0}_{2,n}$.

It is conjectured that all boundaries of the momentum amplituhedron $\mathcal{M}_{n,2,4}$ are all on the hypersurfaces $\langle Y i\, i+1\rangle = 0$ for $i = 1,\ldots,n$. The canonical form we find in this section will make this manifest. We also recall from Proposition 6.1.42 that, for all points inside the momentum amplituhedron $\mathcal{M}_{n,2,4}$, $[\tilde{Y} i\, i+1] > 0$ for all $i = 1,\ldots,n$, as well as $S_{[i,j]} > 0$ for all $i, j \in [n]$.

Let us start by considering the simplest case, *i.e.*, the four-point MHV amplitude. We parametrize points C in the top cell $\mathrm{Gr}^{>0}_{2,4}$ using the positive parameters α_j:

$$C(\alpha) = \begin{pmatrix} 1 & \alpha_2 & 0 & -\alpha_3 \\ 0 & \alpha_1 & 1 & \alpha_4 \end{pmatrix}. \tag{6.2.14}$$

There are various ways to find α's, and a particular choice results in α's depending only on Y and Λ:

$$\alpha_1 = \frac{\langle Y12\rangle}{\langle Y13\rangle},\ \alpha_2 = \frac{\langle Y23\rangle}{\langle Y13\rangle},\ \alpha_3 = \frac{\langle Y34\rangle}{\langle Y13\rangle},\ \alpha_4 = \frac{\langle Y14\rangle}{\langle Y13\rangle}. \tag{6.2.15}$$

The push-forward of the Grassmannian top form through the momentum amplituhedron map is therefore:

$$\mathbf{\Omega}(\mathcal{M}_{4,2,4}) = \bigwedge_{j=1}^{4} \mathrm{dlog}\alpha_j = \mathrm{dlog}\frac{\langle Y12\rangle}{\langle Y13\rangle} \wedge \mathrm{dlog}\frac{\langle Y23\rangle}{\langle Y13\rangle} \wedge \mathrm{dlog}\frac{\langle Y34\rangle}{\langle Y13\rangle} \wedge \mathrm{dlog}\frac{\langle Y14\rangle}{\langle Y13\rangle} \tag{6.2.16}$$

$$= \frac{\langle 1234\rangle^2}{\langle Y12\rangle\langle Y23\rangle\langle Y34\rangle\langle Y41\rangle}\langle Yd^2Y_1\rangle\langle Yd^2Y_2\rangle \,. \tag{6.2.17}$$

If instead we found the α's in terms of $\tilde{Y}$, we would find the following representation for the canonical form:

$$\mathbf{\Omega}(\mathcal{M}_{4,2,4}) = \frac{[1234]^2}{[\tilde{Y}12][\tilde{Y}23][\tilde{Y}34][\tilde{Y}41]}[\tilde{Y}d^2\tilde{Y}_1][\tilde{Y}d^2\tilde{Y}_2]\,. \tag{6.2.18}$$

where, here and in the following, $[ijls]$ and $\langle ijls\rangle$ denote $\langle\tilde{\Lambda}_i,\tilde{\Lambda}_j,\tilde{\Lambda}_l,\tilde{\Lambda}_s\rangle$ and $\langle\Lambda_i,\Lambda_j,\Lambda_l,\Lambda_s\rangle$ respectively. It is easy to check that (6.2.16) and (6.2.18) are related to each other by momentum conservation Definition 6.1.8. Independently of the chosen representation for the canonical form, the canonical function can be evaluated using Definition 6.2.5 and gives the following manifestly parity symmetric answer:

$$\mathbf{\Omega}(\mathcal{M}_{4,2,4}) = \frac{\langle 1234\rangle^2[1234]^2}{\langle Y12\rangle\langle Y23\rangle[\tilde{Y}12][\tilde{Y}23]}\,, \tag{6.2.19}$$

unique up to momentum conservation. Finally, we can extract the amplitude $\mathcal{A}_{4,2}^{\mathtt{tree}}$ using (6.2.7) to get

$$\mathcal{A}_{\mathrm{MHV}_4} = \delta^4(p)\frac{\delta^4(q)\delta^4(\tilde{q})}{\langle 12\rangle_\lambda\langle 23\rangle_\lambda[12]_{\tilde{\lambda}}[23]_{\tilde{\lambda}}}\,, \tag{6.2.20}$$

where $q, \tilde{q}$ are defined in (6.2.1). This formula agrees with the result found in [15].

This calculation can be easily generalized to any MHV amplitude. A particular representation for the volume form reads

$$\mathbf{\Omega}(\mathcal{M}_{n,2,4}) = \bigwedge_{i=2}^{n-1}\left(\mathrm{dlog}\left(\frac{\langle Yi,i+1\rangle}{\langle Y1,i+1\rangle}\right)\wedge\mathrm{dlog}\left(\frac{\langle Y1,i+1\rangle}{\langle Y12\rangle}\right)\right) \tag{6.2.21}$$

$$= \frac{\langle 1\ldots n\rangle^2}{\langle Y12\rangle\langle Y23\rangle\ldots\langle Y1n\rangle}\langle Y\mathrm{d}^2Y_1\rangle\langle Y\mathrm{d}^2Y_2\rangle\ldots\langle Y\mathrm{d}^2Y_{n-2}\rangle\,. \tag{6.2.22}$$

This result agrees with the one we get for the $m=2$ amplituhedron $\mathcal{A}_{n,n-2,2}(\Lambda)\simeq \mathrm{Gr}_{n-2,n}^{\geq 0}$. Finally, the canonical function for MHV_n amplitudes is

$$\Omega(\mathcal{M}_{n,2,4}) = \left(\sum_{i<j} \frac{[12ij]\langle Yij\rangle}{[\tilde{Y}12]}\right)^2 \frac{\langle 1\ldots n\rangle^2}{\langle Y12\rangle\langle Y23\rangle\ldots\langle Y1n\rangle}\,. \tag{6.2.23}$$

The results for $\overline{\text{MHV}}$ amplitudes are the parity conjugate of the previous formulas. In particular, as for the MHV case, we do not need to tile the momentum amplituhedron $\mathcal{M}_{n,n-2}$ since its dimension is already $2n-4$. The boundaries of $\mathcal{M}_{n,n-2}$ are conjectured to lie on the hypersurfaces $[\tilde{Y}ii+1]=0$ for $i=1,\ldots,n$. Moreover, one can show that the analogous of Proposition 6.1.42 holds for for the $\overline{\text{MHV}}$ case: for all points inside $\mathcal{M}_{n,n-2,4}$ we find $\langle Yii+1\rangle > 0$ for $i=1,\ldots,n$ and $S_{[i,j]} > 0$ for $i, j \in [n]$. ◇

Example 6.2.24 (*NMHV*$_6$ *Amplitude*) As a next step, we consider the first example where we need to tile the momentum amplituhedron in order to find the canonical form. The positive Grassmannian $\mathrm{Gr}^{\geq 0}_{3,6}$ is nine-dimensional, while the momentum amplituhedron $\mathcal{M}_{6,3,4}$ is eight-dimensional and therefore the image of the positive Grassmannian through the map $\Phi_{(\Lambda,\tilde{\Lambda})}$ cannot be injective. In order to find the canonical form $\mathbf{\Omega}(\mathcal{M}_{6,3,4})$ we need to therefore find collections of codimension-one cells of $\mathrm{Gr}^{\geq 0}_{3,6}$. There are two possible positroid tilings obtained as images of the following collections of positroid cells:

$$\mathcal{C}_1 = \{S_1, S_3, S_5\}\,, \qquad \mathcal{C}_2 = \{S_2, S_4, S_6\}\,, \tag{6.2.25}$$

where we denoted as $S_i \subset \mathrm{Gr}^{\geq 0}_{3,6}$ the positroid cell where the only vanishing plücker is $p_{[i,i+2]}(C)=0$, with $i \in [6]$. The canonical form can then be written as follows

$$\mathbf{\Omega}(\mathcal{M}_{6,3,4}) = \mathbf{\Omega}^{(2)}_{6,3} + \mathbf{\Omega}^{(4)}_{6,3} + \mathbf{\Omega}^{(6)}_{6,3} = \mathbf{\Omega}^{(1)}_{6,3} + \mathbf{\Omega}^{(3)}_{6,3} + \mathbf{\Omega}^{(5)}_{6,3}\,, \tag{6.2.26}$$

where $\mathbf{\Omega}^{(i)}_{6,3}$ is the pushforward of the logarithmic differential form on the cell S_i.

In the following we focus on $\mathbf{\Omega}^{(1)}_{6,3}$, the other terms can be found by cyclic shifts. We parametrize S_1 using canonical BCFW-bridge coordinates and solve the relations $Y = C(\alpha)^{\perp}\Lambda$ and $\tilde{Y} = C(\alpha)\tilde{\Lambda}$ to find:

$$\alpha_1 = \frac{\langle Y12\rangle}{\langle Y13\rangle}\,, \; \alpha_2 = \frac{\langle Y23\rangle}{\langle Y13\rangle}\,, \; \alpha_3 = \frac{[\tilde{Y}\hat{3}4]}{[\tilde{Y}\hat{1}\hat{3}]}\,, \; \alpha_4 = \frac{[\tilde{Y}64]}{[\tilde{Y}\hat{1}\hat{3}]} \tag{6.2.27}$$

$$\alpha_5 = \frac{[\tilde{Y}6\hat{1}]}{[\tilde{Y}\hat{1}\hat{3}]}\,, \; \alpha_6 = \frac{[\tilde{Y}4\hat{1}]}{[\tilde{Y}\hat{1}\hat{3}]}\,, \; \alpha_7 = \frac{[\tilde{Y}45]}{[\tilde{Y}64]}\,, \; \alpha_8 = \frac{[\tilde{Y}56]}{[\tilde{Y}64]}\,, \tag{6.2.28}$$

where we have denoted the following shifted variables

$$\hat{\tilde{\Lambda}}_1 = \tilde{\Lambda}_1 + \frac{\langle Y23\rangle}{\langle Y13\rangle}\tilde{\Lambda}_2\,, \; \hat{\tilde{\Lambda}}_3 = \tilde{\Lambda}_3 + \frac{\langle Y12\rangle}{\langle Y13\rangle}\tilde{\Lambda}_2\,. \tag{6.2.29}$$

One can notice that the BCFW-bridge variables are just an uplift of the formula found in [15].[7] The push-forward is computed as

$$\boldsymbol{\Omega}_{6,3}^{(1)} = \bigwedge_{i=1}^{8} \mathrm{dlog}\alpha_i\,, \tag{6.2.30}$$

which, using (6.2.27), leads to the following explicit form for the canonical function

$$\Omega_{6,3}^{(1)} = \frac{\left(\sum_{I\in\binom{[1,3]}{2}} \langle YI\rangle[I\,456]\right)^2 \left(\sum_{I\in\binom{[4,6]}{2}} \langle YI\rangle[I\,123]\right)^2}{S_{123}\langle Y12\rangle\langle Y23\rangle[\tilde{Y}45][\tilde{Y}56]\langle Y1|5+6|4\tilde{Y}]\langle Y3|4+5|6\tilde{Y}]}\,. \tag{6.2.31}$$

After using our procedure (6.2.7) for extracting the amplitude, we find that the expression (6.2.31) reduces to the formula found in [15]. While for the denominator it can be easily seen, the numerator requires a more careful analysis.

$$(\langle Y12\rangle[12456] + \langle Y13\rangle[13456] + \langle Y23\rangle[23456])^2 \to \delta^4(q)(\tilde{\eta}_4[56]_{\tilde{\lambda}} + \tilde{\eta}_5[64]_{\tilde{\lambda}} + \tilde{\eta}_6[45]_{\tilde{\lambda}})^2\,, \tag{6.2.32}$$

while the second bracket gives:

$$\left([\tilde{Y}45]\langle 12345\rangle + [\tilde{Y}46]\langle 12346\rangle + [\tilde{Y}56]\langle 12356\rangle\right)^2 \to \delta^4(\tilde{q})\,(\eta_1\langle 23\rangle_\lambda + \eta_2\langle 31\rangle_\lambda + \eta_3\langle 12\rangle_\lambda)^2\,. \tag{6.2.33}$$

Finally, we can write the complete canonical form $\boldsymbol{\Omega}_{6,3}$ by using (6.2.30) and shifting labels:

$$\boldsymbol{\Omega}_{6,3} = \boldsymbol{\Omega}_{6,3}^{(1)} + \boldsymbol{\Omega}_{6,3}^{(1)}\Big|_{i\to i+2} + \boldsymbol{\Omega}_{6,3}^{(1)}\Big|_{i\to i+4}\,, \tag{6.2.34}$$

and similarly for the canonical function. The spurious poles of the type $\langle Yi|j+k|l\tilde{Y}]$ in $\boldsymbol{\Omega}_{6,3}^{(1)}$ cancel in the sum. The form $\boldsymbol{\Omega}(\mathcal{M}_{6,3,4})$ diverges logarithmically on the 15 boundaries of the momentum amplituhedron $\mathcal{M}_{6,3,4}$:

$$\langle Y i\, i+1\rangle = 0\,,\, i = 1,\ldots,6\,, \qquad [\tilde{Y} i\, i+1] = 0\,,\, i = 1,\ldots,6\,, \qquad S_{[i,i+2]} = 0\,, i = 1,2,3\,. \tag{6.2.35}$$

Remark 6.2.36 We can see how the T-duality Conjecture 6.1.70 is reflected in the computation of scattering amplitudes. In the NMHV$_6$ case—i.e., $k=1$ and $n=6$—on one side, the amplitude in momentum space has 3 terms obtained from Eq. (6.2.34). On the other side, the ampliutude in momentum twistor space also has 3 terms, which

[7] The attentive reader will notice that we have a discrepancy of signs w.r.t. [15]. In our formulas they are such that the canonical coordinates are all positive for positive data.

are R-invariants, e.g., $R_{12345} + R_{12356} + R_{13456}$ (cf. Eq. (1.4.19)). Moreover, there are only 2 ways to express these amplitudes using these types of terms, i.e., there are only 2 positroid tilings of $\mathcal{M}_{6,3,4}$ in bijection with the ones of $\mathcal{A}_{6,1,4}$. ◇

6.3 Summary and Outlook

Despite the name, the *amplitu*hedron is more naturally suited to describe the dual Wilson loop—in momentum twistor space—rather than the amplitude itself. In particular, since the latter relies on the Amplitude/Wilson loop duality special to planar $\mathcal{N} = 4$ SYM, a generalization of this geometry to describe scattering amplitudes in other models may be difficult. For example, it would be desirable to find positive geometries encoding scattering amplitudes in less supersymmetric models and beyond the planar sector. On the other hand, a geometric description directly in momentum space—in twistor or spinor helicity space $\{\lambda, \tilde{\lambda}\}$—would overcome this major obstruction.

In this chapter we have introduced a novel geometric object—the *momentum amplituhedron* $\mathcal{M}_{n,k',m}$—which for $k' = k + 2$ and $m = 4$ encodes the tree-level N^kMHV_n scattering amplitudes in $\mathcal{N} = 4$ SYM directly in momentum space. It is the image of the *momentum amplituhedron map* Φ, which is a map from the positive Grassmannian $G^{\geq 0}_{k',n}$ induced by two totally positive matrices $\Lambda^{\perp}$ and $\tilde{\Lambda}$. The canonical form of the momentum amplituhedron $\mathcal{M}_{n,k+2,4}$ encodes the tree-level amplitude $\mathcal{A}_{\text{N}^k\text{MHV}_n}$ if we regard $\Lambda^{\perp}$ and $\tilde{\Lambda}$ as the 'bosonization'[8] of the on-shell variables $\{\lambda, \eta\}$ and $\{\tilde{\lambda}, \tilde{\eta}\}$, respectively, from the non-chiral superpace. This is in the same spirit as when computing scattering amplitudes from the amplituhedron $\mathcal{A}_{n,k,m=4}$: the totally positive matrix Z defining the amplituhedron map $\tilde{Z}$ is regarded as the bosonization of the (super) momentum twistors $\{z, \chi\}$. Our construction is compatible with a conjecture made in [15]. The authors suggested that an 'amplituhedron in momentum space' should have certain sign flips for λ and $\tilde{\lambda}$, and positivity conditions on planar Mandelstam variables. This was motivated by considering the image of the *twistor-string worldsheet* [22] through the Roiban-Spradlin-Volovich (RSV) equations [26].

Our work opens various interesting avenues of investigation. On the physics side, the first question is whether a generalization of our construction to loop amplitudes is possible. There exists a natural extension of tree-level differential forms to loop integrands, as suggested in [15]. Bosonizing those formulas in a similar fashion as for tree level would therefore be a step worth pursuing. Then the underlying positive geometry should bear similarities with the ordinary loop-level amplituhedron. References [23, 24] shows recent developments along these directions. Perhaps the most fascinating question is whether we can extend our construction to other theories. Recently, a momentum amplituhedron construction was developed for ABJM theories [25]. For instance, our work could shed light on positive geometries in twistor

[8] Not to be confused with the procedure used in the context of quantum (field) theories with fermionic degrees of freedom. See Eqs. (6.2.2 and (6.2.3) for the precise definitions.

theories in higher dimensions [27–33]. By dimensional reduction, this could also lead to geometries for scattering amplitudes with massive particles. Moreover, the momentum amplituhedron is formulated directly in spinor helicity variables, which are variables widely used for massless theories in four dimensions (and beyond). This opens the pathway for investigating positive geometries for non-planar, less- or non-supersymmetric theories. For instance, in [34] the *scattering forms* for Yang-Mills and non-linear sigma model were found. These forms do not have logarithmic singularities, which would indicate that there is no underlying positive geometry. However, already for $\mathcal{N} = 4$ SYM one needs to factorize $\delta^4(q)$ to get a logarithmic form on the spinor helicity space, see [15]. Nevertheless, this problem disappears when we consider the forms on the momentum amplituhedron, as we showed in this chapter. We anticipate that similar, but more complicated, behaviour might be possible for less- or non-supersymmetric theories. In this regard, a Grassmannian generalization of the work [35] could lead to the desired 'amplituhedra' for general theories with non-logarithmic forms—enhancing the paradigm of positive geometries.

On the mathematical side, the momentum amplituhedron $\mathcal{M}_{n,k',m}$ in combination with T-duality, opens a path to understand amplituhedra $\mathcal{A}_{n,k,m}$. For example, $\mathcal{M}_{n,k',m=4}$ can be seen as a pair of $m = 2$ amplituhedra intertwined by momentum conservation. Therefore, it should be promising to use our results on $m = 2$ amplituhedra to understand $\mathcal{M}_{n,k+2,m=4}$ and, via T-duality, also $\mathcal{A}_{n,k,m=4}$—which has physical relevance for $\mathcal{N} = 4$ SYM. This, for example, in order to understand the sign stratification, the characterization of positroid tiles, their facets, and connection to cluster algebras (see Chap. 7) for both objects. We recall indeed that, despite the numerous conjectures and broad intuitive understanding emerging from physics, only recently [8] proved that $\mathcal{A}_{n,k,4}$ admits a positroid tiling (from a BCFW recursion). Reference [9] then proved the original conjecture that *all* BCFW collections of cells coming from BCFW recursions give positroid tilings for $\mathcal{A}_{n,k,4}$. Little is know about the set of all tilings and its combinatorics, and about $m > 4$ amplituhedra [36]. We leave these exciting directions for future work.

References

1. N. Berkovits, J. Maldacena, Fermionic T-duality, dual superconformal symmetry, and the amplitude/Wilson loop connection. JHEP **09**, 062 (2008)
2. L.F. Alday, R. Roiban, Scattering amplitudes, wilson loops and the string/gauge theory correspondence. Phys. Rept. **468**, 153–211 (2008)
3. S. Caron-Huot, Notes on the scattering amplitude/Wilson loop duality. JHEP **1107**, 058 (2011)
4. N. Arkani-Hamed, J. Trnka, The amplituhedron. J. High Energy Phys. **10**, 33 (2014)
5. M. Bullimore, L.J. Mason, D. Skinner, Twistor-strings. Grassmannians and leading singularities. JHEP **03**, 070 (2010)
6. N. Arkani-Hamed, F. Cachazo, C. Cheung, J. Kaplan, A duality for the S matrix. JHEP **03**, 020 (2010)
7. N. Arkani-Hamed, J. Bourjaily, F. Cachazo, A. Goncharov, A. Postnikov, J. Trnka, *Grassmannian Geometry of Scattering Amplitudes*. (Cambridge University Press, 2016)

8. C. Even-Zohar, T. Lakrec, R. Tessler, The amplituhedron BCFW triangulation. arXiv:2112.02703 (2021)
9. C. Even-Zohar, T. Lakrec, M. Parisi, M. Sherman-Bennett, J.R. Tessler, L. Williams, Cluster algebras and Tilings for the $m = 4$ amplituhedron. In preparation
10. S.N. Karp, L.K. Williams, Y.X. Zhang, *Decompositions of Amplituhedra* (2017)
11. T. Łukowski, M. Parisi, L.K. Williams, The positive tropical grassmannian, the hypersimplex, and the $m = 2$ amplituhedron. Int. Math. Res. Not. **03**, rnad010 (2023)
12. D. Damgaard, L. Ferro, T. Lukowski, M. Parisi, The momentum amplituhedron. JHEP **08**, 042 (2019)
13. M. Parisi, M. Sherman-Bennett, L. Williams, The $m = 2$ amplituhedron and the hypersimplex: signs, clusters, triangulations, Eulerian numbers, 2023. Accepted and to be published in *Communications of the American Mathematical Society*
14. S.N. Karp, Sign variation, the Grassmannian, and total positivity. J. Combin. Theory Ser. A **145**, 308–339 (2017)
15. S. He, C. Zhang, Notes on scattering amplitudes as differential forms. JHEP **10**, 054 (2018)
16. L. Ferro, T. Łukowski, R. Moerman, From momentum amplituhedron boundaries to amplitude singularities and back. J. High Energy Phys. **7**(18), 201 (2020)
17. N. Arkani-Hamed, J.L. Bourjaily, F. Cachazo, A.B. Goncharov, A. Postnikov, J. Trnka, *Scattering Amplitudes and the Positive Grassmannian.* (Cambridge University Press, 2012)
18. Y. Geyer, L. Mason, M. Parisi, Momentum amplituhedra from the worldsheet. In preparation
19. T. Lukowski, J. Stalknecht, The hypersimplex canonical forms and the momentum amplituhedron-like logarithmic forms. **7** (2021)
20. Y.-T. Huang, Non-chiral S-matrix of $N = 4$ super Yang-Mills. **4** (2011)
21. J.L. Bourjaily, Positroids, plabic graphs, and scattering amplitudes in mathematica. **12** (2012)
22. E. Witten, Perturbative gauge theory as a string theory in twistor space. Commun. Math. Phys. **252**, 189–258 (2004)
23. L. Ferro, T. Lukowski, The Loop Momentum Amplituhedron. J. High Energy Phys., **2023**(5), 1–17 (2023)
24. L. Ferro, R. Glew, T. Lukowski, J. Stalknecht, Prescriptive Unitarity from Positive Geometries. arXiv preprint arXiv:2308.02438 (2023)
25. S. He, C.K. Kuo, Y.Q. Zhang, The momentum amplituhedron of SYM and ABJM from twistor-string maps. J. High Energy Phys., **2022**(2), 1–33
26. R. Roiban, M. Spradlin, A. Volovich, On the tree level S matrix of Yang-Mills theory. Phys. Rev. D **70**, 026009 (2004)
27. M. Heydeman, J.H. Schwarz, C. Wen, M5-brane and D-brane scattering amplitudes. JHEP **12**, 003 (2017)
28. F. Cachazo, A. Guevara, M. Heydeman, S. Mizera, J.H. Schwarz, C. Wen, The S matrix of 6D super Yang-Mills and maximal supergravity from rational maps. JHEP **09**, 125 (2018)
29. M. Heydeman, J.H. Schwarz, C. Wen, S.-Q. Zhang, All tree amplitudes of 6D (2, 0) supergravity: interacting tensor multiplets and the $K3$ moduli space. Phys. Rev. Lett. **122**(11), 111604 (2019)
30. Y. Geyer, L. Mason, Polarized scattering equations for 6D superamplitudes. Phys. Rev. Lett. **122**(10), 101601 (2019)
31. Y. Geyer, L. Mason, Supersymmetric S-matrices from the worldsheet in 10 & 11d. Phys. Lett. B **804**, 135361 (2020)
32. Y. Geyer, L. Mason, D. Skinner, Ambitwistor strings in six and five dimensions. JHEP **08**, 153 (2021)
33. G. Albonico, Y. Geyer, L. Mason, Recursion and worldsheet formulae for 6d superamplitudes. JHEP **08**, 066 (2020)
34. N. Arkani-Hamed, Y. Bai, S. He, G. Yan, *Scattering Forms and the Positive Geometry of Kinematics, Color and the Worldsheet* (2017)
35. P. Benincasa, M. Parisi, Positive geometries and differential forms with non-logarithmic singularities. Part I. J. High Energy Phys. **8**(44), 023 (2020)
36. N. Arkani-Hamed, C. Even-Zohar, M. Parisi, M. Sherman-Bennett, R.J. Tessler, T. Lakrec, L. Williams, The $m = 6$ amplituhedron. In preparation

Chapter 7
Cluster Algebras and Amplituhedra

Introduced by Fomin and Zelevinsky in 2002 [1], *cluster algebras* are incredibly rich structures which connects many areas in mathematics, such as algebraic geometry and combinatorics. We refer to [2–4] for the basics of cluster algebras and cluster varieties. Moreover, in a seminal paper [5], Scott showed that (the coordinate rings of) the Grassmannian $\mathrm{Gr}_{k,n}$ has a remarkable cluster structure.

In physics, cluster algebras have been recently playing a remarkable role in both understanding the singularities of scattering amplitudes and advancing cutting-edge computations. In 2013 Golden-Goncharov-Spradlin-Vergu-Volovich [6] first established that singularities of scattering amplitudes of planar $\mathcal{N} = 4$ SYM at loop level can be described using cluster algebras. In particular, a large class of loop amplitudes can be expressed in terms of *multiple polylogarithms* whose branch points are encoded in the so-called *symbol alphabet*. Remarkably, elements of this alphabet were observed to be cluster variables of the cluster algebra associated to the Grassmannian $\mathrm{Gr}_{4,n}$. This enabled the powerful program of *cluster bootstrap* which pushed both the computation and the understanding of the mathematical structure of scattering amplitudes beyond the frontiers, see [7] for a recent review.

In 2017 Drummond–Foster–Gurdogan [8] enhanced the connection with cluster algebras by observing phenomena they called *cluster adjacencies*, related to compatibility of cluster variables. Shortly thereafter, they considered ‘BCFW terms’ in tree-level $\mathcal{N} = 4$ SYM amplitudes—rational functions coming from the BCFW recursions—and conjectured that their poles correspond to collections of compatible cluster variables of the $\mathrm{Gr}_{4,n}$ cluster algebra [9]. In [10], this conjecture was extended to all (rational) *Yangian invariants*—the ‘building blocks’ of tree-level amplitudes and leading singularities of planar $\mathcal{N} = 4$ SYM. Since then, cluster algebras have been taking more space on the mathematical canvas of physics. In particular, its connection with tropical geometry has revealed to be fruitful for bootstrapping loop amplitudes in $\mathcal{N} = 4$ SYM [11–14]. We can also find recent works on cluster structures and plabic graphs [15–18], *wall crossing* [19], and *tensor diagrams* [20]. More cluster and tropical structures have also been explored in Feynman integrals [21, 22].

M. Parisi, *Combinatorial Aspects of Scattering Amplitudes*, Springer Theses,
https://doi.org/10.1007/978-3-031-41069-7_7

Despite the unveiling of so many cluster structures in scattering amplitudes of $\mathcal{N} = 4$ SYM, their relation to the paradigm of positive geometries—in particular to amplituhedra—is still unknown. Do cluster phenomena in scattering amplitudes have a purely geometric origin? If so, where are the cluster structures hiding inside the geometry? In turn, can the geometry suggest new ones? In this chapter we start tackling these questions by exploring the connection between amplituhedra and cluster algebras.

Summary of the Chapter. This chapter is based on the following works by the author: [23, Sects. 6,8], [24], and [25].

In Sect. 7.1 we discuss phenomena of *cluster adjacency* in connection with the amplituhedron, Yangian invariants, and scattering amplitudes. We formulate and generalize the cluster adjacency conjecture in terms of the amplituhedron $\mathcal{A}_{n,k,m}$ (Sect. 7.1.1). In Sect. 7.1.2 we prove this conjecture for the $m = 2$ amplituhedron $\mathcal{A}_{n,k,2}$, using results of Sect. 4.2.2. Via T-duality, we also prove its hypersimplex-analogue for $\Delta_{k+1,n}$. Finally, we provide a conjectural formula for all $m = 2$ Yangian invariants which manifest cluster adjacency. In Sect. 7.1.3 we explore the $m = 4$ cluster adjacency conjecture—relevant for scattering amplitudes in $\mathcal{N} = 4$ SYM—from the amplituhedron $\mathcal{A}_{n,k,4}$. In particular, we obtain the list of poles of all Yangian Invariants in the N^2MHV sector from the facets of the respective Grasstopes. We show these are expressed in terms of compatible cluster variables of $\text{Gr}_{4,n}$.

In Sect. 7.2, we will prove a more geometric statement, which illustrates a new phenomenon in the setting of amplituhedra. Using twistor coordinates and results from Sect. 2.3, we will associate a cluster variety to each positroid tile in $\mathcal{A}_{n,k,2}(Z) \subset \text{Gr}_{k,k+2}$, and we will show that the positroid tile is the totally positive part of that cluster variety. We then have the interesting phenomenon that the $2k$-dimensional amplituhedron $\mathcal{A}_{n,k,2}(Z)$ can be tiled into $\binom{n-2}{k}$ $2k$-dimensional positroid tiles, each of which is the totally positive part of a cluster variety. Moreover, there are many ways to tile the amplituhedron (cf. Sect. 5.1).

In Sect. 7.3 we review how the singularity structure of scattering amplitudes can be understood in terms of amplituhedra. We start with Sect. 7.3.1 by introducing the loop amplituhedron—the positive geometry for loop integrands. We then review what Leading Singularities (Sect. 7.3.2) and Landau Singularities (Sect. 7.3.3) are and how they can be obtained from amplituhedra.

In Sect. 7.4 we explore new connections between cluster algebras and singularities of scattering amplitudes. In Sect. 7.4.1 we phrase a new conjecture about cluster structures relating Leading and Landau singularities—*LL-cluster adjacency*. In Sect. 7.4.2 we explore this conjecture for one-loop amplitudes. We prove LL-cluster adjacency for NMHV amplitudes and test it for N^2MHV up to 9-points amplitudes. Finally, in Sect. 7.4.3 we discuss possible implication of LL-cluster adjacency for bootstrapping loop amplitudes. We conclude with a summary and future directions opened by our work in Sect. 7.4.4.

7.1 Cluster Adjacency and the Amplituhedron

There are various different but related cluster-adjacency statements for scattering amplitudes in $\mathcal{N} = 4$ super Yang-Mills:

Symbol letters. Two words appear next to each other in the symbol of a *BDS-like normalised* amplitude only if they are *compatible* cluster variables[1] of the $\mathrm{Gr}_{4,n}$ cluster algebra [8]—i.e. there exists a cluster of $\mathrm{Gr}_{4,n}$ which contains both cluster variables. For all known *integrable words* with physical initial entries, this requirement appears to be equivalent to the *extended Steinmann conditions* of [26, 27].

BCFW terms. In [9] it is conjectured that BCFW representations of tree amplitudes in $\mathcal{N} = 4$ SYM are linear combinations of terms whose poles are compatible cluster variables. The simplest case of this statement is for NMHV tree amplitudes, that are sums of R-invariants:

$$\mathcal{A}_{NMHV} = \sum_{1<i<j<n} R_{1\,i\,i+1\,j\,j+1}\,, \tag{7.1.1}$$

and the adjacency for the latter has was proved in [9] through a procedure in which one starts from the initial cluster of $\mathrm{Gr}_{4,n}$ and arrives at a cluster containing the poles of $R_{1\,i\,i+1\,j\,j+1}$ through a sequence of (partial) cyclic rotations.

This observation, in particular the simple proof of the cluster-adjacency of R-invariants, motivates the question of how far this property extends. In [10] it was conjectured that all (rational) Yangian invariants satisfy such cluster adjacency properties.

(Rational) Yangian invariants. The natural question of whether the manifestation of cluster adjacency in BCFW terms extends to more general Yangian Invariants was asked in [9]. The authors of [10] conjectured that such a question has indeed an affirmative answer for rational Yangian invariants, providing a lot evidence with the use of Sklyanin brackets.

R-invariants and NMHV final entries. Finally, the fourth statement of cluster adjacency concerns NMHV loop amplitudes, which are sums of iterated integrals whose coefficients are R-invariants. Schematically they have the form

$$\mathcal{A}^{(L)}_{NMHV} = \sum_{\alpha, i_1,\ldots,i_{2L}} R_\alpha\, c_{1,\ldots,2L}\ \phi_{i_1} \otimes \cdots \otimes \phi_{i_{2L}}\,, \tag{7.1.2}$$

where the index α enumerates all relevant R-invariants, L is the loop order, and the indices i_k enumerate letters, the rational ones of which are cluster variables of the $\mathrm{Gr}_{4,n}$ cluster algebra. The observation which holds for all known such amplitudes is that the final entries of the symbols of the polylogarithms multiplying the R-

[1] They are referred to as '$\mathcal{A}$-coordinates' in the physics literature.

invariants are cluster-adjacent to all of the poles of the R-invariant that multiplies them. Note that in general one needs to write out the amplitude with a redundant set of R-invariants that satisfy linear 6-term identities in order to make manifest this cluster-adjacency property.

7.1.1 Cluster Adjacency for Amplituhedra $\mathcal{A}_{n,k,m}$

Scattering amplitudes at tree level can be expressed as a sum of certain building blocks $\{\mathcal{Y}_a\}$ which are rational functions in the kinematics called *Yangian invariants*, e.g. via the BCFW recursions. From a geometric perspective, this corresponds to considering a positroid tiling $\{Z_a\}$ of the amplituhedron in positroid tiles Z_a—(the closure of the) image $\tilde{Z}(S_a)$ of the positroid cell $S_a \subseteq \mathrm{Gr}_{k,n}^{\geq 0}$ where the $\tilde{Z}$ map is injective (see Sect. 2.1.2). In particular, a Yangian invariant $\mathcal{Y}_a$ can be computed from the canonical function $\Omega(Z_a)$ of the associated positroid tile Z_a (see Appendix B). Claims about singularities of $\mathcal{Y}_a$ correspond to claims about facets (or faces) of positroid tiles Z_a. Therefore, tree-level cluster adjacency—i.e. poles of $\mathcal{Y}_a$ are cluster variables which can be found in a common cluster of $\mathrm{Gr}_{4,n}$—says something deep about the geometry of Z_a! In the following we will reformulate cluster adjacency conjectures in terms of the geometry of the amplituhedron $\mathcal{A}_{n,k,m}(Z)$ and the facets of its positroid tiles.

Let us start from the $m = 2$ amplituhedron. To each positroid tile $Z_{\hat{G}(T)}$ of $\mathcal{A}_{n,k,m=2}$, one can associate a rational function in the twistor coordinates which is called a *Yangian invariant*.[2] A defining property of this function is that it has a simple pole at $\langle Yij\rangle = 0$ if and only if there is a facet of $Z_{\hat{G}(T)}$ lying on the hypersurface $\{\langle Yij\rangle = 0\}$. Let us consider the collection $\{\langle Yij\rangle\}_{\hat{G}(T)}$ of twistor coordinates corresponding to such poles, and identify it via Proposition 2.2.3, with the corresponding collection of Plücker coordinates $\{p_{ij}(z)\}_{\hat{G}(T)}$ in the Grassmannian $\mathrm{Gr}_{2,n}(\mathbb{C})$ (with $z = Y^{\perp}Z^T$). These Plücker coordinates are cluster variables of the type A_{n-3} cluster algebra associated to $\mathrm{Gr}_{2,n}(\mathbb{C})$ [29]. In this cluster algebra, p_{ab} and p_{cd} are *compatible* cluster variables if the arcs $a \to b$ and $c \to d$ in the polygon $\mathbf{P}_n$ do not cross. The $m = 2$ *cluster adjacency conjecture* of Lukowski–Parisi–Spradlin–Volovich [24] says that the cluster variables associated to the facets of each positroid tile of $\mathcal{A}_{n,k,2}$ represent compatible cluster variables for $\mathrm{Gr}_{2,n}(\mathbb{C})$. We generalize this conjecture as follows.

Conjecture 7.1.3 *Let $Z_{\hat{G}(T)}$ be a positroid tile of $\mathcal{A}_{n,k,2}(Z)$. Each facet lies on a hypersurface $\langle Yij\rangle = 0$, and the corresponding collection of Plücker coordinates $\{p_{ij}\}_{\hat{G}(T)}$ labeling facets is a collection of compatible cluster variables for the Grassmannian* $\mathrm{Gr}_{2,n}(\mathbb{C})$.

Moreover, if p_{hl} is compatible with $\{p_{ij}\}_{\hat{G}(T)}$, then the twistor coordinate $\langle Yhl\rangle$ has a fixed sign on the open positroid tile $Z^{\circ}_{\hat{G}(T)}$.

We will prove Conjecture 7.1.3 in Theorem 7.1.8.

---

[2] Within the framework of *positive geometries*, this is the *canonical function* of $Z_{\hat{G}(T)}$ [28].

We now explain how to generalize this conjecture for other m. The relevant cluster algebra is the homogeneous coordinate ring of $\mathrm{Gr}_{m,n}(\mathbb{C})$ [5]. Each cluster variable can be written as a polynomial $Q(p_I)$ in the $\binom{n}{m}$ Plücker coordinates. Meanwhile, each facet of a positroid tile Z_π of $\mathcal{A}_{n,k,m}$ lies on a hypersurface defined by the vanishing of some (often non-linear) polynomial $Q(\langle YZ_I\rangle)$ in the $\binom{n}{m}$ twistor coordinates, where $\langle YZ_I\rangle$ is shorthand for $\langle YZ_{i_1}\dots Z_{i_m}\rangle$ with $I = \{i_1,\dots,i_m\}$. Then the *cluster adjacency conjecture* for $\mathcal{A}_{n,k,m}$ is the following.

Conjecture 7.1.4 (Cluster adjacency for $\mathcal{A}_{n,k,m}(Z)$) *Let Z_π be a positroid tile of the amplituhedron $\mathcal{A}_{n,k,m}(Z)$. Then there is a set*

$$\mathrm{Facet}(Z_\pi) := \{Q(p_I) \mid \textit{a facet of } Z_\pi \textit{ lies on the hypersurface } Q(\langle YZ_I\rangle) = 0\},$$

such that: where Q is an irreducible polynomial in the $\binom{n}{m}$ Plücker coordinates. Then

1. *Each $Q \in$ Facet(Z_π) is a cluster variable for* $\mathrm{Gr}_{m,n}(\mathbb{C})$.
2. Facet(Z_π) *consists of compatible cluster variables.*
3. *If $\tilde{Q}$ is a cluster variable compatible with* Facet(Z_π), *the polynomial $\tilde{Q}(\langle YZ_I\rangle)$ in twistor coordinates has a fixed sign on the open positroid tile Z_π°.*

Remark 7.1.5 We note that positroid tiles for $m = 4$ are not yet characterized.[3] In general, the polynomials appearing in the sets Facet(Z_π) are unknown. Moreover, for $n \geq 8$, there is no classification of the cluster variables of $\mathrm{Gr}_{4,n}$.

Remark 7.1.6 In a work in preparation [30], the author together with some collaborators tackle Conjecture 7.1.4 for the $m = 4$ amplituhedron.

There is a question which ties together cluster algebras, positroid tiles and tilings:

Question 7.1.7 *Let choose $\mathcal{F}$ in Remark 2.2.13 to be the union of* Facet(Z_π) *over all positroid tiles Z_π of $\mathcal{A}_{n,k,m}(Z)$, where the polynomials are regarded as functions of twistor coordinates. What are the realisable chambers $\mathcal{A}^{\mathcal{F},\sigma}_{n,k,m}$, with $\sigma \in \{+,-\}^N$?*

This would provide insights on the common refinement of all positroid tilings of $\mathcal{A}_{n,k,m}(Z)$. We note that, if Conjecture 7.1.4 holds, $\mathcal{F}$ consists of cluster variables for $\mathrm{Gr}_{m,n}$.

7.1.2 $m = 2$ Cluster Adjacency and Yangian Invariants

Aspects of the $\mathrm{Gr}_{4,n}$ Grassmannian cluster algebras have been found to play several still rather mysterious roles in the mathematical structure of scattering amplitudes in planar $\mathcal{N} = 4$ Yang-Mills theory. A simple toy model which serves as a nice

[3] Conjecturally they are images of positroid cells with *intersection number* one [25], which correspond to 'rational' Yangian invariants [10].

playground for studying features of this cluster structure is the $m = 2$ version of the theory, where the momentum twistors describing the kinematic scattering data are restricted to lie in a $\mathbb{P}^1$ subspace[4] of the usual $\mathbb{P}^3$ (cf. Sect. 1.4). The associated $\mathrm{Gr}_{2,n} \cong A_{n-3}$ cluster algebra[5] is completely understood. The positive geometry associated to these '$m = 2$ amplitudes' is the $m = 2$ amplituhedron $\mathcal{A}_{n,k,m=2}$, which has been central in this work. In this section we explore the cluster adjacency conjecture for (rational) Yangian invariants [9, 10]—poles of every rational Yangian invariant are given by cluster coordinates in a common cluster. The full $\mathcal{N} = 4$ Yang-Mills theory has a whole zoo of n-particle N^{k}MHV Yangian invariants (see e.g. Chapter 12 of [32]). In contrast, in the $m = 2$ toy model, by using the amplituhedron formulation of scattering amplitudes, we are able to conjecture an explicit formula for all Yangian invariants for any n and k. Following the results from Chap. 2, each N^{k}MHV invariant is labelled by a (k, n)-unpunctured plabic tiling—i.e. a collection of k non-crossing triangles in an n-gon, with poles corresponding to some edges of the triangles. The result manifestly satisfies cluster adjacency with respect to the $\mathrm{Gr}_{2,n}$ cluster algebra.

We now prove Conjecture 7.1.3, which extends the $m = 2$ *cluster adjacency conjecture* of Lukowski–Parisi–Spradlin–Volovich [24].

Theorem 7.1.8 (Cluster adjacency for $\mathcal{A}_{n,k,2}$) *Let* $Z_{\hat{G}(T)}$ *be a positroid tile of* $\mathcal{A}_{n,k,2}(Z)$.

Set $\mathrm{Facet}(Z_{\hat{G}(T)}) := \{p_{ij} \mid$ *there is a facet of* $Z_{\hat{G}(T)}$ *on the hypersurface* $\langle Yij\rangle = 0.\}$. *Then:*

1. $\mathrm{Facet}(Z_{\hat{G}(T)})$ *consists of compatible cluster variables for* $\mathrm{Gr}_{2,n}$.
2. *If* $p_{h\ell}$ *is compatible with* $\mathrm{Facet}(Z_{\hat{G}(T)})$, *then* $\langle Yh\ell\rangle$ *has a fixed sign on* $Z^{\circ}_{\hat{G}(T)}$.

Proof The first part follows directly from Theorem 4.2.10 as $Z_{\hat{G}(T)}$ has a facet on $\{\langle Yij\rangle = 0\}$ if and only if $i \to j$ is a facet defining arc in T, and the facet-defining arcs do not cross. The second part follows from Theorem 4.2.2. □

Using Theorems 4.2.2 and 4.2.10, we can translate the cluster adjacency theorem for the $m = 2$ amplituhedron into a cluster adjacency theorem for the hypersimplex.

---

[4] Note that this is quite different from restricting to two space-time dimensions.

[5] This algebra has also been found to govern the structure of $\mathcal{N} = 4$ Yang-Mills amplitudes in the multi-Regge limit [31].

Theorem 7.1.9 (Cluster adjacency for $\Delta_{k+1,n}$) *Let $\Gamma_{G(\mathcal{T})}$ be a positroid tile of $\Delta_{k+1,n}$.*

Set $\mathrm{Facet}(\Gamma_{G(\mathcal{T})}) := \{p_{ij} \mid$ *there is a facet of* $\Gamma_{G(\mathcal{T})}$ *on the hyperplane* $x_{[i,j-1]} = a_{i,j}\}$,

where $a_{i,j}$ are some non-negative integers. Then:

1. $\mathrm{Facet}(\Gamma_{G(\mathcal{T})})$ *consists of compatible cluster variables for* $\mathrm{Gr}_{2,n}$.
2. *If $p_{h\ell}$ is compatible with* $\mathrm{Facet}(\Gamma_{G(\mathcal{T})})$*, then* $x_{[h,\ell-1]} > \mathrm{area}(h \to \ell)$ *in* $\Gamma^{\circ}_{G(\mathcal{T})}$.

Yangian Invariants for $\mathcal{A}_{n,k,2}$. *Yangian invariants* are basic building blocks for many amplitude-related quantities of interest (cf. Sect. 2.1.2). In this section we conjecture a formula for all Yangian invariants for $m = 2$. They can be directly extracted from the canonical functions $\Omega(Z_\pi)$ of the corresponding positroid tiles Z_π of the $m = 2$ amplituhedron.[6] These are the building blocks for positroid tilings and have been completely classified in Theorem 2.3.25.

Given a (k, n)-unpunctured plabic tiling $\overline{\mathcal{T}}$, we denote its black polygons as $P_1, \ldots, P_r$ and write $\overline{\mathcal{T}} = \{P_1, \ldots, P_r\}$. For each P_i, we denote $\mathrm{V}(P_i) = \{a_1 < \ldots < a_s\} \subseteq [n]$ the labels of its vertices and $\mathrm{E}(P_i) = \{\{a_1 < a_2\}\} \subseteq \binom{[n]}{2}$ the collection of edges of P_i labelled via the pairs of adjacent vertices.

Let $J \subseteq [k]$ and let $Y = Y_1 \wedge \ldots \wedge Y_k$ be an element of $\mathrm{Gr}_{k,k+2}$, we denote $Y_J^{\perp} \subseteq Y$ as the $(k - |J|)$-plane $Y_J^{\perp} = \epsilon_{J,[k]\setminus J} Y_{[k]\setminus J}$, where $Y_{[k]\setminus J} = \wedge_{\bar{j} \in [n]\setminus j} Y_{\bar{j}} \in \mathrm{Gr}_{k-|J|,k+2}$. For $I \subseteq [k]$ such that $|J| = |I| - 2$, we then have:

$$\langle Y_J^{\perp} Z_I \rangle = \epsilon_{J,[k]\setminus J} \langle Y_{[k]\setminus J} Z_I \rangle. \tag{7.1.10}$$

Definition 7.1.11 Let $\overline{\mathcal{T}} = \{P_1, \ldots, P_r\}$ be a (k, n)-unpunctured plabic tiling. For each black polygon P_i, let $I_i = \mathrm{V}(P_i)$. Then we define

$$\langle Y\, I_1 \cap \ldots \cap I_r \rangle := \sum_{J_1,\ldots,J_r} \epsilon_{J_1,\ldots,J_r} \langle Y_{J_1}^{\perp} Z_{I_1} \rangle \ldots \langle Y_{J_r}^{\perp} Z_{I_r} \rangle\,, \tag{7.1.12}$$

where the sum is over partitions of $[k]$ into $\{J_1, \ldots, J_r\}$, with $|J_i| = |I_i| - 2$.

Remark 7.1.13 The partitions above are well defined as $\overline{\mathcal{T}}$ is a (k, n)-unpunctured plabic tiling. Therefore, we need k triangles to triangulate all the black polygons $P_1, \ldots, P_r$, i.e. $k = \sum_{i=1}^{r} (\mathrm{V}(P_i) - 2) = \sum_{i=1}^{r} |J_i|$. Moreover, even if (7.1.10) depends on the choice of vectors $Y_1, \ldots, Y_k$ we used to express Y, the set of expressions in (7.1.12) depends only on Y (if we consider them in projective space).

Conjecture 7.1.14 (All $m = 2$ Yangian invariants) *Let $\overline{\mathcal{T}} = \{P_1, \ldots, P_r\}$ be a (k, n)-unpunctured plabic tiling. Then the Yangian invariant $\Omega(Z_{\overline{\mathcal{T}}})$ associated to the positroid tile $Z_{\overline{\mathcal{T}}}$ of $\mathcal{A}_{n,k,2}(Z)$ is:*

---

[6] With a slight abuse of terminology, we will also refer to $\Omega(Z_\pi)$ as 'Yangian invariants'—following the original paper [24].

$$\Omega(Z_{\overline{T}}) = \frac{\langle Y, \mathrm{V}(P_1) \cap \ldots \cap \mathrm{V}(P_r)\rangle^2}{\prod_{i=1}^r \prod_{J \in \mathrm{E}(P_i)} \langle Y Z_J \rangle}. \tag{7.1.15}$$

Moreover, all poles of $\Omega(\Omega(Z_{\overline{T}}))$ are simple and are exactly in $\langle Y Z_J \rangle = 0$, where $J \in E(P_i)$, with $i \in [r]$. In other words, the numerator of (7.1.15) does not have zeros in correspondence of those hypersurfaces.

Remark 7.1.16 In the case of P_i be triangles for all $i \in [r]$, this formula agrees with [28].[7] Moreover, let us now consider a (k, n)-unpunctured plabic tilings $\overline{T} = \{P_1, \ldots, P_r\}$ and subdivide the polygon P_s further into two polygons $\{P_a, P_b\}$ which share an edge $\{i, j\}$. It can be proved by induction, that the right hand side of (7.1.15) with the set of polygons $\{P_1, \ldots, P_r\} \setminus \{P_s\} \cup \{P_a, P_b\}$ still gives $\Omega_{\overline{T}}(Y; Z)$. In particular, the pole arising from the factor $\langle Y Z_{ij}\rangle^2$ in the denominator is cancelled by the fact the numerator develops a double zero in $\langle Y Z_{ij}\rangle = 0$. In other words, the formula (7.1.15) works with any set of polygons $\{\tilde{P}_1, \ldots, \tilde{P}_\ell\}$ such that appropriate subsets subdivide the polygons $\{P_1, \ldots, P_r\}$ respectively.

Example 7.1.17 ($k = 1$). $\mathcal{A}_{n,1,2}(Z)$ is an n-gon and its positroid tiles $Z_{\overline{T}}$ are triangles with vertices Z_a, Z_b, Z_c, with $\mathcal{T} = \{a < b < c\}$. Then the Yangian invariants are:

$$\Omega(Z_{\mathcal{T}}) = \frac{\langle Z_{abc}\rangle^2}{\langle Y Z_{ab}\rangle \langle Y Z_{bc}\rangle \langle Y Z_{ca}\rangle}. \tag{7.1.18}$$

◇

Example 7.1.19 ($k = 2$). $\mathcal{A}_{n,2,2}(Z)$ has positroid tiles $Z_{\overline{T}}$ of two types: when the unpunctured plabic tiling is made of two triangles $\overline{T}_1 = \{\{a_1 < b_1 < c_1\}, \{a_2 < b_2 < c_2\}\}$, or when it is made of just a quadrilateral $\overline{T}_2 = \{\{a < b < c < d\}\}$. Then the Yangian invariants are:

$$\Omega(Z_{\overline{T}_1}) = \frac{\langle Y(a_1 b_1 c_1) \cap (a_2 b_2 c_2)\rangle^2}{\langle Y a_1 b_1\rangle \langle Y b_1 c_1\rangle \langle Y c_1 a_1\rangle \langle Y a_2 b_2\rangle \langle Y b_2 c_2\rangle \langle Y c_2 a_2\rangle}, \tag{7.1.20}$$

$$\Omega(Z_{\overline{T}_2}) = \frac{\langle Z_{abcd}\rangle^2}{\langle Y a, b\rangle \langle Y b, c\rangle \langle Y c, d\rangle \langle Y d, a\rangle}. \tag{7.1.21}$$

◇

In this section we showed that the connection between cluster adjacency and Yangian invariants can be made manifest for $m = 2$ and originates from the geometry of each positroid tile inside the $m = 2$ amplituhedron. In Sect. 7.2 we go at the roots and find a cluster structure inside the geometry itself. In particular, we discover that a positroid tile is the positive part of a cluster variety (of type $A_{n_1} \times \cdots \times A_{n_r}$). Moreover, this cluster structure seems different from the ones we would naively derive from $\mathrm{Gr}_{k,n}$ or $\mathrm{Gr}_{2,n}$. The associated cluster variables are indeed positive ratios of twistor coordinates which have definite sign (non necessarily positive) inside the positroid tile.

[7] Notice a typo in the numerator of formula (7.52) in [28]: the power should be 2 instead of k.

7.1.3 $m = 4$ Cluster Adjacency and Yangian Invariants

We recall that in [10] the cluster adjacency conjecture of [9] was generalised for all *rational*[8] Yangian invariants of $\mathcal{N} = 4$ SYM. In particular, in [9] cluster adjacency between Yangian invariants was checked up to 8-points N^2MHV by looking at Yangian invariants appearing into a specific representation of the amplitude. This is not an exhaustive check since, starting from 8-points N^2MHV, one in general finds Yangian invariants which are not related by cyclic symmetry to any of the Yangian invariants appearing in a fixed representation. Whereas in [10], *pair-wise* cluster adjacency between poles of Yangian invariants was checked for all $k \leq 2$ and many $k = 3$ Yangian invariants, by employing Sklyanin Poisson brackets. In this section we prove that all (rational) N^2MHV Yangian invariants are cluster adjacent by finding explicit examples of corresponding clusters. Moreover, we provide the list of their poles in terms of cluster variables of the $\text{Gr}_{4,n}$ cluster algebra—polynomials in Plückers of momentum twistors (see the files `yik2n_.m` attached to [25], for which we used the Mathematica package `positroids.m` [33]).

In the following, we will present[9] the list of N^2MHV_n Yangian invariants. We will only focus on a representative Yangian invariant for each cyclic class, i.e. all the others can be obtained by cyclically shifting labels. Moreover, we consider those Yangian invariants $\mathcal{Y}_\pi$ that can not be reduced to other Yangian invariants with a lower number of particles. More precisely, $\mathcal{Y}_\pi$ is associated to a positroid cell S_π whose elements are represented by matrices which don't have any zero column (i.e. π is a loopless permutation). As observed in [34], there are only 14 such types of Yangian invariants for N^2MHV_n scattering amplitudes in $\mathcal{N} = 4$ SYM. Equivalently, conjecturally there are 14 types of positroid tiles for $\mathcal{A}_{n,k=2,m=4}$ amplituhedra. Following the physics notation, we will use $\langle ijls \rangle$ to denote plücker coordinates $p_{ijls}(z)$ in $\text{Gr}_{4,n}$, with $i, j, l, s \in [n]$. Moreover, we will use the following conventions:

$$
\begin{aligned}
\langle i \rangle &:= \langle i, i+1, i+2, i+3 \rangle \qquad (7.1.22)\\
\langle i \bar{j} \rangle &:= \langle i, j-1, j, j+1 \rangle \\
\langle a(bc)(de)(fg) \rangle &:= \langle abcd \rangle \langle aefg \rangle - \langle abce \rangle \langle adfg \rangle \\
\langle I(ab) \cap (cde) \rangle &:= \langle Ia \rangle \langle bcde \rangle - \langle Ib \rangle \langle acde \rangle \\
\langle I(abc) \cap (def) \rangle &:= \langle Iab \rangle \langle cdef \rangle - \langle Iac \rangle \langle bdef \rangle + \langle Ibc \rangle \langle adef \rangle,
\end{aligned}
$$

where $a, b, c, d, e, f, g \in [n]$ and $I \in \binom{[n]}{2}$. Below we will also denote as ∂R_J the set of cluster variables associated to poles of an R-invariant R_J with labels $J = \{j_1, \ldots, j_5\}$, i.e. the collection $\{\langle J \setminus \{j_1\} \rangle, \ldots, \langle J \setminus \{j_5\} \rangle\}$.

N^2MHV_6 Yangian Invariants. For $n = 6$, there is only one Yangian invariant $\mathcal{Y}_1$ whose poles are the frozen variables $\{\langle i \rangle\}_{i \in [6]}$. Therefore, it trivially satisfies cluster adjacency:

---

[8] I.e. with *intersection number* equals to one, in the terminology of [32, 33].

[9] The labelling of Yangian invariants differs from [25].

$$\text{Facet}(\mathcal{Y}_1) = \{\langle 1\rangle, \langle 2\rangle, \langle 3\rangle, \langle 4\rangle, \langle 5\rangle, \langle 6\rangle\}$$

N^{2}MHV$_7$ Yangian Invariants. For $n = 7$, there are 3 cyclic classes, but only 2 of them are of a new type, i.e. the other is just a relabelling of $\mathcal{Y}_1$.

$$\begin{aligned}\text{Facet}(\mathcal{Y}_2) &= \{\langle 1\bar{4}\rangle, \langle 5(67)(12)(34)\rangle, \langle 3(45)(67)(12)\rangle, \langle 1\rangle, \langle \bar{4}7\rangle, \langle 2\rangle, \langle 3\rangle, \langle 4\rangle\}\\ \text{Facet}(\mathcal{Y}_3) &= \{\langle 3471\rangle, \langle 7(12)(34)(56)\rangle, \langle \bar{2}7\rangle, \langle 3467\rangle, \langle 1\rangle, \langle \bar{4}7\rangle, \langle 3\rangle, \langle \bar{3}7\rangle, \langle 4\rangle\}\end{aligned}$$

N^{2}MHV$_8$ Yangian Invariants. For $n = 8$, there are 24 cyclic classes. 4 of them are of $n = 6$ type, 14 are of $n = 7$ type, and 6 are of a new types:

$$\begin{aligned}\text{Facet}(\mathcal{Y}_4) &= \partial R_{123,(45)\cap\bar{7},8} \cup \partial R_{45678}\\ \text{Facet}(\mathcal{Y}_5) &= \partial R_{81234} \cup \partial R_{45678}\\ \text{Facet}(\mathcal{Y}_6) &= \partial R_{1234,\bar{5}\cap(78)} \cup \partial R_{45678}\\ \text{Facet}(\mathcal{Y}_7) &= \{\langle 1\bar{4}\rangle, \langle 1245\rangle, \langle 123(45)\cap\bar{7}\rangle, \langle 1\rangle, \langle 4578\rangle, \langle 2\rangle, \langle \bar{5}8\rangle, \langle 4\rangle, \langle 5\rangle\}\\ \text{Facet}(\mathcal{Y}_8) &= \{\langle 6(13)(45)(78)\rangle, \langle 6(12)(45)(78)\rangle, \langle 123, \bar{5}\cap(78)\rangle, \langle 4\bar{7}\rangle, \langle 123(45)\cap\bar{7}\rangle,\\ &\quad \langle \bar{5}8\rangle, \langle 6(23)(45)(78), \langle 4\rangle, \langle 5\rangle\}\\ \text{Facet}(\mathcal{Y}_9) &= \{\langle 1(34)(56)(78)\rangle, \langle 7\rangle, \langle 6(12)(34)(78)\rangle, \langle 3(12)(56)(78)\rangle, \langle 1\rangle,\\ &\quad \langle 8(12)(34)(56)\rangle, \langle 5(12)(34)(78)\rangle, \langle 3\rangle, \langle 2(34)(56)(78)\rangle, \langle 7(12)(34)(56)\rangle,\\ &\quad \langle 5\rangle, \langle 4(12)(56)(78)\rangle\}\end{aligned}$$

Remark 7.1.22 In the physics literature, it is common to write Yangian invariants as products of R-invariants. For example, the Yangian invarian associated to $\mathcal{Y}_7$ can be written as a product of R_{12345} and $R_{678,(123)\cap(45),5}$. However, this might obscure its actual poles. Observing that

$$\langle (123)\cap(45)567\rangle = \langle 1235\rangle\langle 4567\rangle, \tag{7.1.23}$$

and that $\langle 1235\rangle$ is also a pole of R_{12345}, it seems this pole will be double in the product. As we appeal purely to the geometry of the amplituhedron, we will be able to detect only the actual poles of Yangian invariants, e.g. in this case $\langle 1235\rangle$ is spurious due to zeros of the numerator.

All of these Yangian invariants have poles that are polynomial in momentum twistor brackets. These polynomials are all cluster variables of $\text{Gr}_{4,8}$. We verified that cluster variables corresponding to all the poles of these Yangian invariants are cluster compatible, with a single exception. Namely, the Yangian invariant $\mathcal{Y}_9$, which corresponds to the *four-mass box*, contains non compatible poles, i.e. it violates cluster adjacency. We will comment on this in Sect. 7.1.3.

The $\text{Gr}_{4,n}$ cluster algebras are infinite for $n \geq 8$ but recent understanding [11, 13, 35] suggest natural truncations of these in terms of *positive tropical Grassmannians* or their generalisations. For $n = 8$ there have been three such constructions, by considering all tropicalised Plücker coordinates, only a parity-invariant subset thereof, or the parity completion of the set of Plücker coordinates. These, as polytopes, have 274, 260 and 548 vertices, respectively.

One may also wonder if and which of the proposed truncations of the infinite cluster algebras via tropical fans do accommodate the adjacencies we found for 8 points N^{2}MHV yangian invariants. We find that all Yangian invariants, except $\mathcal{Y}_9$, are also cluster-adjacent in the more restrictive sense of tropical fans. In particular, the corresponding g-vectors of their unfrozen poles always form a cone of the tropical fan with 274 vertices, obtained by tropicalising the maximal parity-invariant subset of the Plücker coordinates.

N^{2}MHV$_9$ Yangian Invariants. There are 108 cyclic classes. 10 of them are of $n = 6$ type, 56 are of $n = 7$ type, 38 are of $n = 8$ type, and 4 new types:

$$\text{Facet}(\mathcal{Y}_{10}) = \partial R_{12349} \cup \partial R_{56789} \tag{7.1.24a}$$

$$\text{Facet}(\mathcal{Y}_{11}) = \partial R_{1234,(567)\cap(89)} \cup \partial R_{56789} \tag{7.1.24b}$$

$$\text{Facet}(\mathcal{Y}_{12}) = \partial R_{1234,(56)\cap(789)} \cup \partial R_{56789} \tag{7.1.24c}$$

$$\begin{aligned}\text{Facet}(\mathcal{Y}_{13}) = \{&\langle 5\,6\,\bar{2} \cap \bar{8}\rangle, \langle 4\,6\,\bar{2} \cap \bar{8}\rangle, \langle 4\,5\,\bar{2} \cap \bar{8}\rangle, , \\ &\langle 2\,3\,\bar{5} \cap \bar{8}\rangle, \langle 1\,3\,\bar{5} \cap \bar{8}\rangle, \langle 1\,2\,\bar{5} \cap \bar{8}\rangle, \\ &\langle 8\,9\,\bar{2} \cap \bar{5}\rangle, \langle 7\,9\,\bar{2} \cap \bar{5}\rangle, \langle 7\,8\,\bar{2} \cap \bar{5}\rangle\,\}\end{aligned} \tag{7.1.24d}$$

All 108 of these objects are cluster adjacent in $\text{Gr}_{4,9}$, except the two, which are of the $n = 8$ four-mass box type.

Example 7.1.25 (*The "Spurion"*). We also present the cluster that contains the poles of a Yangian invariant which is particularly interesting, namely $\mathcal{Y}_{14}$ in (7.1.24d). Informally, it is called the *spurion*,[10] since it does not contain any physical pole. Therefore, it can not appear in any of the BCFW representations of the amplitude $\mathcal{A}_{9,2}$. Nevertheless, from the perspective of the amplituhedron, it is a positroid tile and can be part of a tiling, giving a representation of $\mathcal{A}_{9,2}$ not obtainable with standard BCFW. See [30] for an explicit example.

This cluster has the quiver diagram displayed in Fig. 7.1, where we abbreviated the relevant cluster variables as

$$\begin{aligned}
a_1 &= \langle 5\,6\,\bar{2} \cap \bar{8}\rangle & a_2 &= \langle 4\,6\,\bar{2} \cap \bar{8}\rangle & a_3 &= \langle 4\,5\,\bar{2} \cap \bar{8}\rangle \\
a_4 &= \langle 2\,3\,\bar{5} \cap \bar{8}\rangle & a_5 &= \langle 1\,3\,\bar{5} \cap \bar{8}\rangle & a_6 &= \langle 1\,2\,\bar{5} \cap \bar{8}\rangle \\
a_7 &= \langle 8\,9\,\bar{2} \cap \bar{5}\rangle & a_8 &= \langle 7\,9\,\bar{2} \cap \bar{5}\rangle & a_9 &= \langle 7\,8\,\bar{2} \cap \bar{5}\rangle\,,
\end{aligned} \tag{7.1.26}$$

while the remaning irrelevant ones are

[10] We would like to thank Jacob Bourjaily for suggesting this nickname.

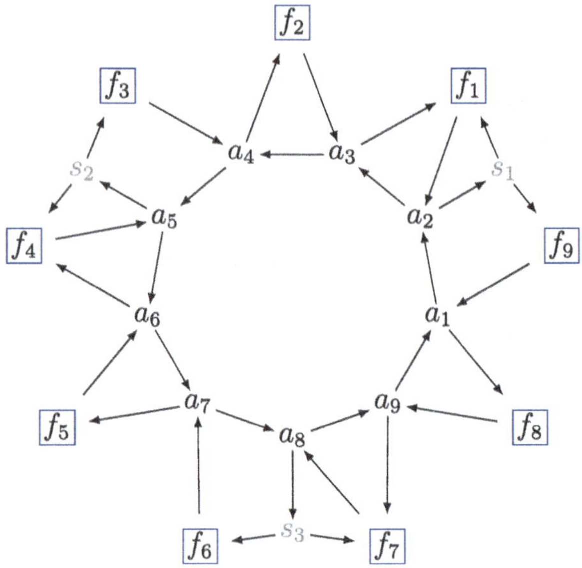

Fig. 7.1 Quiver diagram for a $\mathrm{Gr}_{4,9}$ showing cluster adjacency of the poles of $\mathcal{Y}_{101}$ for $n = 9$

$$s_1 = \langle 1456\rangle \qquad s_2 = \langle 1237\rangle \qquad s_3 = \langle 4789\rangle\,. \tag{7.1.27}$$

These three irrelevant cluster variables can be freely mutated and therefore one may say that this Yangian invariant corresponds to a cube in the cluster polytope. Note also the $\mathbb{Z}_3$ symmetry of the spurion is reflected in this cluster. ◇

N²MHV$_{10}$ Yangian Invariants. For $n = 10$, there are 395 cyclic classes of Yangian invariants. 22 of them are of $n = 6$ type, 168 are of $n = 7$ type, 174 are of $n = 8$ type, 30 are of $n = 9$ type and only 1 is of $n = 10$:

$$\mathrm{Facet}(\mathcal{Y}_{14}) = \partial R_{12345} \cup \partial R_{6789,10}\,. \tag{7.1.28}$$

As observed in [32], there are no new types of Yangian invariants beyond $n = 10$. In Fig. 7.2, we represent a cluster in $\mathrm{Gr}_{4,10}$ cluster algebra which contains all poles of $\mathcal{Y}_{14}$. We observe that all poles of R_{12345} are in the left-most position, whereas all the poles of $R_{6789,10}$ are in the right-most position.

With this, we proved cluster adjacency for *all* N²MHV Yangian Invariants (conjecturally) corresponding to positroid tiles of $\mathcal{A}_{n,k=2,4}$. Moreover, we observe that Yangian invariants of the four-mass box type, which are *not* positroid tiles, do not satisfy cluster adjacency. The corresponding cluster algebras are infinite, and one may wonder how one can check this conclusively. We comment on this in the next section.

Yangian invariants of "Four-mass Box" type. The $n = 8$ Yangian invariant $\mathcal{Y}_9$ corresponds to a four-mass box cut. These types of Yangian invariants fall in the

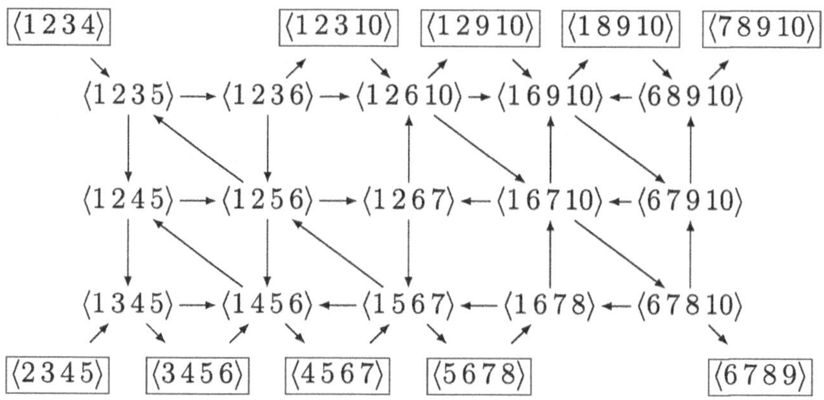

Fig. 7.2 The cluster in $\mathrm{Gr}_{4,10}$ demonstrating the adjacency of the Yangian invariant $\mathcal{Y}_1$ for $n = 10$

category of Yangian invariants with *intersection number* higher than one, and they have always been excluded in cluster-adjacency analysis, e.g. in [10].

In particular, for the case of $\mathcal{Y}_9$, points in the amplituhedron can have 2 preimages[11] in the associated cell S_9 in $\mathrm{Gr}_{k,n}^{\geq 0}$. From an algebraic perspective, this corresponds to the fact that $\mathcal{Y}_9$ can be expressed as the sum of two contributions:

$$\mathcal{Y}_9 = \mathcal{Y}_9^{(1)} + \mathcal{Y}_9^{(2)}, \tag{7.1.29}$$

each of which contains square-roots, however the sum is rational. Moreover, only the actual poles of the sum in Eq. (7.1.29) will corresponds to facets of $\mathcal{Y}_9$. We reported these poles in Eq. (7.1.22) and checked that they are *not* cluster adjacent.[12] The infinite nature of the $\mathrm{Gr}_{4,8}$ cluster algebra might make it difficult to perform exhaustive checks. Nevertheless, we explain below how one proves that this Yangian invariant is not cluster adjacent.

We first find a cluster containing only two of the poles of the Yangian invariant $\mathcal{Y}_9$, e.g.

$$\langle 6(12)(34)(78)\rangle, \qquad \langle 8(12)(34)(56)\rangle\,. \tag{7.1.30}$$

Then, we consider the sub-algebra obtained by freezing these two cluster variables and mutating all the other in the cluster. We can then try to find the other poles of $\mathcal{Y}_9$. The sub-algebra turns out to be infinite. However, after a relatively small number of mutations, one finds that all mutations are exhausted apart from those corresponding to the 1-dimensional infinite sub-affine A_2 sequences. Each mutation in these sequences produces Plücker polynomials of increasing degree, exhausting

[11] It is an open problem to understand when preimages of cells with intersection number higher than one lie inside the corresponding positroid cells.

[12] $\mathcal{Y}_9$, as other Yangian invariants with intersection number higher than one, can be rewritten as a sum of other rational Yangian invariants, each of which we showed to satisfy cluster adjacency.

the possibility of generating any of the poles of Eq. (7.1.22). One might wonder[13] whether these mutations cover all possible clusters containing the two cluster variables (7.1.30) we began with. The answer is positive thanks to [36, Theorem 6.2]. This indeed guarantees that for *any* skew-symmetrisable cluster algebra, the seeds whose clusters contain a given collection of cluster variables form a connected subgraph of the exchange graph of the cluster algebra.

7.2 Cluster Varieties in the Amplituhedron

In this section we advance our understanding on why cluster structures emerge from geometry. We focus on the $m = 2$ amplituhedron, but we expect similar structures to be present for general amplituhedra. In particular, we associate a cluster variety $\mathcal{V}_{\overline{\mathcal{T}}}$ in $\mathrm{Gr}_{k,k+2}(\mathbb{C})$ to each positroid tile $Z_{\hat{G}(\overline{\mathcal{T}})}$ of the $m = 2$ amplituhedron $\mathcal{A}_{n,k,m=2}$. The *seed tori* of $\mathcal{V}_{\overline{\mathcal{T}}}$ are in bijection with triangulated unpunctured plabic tilings $\mathcal{T}$ represented by the unpunctured tiling $\overline{\mathcal{T}}$. Moroever, we show that the positroid tile $Z^{\circ}_{\hat{G}(\overline{\mathcal{T}})}$ is exactly the *totally positive part* of $\mathcal{V}_{\overline{\mathcal{T}}}$.

Fix a (k, n)-unpunctured plabic tiling $\overline{\mathcal{T}}$, with black polygons $P_1, \ldots, P_r$. For each black polygon P_i, fix an arc $h_i \to j_i$ with $h_i < j_i$ in the boundary of P_i. We call this the *distinguished boundary arc* of P_i. We will build $\mathcal{V}_{\overline{\mathcal{T}}}$ by defining seeds in the field of rational functions on $\mathrm{Gr}_{k,k+2}(\mathbb{C})$.

Definition 7.2.1 (*Cluster variables*) Let $a \to b$ with $a < b$ be an arc which is contained in a black polygon P_i and is not the distinguished boundary arc $h_i \to j_i$. We define

$$x_{ab} := \frac{(-1)^{\mathrm{area}(a\to b)}\langle Yab\rangle}{(-1)^{\mathrm{area}(h_i\to j_i)}\langle Yh_i j_i\rangle}.$$

This is a rational function on $\mathrm{Gr}_{k,k+2}(\mathbb{C})$ and is regular away from the hypersurface $\{\langle Yh_i j_i\rangle = 0\}$.

Definition 7.2.2 (*Seed of a tile*) Let $\mathcal{T}$ be a triangulated unpunctured plabic tiling represented by $\overline{\mathcal{T}}$. The *quiver* $Q_{\mathcal{T}}$ is obtained as follows:

- Place a frozen vertex on each non-distinguished boundary arc of $P_1, \ldots, P_r$ and a mutable vertex on every other black arc of $\mathcal{T}$.
- If arcs $a \to b$, $b \to c$, $c \to a$ form a triangle, we put arrows in Q between the corresponding vertices, going clockwise around the triangle.

We label the vertex of $Q_{\mathcal{T}}$ on arc $a \to b$ of $\mathcal{T}$ with the function x_{ab}. The collection of vertex labels is the *(extended) cluster* $\mathbf{x}_{\mathcal{T}}$. The pair $(Q_{\mathcal{T}}, \mathbf{x}_{\mathcal{T}})$ is the *seed* $\Sigma_{\mathcal{T}}$.

Note that there are no frozen variables corresponding to the distinguished boundary arcs. Also note the cluster $\mathbf{x}_{\mathcal{T}}$ has size $2k$. See Fig. 7.3 for an example.

Now we show that each seed gives a *seed torus* in $\mathrm{Gr}_{k,k+2}(\mathbb{C})$.

---

[13] We are grateful to Andrew McLeod for raising this question.

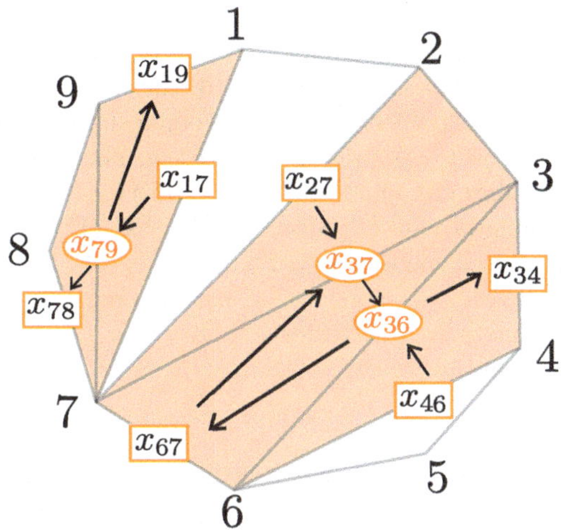

Fig. 7.3 In orange, a triangulated unpunctured plabic tiling $\mathcal{T}$. In black, the seed $\Sigma_{\mathcal{T}}$. The frozen variables are inside rectangles and the mutable variables are encircled in orange. The distinguished boundary arcs are $2 \to 3$ and $8 \to 9$

Proposition 7.2.3 *Let $\mathcal{T}$ be a triangulated unpunctured plabic tiling represented by $\overline{\mathcal{T}}$. Consider the Zariski-open subset*

$$\mathcal{V}_{\mathcal{T}} := \left\{ Y \in \mathrm{Gr}_{k,k+2}(\mathbb{C}) : \prod_{a \to b \text{ black arc of } \mathcal{T}} \langle Yab \rangle \neq 0 \right\}.$$

This is birational to an algebraic torus of dimension $2k$, with field of rational functions $\mathbb{C}(\mathbf{x}_{\mathcal{T}})$, the field of rational functions in the cluster $\mathbf{x}_{\mathcal{T}}$.

Proof The main idea is that Corollary 2.3.40—which gave a bijection between

$$Z^{\circ}_{\hat{G}(\overline{\mathcal{T}})} = \{Y \in \mathrm{Gr}_{k,k+2}(\mathbb{R}) : \text{for all arcs } i \to j \text{ of } \mathcal{T} \text{ with } i < j, (-1)^{\mathrm{area}(i \to j)} \langle Yij \rangle > 0\}$$

and $(\mathbb{R}_{>0})^{2k}$—extends directly to give a birational morphism from $\mathcal{V}_{\mathcal{T}}$ to $(\mathbb{C}^*)^{2k}$. When we let the edge weights α_i, β_i, γ_i (used to define matrix M in (2.3.8)) range over all nonzero complex numbers, the set of $k \times n$ matrices we get sweeps out the *open Deodhar stratum*[14] $D_{\mathcal{T}}$ as opposed to the positroid cell $S_{\hat{G}(\mathcal{T})}$. That is, the stratum $D_{\mathcal{T}} \subset \mathrm{Gr}_{k,n}(\mathbb{C})$ consists of subspaces represented by the matrices $M_{\mathcal{T}}(\boldsymbol{\alpha}, \boldsymbol{\beta}, \boldsymbol{\gamma})$ of (2.3.8), where $(\boldsymbol{\alpha}, \boldsymbol{\beta}, \boldsymbol{\gamma})$ vary over $(\mathbb{C}^*)^{3k}$ rather than $(R_{>0})^{3k}$.

Let us define the map

$$\mathcal{V}_{\mathcal{T}} \to (\mathbb{C}^*)^{2k}$$
$$Y \mapsto \mathbf{x}_{\mathcal{T}}(Y).$$

[14] Parameterizations of Deodhar strata in flag varieties are given in [37]; in the Grassmannian, these can be equivalently parameterized using weighted networks, as shown in [38].

To show that the map is injective, pick $Y, Y' \in \mathcal{V}_{\mathcal{T}}$ which map to the same point. By applying the proof of Theorem 2.3.28 to Y and Y', we obtain matrices C and C' whose entries are twistor coordinates of Y, Y'. These matrices are full rank since they represent points in the open Deodhar stratum $D_{\mathcal{T}}$. So by the proof of Theorem 2.3.28, CZ and $C'Z$ represent Y and Y', respectively. But we can rescale the rows of C so that the entry $\langle Yab \rangle$ becomes $x_{ab}(Y)$ (or 1 if $a \to b$ is a distinguished boundary arc). Similarly with C'. Since $\mathbf{x}_{\mathcal{T}}(Y) = \mathbf{x}_{\mathcal{T}}(Y')$, we have that C, C' represent the same subspace and thus $Y = Y'$.

In the other direction, consider some $2k$-tuple of nonzero complex numbers $\mathbf{q}_{\mathcal{T}}$. Define a weight vector (α, β, γ) for $\hat{G}(\mathcal{T})$, where for a triangle $\{a_i, b_i, c_i\}$, the weights are

$$\alpha_i = q_{b_i c_i}, \quad \beta_i = q_{a_i c_i}, \quad \gamma_i = q_{a_i b_i}.$$

(As usual, if $a \to b$ is a distinguished boundary arc, we take $x_{ab} = 1$.) Let $C := M_{\mathcal{T}}(\alpha, \beta, \gamma)$.

The matrix C lies in the Deodhar stratum $D_{\mathcal{T}}$ and so has full rank. Let $Y := CZ$. Consider an arc $a \to b$ of $\mathcal{T}$ which is in a black polygon P. From the proof of Proposition 2.3.19, we have

$$\langle Yab \rangle = (-1)^{\text{area}(a \to b)} q_{ab} \cdot \mathcal{Q}_P$$

where $\mathcal{Q}_P$ is a polynomial with positive coefficients in the q_{ij}'s and the minors of Z depending only on the polygon P. $\mathcal{Q}_P$ is generically nonzero, in which case it is easy to check that $x_{ab}(Y) = q_{ab}$. Moreover, in this case, Y is a full-rank matrix, since it has at least one non-zero twistor coordinate.

Now, suppose $\mathbf{q}_{\mathcal{T}}$ lies in the open subset of $(\mathbb{C}^*)^{2k}$ where the polynomials $\mathcal{Q}_P$ are nonzero for all polygons P. Then Y, as defined above, lies in $\mathcal{V}_{\mathcal{T}}$ and maps to $\mathbf{q}_{\mathcal{T}}$. □

Next, we verify that the seeds given by different triangulated unpunctured tilings are related by mutation.

Proposition 7.2.4 *Let $\mathcal{T}$ be a triangulated unpunctured tiling represented by $\overline{\mathcal{T}}$ and let $a \to b$ correspond to a mutable vertex of $\Sigma_{\mathcal{T}}$. Let $\mathcal{T}'$ be related to $\mathcal{T}$ by flipping the arc $a \to b$. Then $\Sigma_{\mathcal{T}}$ and $\Sigma_{\mathcal{T}'}$ are related by mutation at x_{ab}.*

The seeds which can be obtained from $\Sigma_{\mathcal{T}}$ by an arbitrary sequence of mutations are exactly the seeds $\Sigma_{\mathcal{T}'}$ where $\mathcal{T}'$ is represented by $\overline{\mathcal{T}}$.

In light of Proposition 7.2.4, we can make the following definition.

Definition 7.2.5 Let $\mathcal{T}$ be a triangulated unpunctured plabic tiling and $\overline{\mathcal{T}}$ the plabic tiling obtained by forgetting the triangulation. We let $\mathcal{A}(\overline{\mathcal{T}})$ denote the cluster algebra $\mathcal{A}(Q_{\mathcal{T}}, \mathbf{x}_{\mathcal{T}})$.

Proof of Proposition 7.2.4 On the level of quivers, the first statement follows immediately from the well-known combinatorics of type A cluster algebras.

Say the arc $a \to b$ is in triangles $\{a < u < b\}$ and $\{a < b < v\}$ in $\mathcal{T}$, so $a \to b$ is flipped to $u \to v$ (the argument is analogous if instead $v < a$). We need to check that, in the field of rational functions on $\mathrm{Gr}_{k,k+2}(\mathbb{C})$, we have

$$x_{ab}x_{uv} = x_{au}x_{bv} + x_{av}x_{ub}$$

(where $x_{h_i j_i}$ is defined to be 1). This follows easily from the 3-term Plücker relations for the corresponding twistor coordinates.

The second statement follows immediately from the combinatorics of flipping diagonals in triangulations. □

Together, Propositions 7.2.3 and 7.2.4 tell us that the union of the seed tori is a cluster variety in $\mathrm{Gr}_{k,k+2}(\mathbb{C})$.

Theorem 7.2.6 *Let $\overline{\mathcal{T}}$ be a (k, n)-unpunctured plabic tiling. Then*

$$\mathcal{V}_{\overline{\mathcal{T}}} := \bigcup_{\mathcal{T}} \mathcal{V}_{\mathcal{T}}$$

is a cluster variety in $\mathrm{Gr}_{k,k+2}(\mathbb{C})$, where the union is over triangulated unpunctured plabic tilings represented by $\overline{\mathcal{T}}$. We call $\mathcal{V}_{\overline{\mathcal{T}}}$ the amplituhedron (cluster) variety[15] *of $Z_{\hat{G}(\mathcal{T})}$.*

Moreover, the positive part

$$\mathcal{V}^{\geq 0}_{\overline{\mathcal{T}}} := \{Y \in \mathcal{V}_{\overline{\mathcal{T}}} : x_{ab}(Y) > 0 \textit{ for all cluster variables } x_{ab}\}$$

is equal to the positroid tile $Z^{\circ}_{\hat{G}(\overline{\mathcal{T}})}$.

Proof The first statement follows directly from the definition of cluster variety and Propositions 7.2.3 and 7.2.4.

For the second statement, note that by Theorem 2.3.28, points of $Z^{\circ}_{\hat{G}(\overline{\mathcal{T}})}$ are in the positive part $\mathcal{V}^{\geq 0}_{\overline{\mathcal{T}}}$. To see the opposite inclusion, take a point Y in the positive part and choose a triangulated unpunctured tiling $\mathcal{T}$ represented by $\overline{\mathcal{T}}$. Run the proof of Theorem 2.3.28 to obtain a matrix C as in Proposition 2.3.19, whose entries are $\langle Yab \rangle$ for arcs $a \to b$ of $\mathcal{T}$. This matrix is full rank, and CZ represents Y. We just need to show that C represents an element of $S_{\hat{G}(\mathcal{T})}$.

If row i of C corresponds to a triangle in $\mathcal{T}$ lying in polygon P_i, rescale row i by $(-1)^{\mathrm{area}(h_i \to j_i)}/\langle Y h_i j_i \rangle$. Call the resulting matrix C'. Because $x_{ab} > 0$ for all arcs $a \to b$ of $\mathcal{T}$, the entry $\langle Yab \rangle$ of C has been rescaled to a real number with sign $(-1)^{\mathrm{area}(a \to b)}$. By the same argument as the last paragraph of Theorem 2.3.28, C' (and thus C) represents an element of $S_{\hat{G}(\mathcal{T})}$. □

Theorem 7.2.7 *The cluster algebra $\mathcal{A}(\overline{\mathcal{T}})$ equals the upper cluster algebra $\overline{\mathcal{A}}(\overline{\mathcal{T}})$. If the unpunctured plabic tiling $\overline{\mathcal{T}}$ consists of polygons $P_1, \ldots, P_r$, where P_i has n_i*

[15] This is closely related to the amplituhedron variety defined in [39].

vertices, then $\mathcal{A}(\overline{\mathcal{T}})$ *is a finite type cluster algebra of Cartan-Killing type* $A_{n_1-2} \times \cdots \times A_{n_r-2}$.

Proof The quiver we are associating to each unpunctured plabic tiling is a disjoint union of the quivers associated to triangulated polygons, or equivalently to $\mathbb{C}[\mathrm{Gr}_{2,n}]$. It is well known that the quiver associated to a triangulated n-gon has Cartan-Killing type A_{n-2} [40]. This implies that $\mathcal{A}(\overline{\mathcal{T}}) = \mathcal{A}(Q_{\mathcal{T}}, \mathbf{x}_{\mathcal{T}})$ has type $A_{n_1-2} \times \cdots \times A_{n_r-2}$.

Because our quivers are just a disjoint union of type A quivers, in particular our cluster algebra has an acyclic seed. Moreover each cluster algebra associated to the n-gon has an exchange matrix with full $\mathbb{Z}$-rank, see [41, Proof of Theorem 5.3.2], so the same is true of $\mathcal{A}(\overline{\mathcal{T}})$.

Using [42, Proposition 1.8 and Remark 1.22], the fact that $\mathcal{A}(\overline{\mathcal{T}})$ has an acyclic seed and also has a full rank exchange matrix implies that the upper cluster algebra $\overline{\mathcal{A}}(\mathcal{T})$ equals the cluster algebra $\mathcal{A}(\overline{\mathcal{T}})$. □

Remark 7.2.8 Given what we've proved, one can make an argument as in the proof of [42, Theorem 2.10] that $\mathcal{A}(\overline{\mathcal{T}})$ is the coordinate ring of the amplituhedron variety $\mathcal{V}_{\overline{\mathcal{T}}}$ and also the closely related variety

$$V_{\overline{\mathcal{T}}} := \{Y \in \mathrm{Gr}_{k,k+2}(\mathbb{C}) \mid \langle Yij \rangle \neq 0 \text{ for } h \to j \text{ a boundary arc of a black polygon of } \overline{\mathcal{T}}.\}.$$

In this section we showed that each positroid tile of the $m = 2$ amplituhedron is the positive part of a cluster variety. The cluster variables associated to the respective cluster algebras are positive ratios of twistor coordinates with definite sign inside the positroid tile. Interestingly, as the twistor coordinates for $\mathcal{A}_{n,k,m=2}$ can be seen as plücker coordinates of $\mathrm{Gr}_{2,n}$ (e.g. via the B-amplituhedron picture, see Proposition 2.2.3), such cluster structure is not the one emerging simply from the standard positive Grassmannian $\mathrm{Gr}_{2,n}^{>0}$. We expect positroid tiles of general amplituhedra $\mathcal{A}_{n,k,m}$ can be regarded as (the positive part of) some cluster varieties, and connected with cluster structures of $\mathrm{Gr}_{m,n}$. See [30] for recent developments. We leave this as a future research direction.

7.3 Leading and Landau Singularities from Amplituhedra

Constructing scattering amplitudes from the knowledge of their singularities, i.e. their poles and branch-cut structure, is an approach with a long history [43], which has proven to be particularly effective for scattering amplitudes in $\mathcal{N} = 4$ super Yang-Mills theory.

Singularities of scattering amplitudes at tree-level are given by multi-particle factorisation channels, which correspond to *Mandelstam invariants*, and are constructed from subsets of the momenta of the particles in the scattering process. Whereas, loop amplitudes exhibit more complicated singularities, leading to logarithmic divergences. In cases where loop amplitudes are expressed as (multiple) polylogarithms, the collection of these logarithmic singularities is called the symbol alphabet. When expressed in terms of momentum twistors, many (all for $n \leq 7$, where n is the number of particles) of these are simply polynomials in the Plücker coordinates in $\mathrm{Gr}_{4,n}$. Moreover, their vanishing loci correspond to special configurations of momentum twistors in $\mathbb{CP}^3$.

On one side, we have seen the emergence of positive geometries [28] as an overarching framework to geometrise scattering amplitudes and their analytic structure, at tree-level and for loop integrands in several theories, among which is $\mathcal{N} = 4$ SYM (cf. the *amplituhedron* in Chap. 2 and the *momentum amplituhedron* in Chap. 6).

On the other, we have witnessed an increasing appearance of cluster algebra structures in scattering amplitudes, especially in capturing singularities of (integrated) loop amplitudes in $\mathcal{N} = 4$ SYM. This started in 2013 with the conjecture made by Golden et al in [6] that the symbol letters of six- and- seven-particle loop amplitudes are cluster variables of the $\mathrm{Gr}_{4,n}$ cluster algebra. A few years later in [8] it was conjectured that these letters satisfy remarkable cluster properties, called *cluster adjacency*. See Sect. 7.1). In terms of the symbol, they dictate which letters can appear consecutively. Moreover, shortly after these adjacencies were also observed at tree-level, in connection with symbol entries [9]. See also [44] for a recent work on the cluster-adjacency of one-loop amplitudes. The guidance of cluster algebras has unlocked the possibility of developing a powerful bootstrap programme which allowed to perform computations that otherwise would have been beyond reach [45–54]. At the same time, they shed more light on the mathematical structures describing singularities of scattering amplitudes and motivate the existence of a possible geometric origin.

One manifestation of the cluster-algebraic phenomena is an observation that building blocks of a BCFW representation of the tree-level amplitude, which are *Yangian invariants*, are *cluster adjacent* [9]. In other words, all poles of each of them are expressed by a collection of cluster variables of the $\mathrm{Gr}_{4,n}$ cluster algebra that can be found together in a common cluster. Moreover, this conjecture was generalised in [10], for all (rational) Yangian invariants of $\mathcal{N} = 4$ SYM. In geometric terms, poles of (rational) Yangian invariants correspond to facets of *positroid tiles* of the amplituhedron. Furthermore, different representations of scattering amplitudes, obtained from identities among Yangian invariants, correspond to different *positroid tilings* of the same geometric space.

One of the first steps towards an amplituhedronic understanding of cluster phenomena was taken in section Sect. 7.1.2, where a toy model for tree-level cluster adjacency of $\mathcal{N} = 4$ SYM was considered. We showed that Yangian invariants of the $m = 2$ amplituhedron are cluster adjacent with respect to the well-known $\mathrm{Gr}_{2,n} \simeq A_{n-3}$ cluster algebra. The $m = 2$ amplituhedron is often considered as a toy model for the physical $m = 4$ case, moreover it also governs the geometry of one-loop MHV integrands [55] and it has some relevance for the NMHV ones as well [56]. By exploiting the geometry of the $m = 2$ amplituhedron, we provided an explicit expression of all Yangian invariants was, where cluster adjacency of their poles is manifest (Conjecture 7.1.14).

The interest in understanding how cluster algebras encode the analytic properties of scattering amplitudes led physicists to explore the connection between cluster algebras and the *positive tropical Grassmannian*, originally introduced in [57]. See for examples [11, 13, 35], for applications in $\mathcal{N}=4$ SYM. Remarkably, the very same positive tropical Grassmannian has been found to regulate the combinatorics of (regular) *tilings* (and, more generally, dissections) of the $m=2$ amplituhedron (see Sect. 5.2.2). This raises the question of whether there is a deeper connection between the latter object and cluster algebras themselves.

A remarkable instance of how geometry encodes singularities of scattering amplitudes in $\mathcal{N}=4$ SYM is the fact that all *Leading Singularities* of the theory, at any loop order, can be computed by a contour integral over the space of k-planes in n dimensions, called *Grassmannian* [58, 59]. Leading Singularities are the singularities of the integrand of a loop amplitude with maximal codimension in loop momenta. The geometrisation was pushed even further via [32] and, a year after, the authors of [60] defined the *loop amplituhedron*, whose boundaries encode singularities of the integrand, among which are the Leading Singularities corresponding to maximal cuts.

The application of this geometric approach to *Landau Singularities* [61–63] is another example of its utility to obtain a better understanding of the structure of singularities of scattering amplitudes. The Landau analysis allows to connect singularities of the *integrand*, described geometrically from boundaries of loop amplituhedra, to the ones of the *integrated* amplitudes. Among all Landau singularities, there are in general many spurious ones coming from summing over Feynman diagrams. On the other hand, the amplituhedron can tell which are the true singularities of the integrand, and therefore select the true Landau singularities of the loop amplitude.

In Sect. 7.4, using an amplituhedron-based approach, we will discover new cluster structures involving both Landau Singularities and Leading Singularities of $\mathcal{N}=4$ SYM. In this section we review the main ingredients needed for this purpose. In Sect. 7.3.1 we recall the definition of the *loop amplituhedron* and in Sect. 7.3.2 we introduce the concept of *Leading Singularities* and how one can obtain them from special boundaries of the loop amplituhedron. In Section Sect. 7.3.3 we present the definition of *Landau singularities* and how the loop amplituhedron can select the non-spurious ones. For both Leading and Landau singularities we present in the respective sections examples at one-loop which will be relevant for our work.

7.3.1 Loop Amplituhedra

In this section, we introduce the *loop amplituhedron* $\mathcal{A}_{n,k,m;\ell^L}$ [60, 64] which—for $m=4$ and $\ell=2$—is the geometry encoding the integrands of loop amplitudes of planar $\mathcal{N}=4$ SYM [65]. In analogy with the tree-level amplituhedron $\mathcal{A}_{n,k,m}$ (cf. Chap. 2), $\mathcal{A}_{n,k,m;\ell^L}$ is the image under a map induced by a totally positive matrix Z. The domain of the map is a generalization of the *positive Grassmannian* $\mathrm{Gr}_{k,n}^{\geq 0}$ called *L-loop*[16] *positive Grassmannian* $\mathrm{Gr}_{k,n;\ell^L}^{\geq 0}$.

[16] This is not the standard *loop Grassmannian* in the math literature [66].

Definition 7.3.1 (*L-Loop Grassmannian*) Fix $k \leq n$, L positive integers, and let $\underline{k} = (k_1, \ldots, k_L)$ be a vector of positive integers with $k_1 + \ldots + k_L \leq n - k$. Then we define the *L-loop Grassmannian* $\mathrm{Gr}_{k,n;\underline{k}}$ to be the set of points $V = \{V_S\}_{S \subseteq [L]}$ which are collections of linear subspaces V_S of dimension $k + \sum_{s \in S} k_s$, and such that $V_S \subset V_{S'}$ if $S \subset S'$, with $S, S' \subseteq [L]$.

Definition 7.3.2 (*The positive L-loop Grassmannian*) Fix $k, n, L, \underline{k}$ as in Definition 7.3.1. Let us consider a matrix C in $\mathrm{Mat}_{k,n}$ and the matrices $D^{(1)}, \ldots, D^{(L)}$, where $D^{(\ell)}$ in $\mathrm{Mat}_{k_\ell,n}$, $\ell \in [L]$. Given $S \subseteq [L]$, let us denote as P_S the matrix obtained by staking C with the matrices $(D^{(s)})_{s \in S}$, i.e. P_S is in $\mathrm{Mat}_{k+k_S,n}$, with $k_S = \sum_{s \in S} k_s$. We say that $P = P_{[L]}$ is *positive* if for each $S \subseteq [L]$, P_S is a totally positive matrix, i.e. P_S is in $\mathrm{Mat}^{>0}_{k+k_S,n}$. Then we define the *(totally) positive L-loop Grassmannian* $\mathrm{Gr}^{>0}_{k,n;\underline{k}}$ to be the set of points $V = \{V_S\}_{S \subseteq [L]}$ in $\mathrm{Gr}_{k,n;\underline{k}}$ for which there exists a matrix P as above such that for each $S \subseteq [L]$, V_S is the row-span of P_S. We say that the matrix P represents the element V in $\mathrm{Gr}^{>0}_{k,n;\underline{k}}$. Finally, we define the *(totally) non-negative L-loop Grassmannian* $\mathrm{Gr}^{\geq 0}_{k,n;\underline{k}}$ to be the closure of $\mathrm{Gr}^{>0}_{k,n;\underline{k}}$ in $\mathrm{Gr}_{k,n;\underline{k}}(\mathbb{R})$.

Definition 7.3.3 (*The L-loop Amplituhedron*) Choose positive integers $k < n$, and m, ℓ, L such that $\ell \leq k + m \leq n$ and $\ell L \leq n - k$, and let $Z \in \mathrm{Mat}^{>0}_{n,k+m}$ thought as a linear map $Z : \mathbb{R}^n \to \mathbb{R}^{m+k}$. Let ℓ^L denote $\underline{k} = (\ell, \ldots, \ell)$. Then Z induces a map $\tilde{Z} : \mathrm{Gr}^{\geq 0}_{k,n;\ell^L} \to \mathrm{Gr}_{k,k+m;\ell^L}$ defined by

$$\tilde{Z}(\{V_S\}_{S \subseteq [L]}) := \{Z(V_S)\}_{S \subseteq [L]}.$$

If P is a matrix representing an element of $\mathrm{Gr}^{\geq 0}_{k,n;\ell^L}$, then $\tilde{Z}(P)$ is defined to be the element of $\mathrm{Gr}_{k,k+m;\ell^L}$ represented by the matrix PZ. The *L-loop amplituhedron* $\mathcal{A}_{n,k,m;\ell^L}(Z)$ is defined to be the image $\tilde{Z}(\mathrm{Gr}^{\geq 0}_{k,n;\ell^L})$ inside $\mathrm{Gr}_{k,k+m;\ell^L}$.

If P is obtained by stacking the matrices $(C, D^{(1)}, \ldots, D^{(L)})$ as in Definition 7.3.2, then PZ is $(CZ, D^{(1)}Z, \ldots, D^{(L)}Z)$, which we denote as $(Y, \mathcal{L}^{(1)}, \ldots, \mathcal{L}^{(L)})$ or $(Y, \mathcal{L})$, with $\mathcal{L} = (\mathcal{L}^{(1)}, \ldots, \mathcal{L}^{(L)})$ referred to as the *loop momenta*. Moreover, when $m = 4$ and $\ell = 2$, we denote $\mathcal{A}_{n,k,4;2^L}(Z)$ as $\mathcal{A}^{(L)}_{n,k}$—whose canonical form encodes the integrand of the L-loop $\mathrm{N}^k\mathrm{MHV}_n$ scattering amplitudes of planar $\mathcal{N} = 4$ SYM. We note that when $L = 0$ the loop amplituhedron $\mathcal{A}_{n,k,4;2^0}(Z)$ equals the tree amplituhedron $\mathcal{A}_{n,k,4}(Z)$.

As for the tree-level amplituhedron $\mathcal{A}_{n,k,m}$, there is a conjectural characterization of the loop amplituhedron $\mathcal{A}^{(L)}_{n,k}$ in terms signs of its twistor coordinates [67]. Given a point $(Y, \mathcal{L}^{(1)}, \ldots, \mathcal{L}^{(L)})$ in $\mathcal{A}^{(L)}_{n,k}$, we will consider *twistor coordinates* as the $(k+4) \times (k+4)$ determinants of the type $\langle Y, i, j, \ell, s \rangle$, $\langle Y \mathcal{L}^{(\ell)} i, j \rangle$, $\langle Y \mathcal{L}^{(\ell_1)} \mathcal{L}^{(\ell_2)} \rangle$, obtained by stacking Y with some $\mathcal{L}^{(\ell)}$, $\ell \in [L]$ and rows Z_i, $i \in [n]$ of the matrix Z.

Conjecture 7.3.4 (Sign Characterization for $\mathcal{A}^{(L)}_{n,k}$) *Let k, n, L and Z as in Definition 7.3.3. Let $\mathcal{J}^{\circ\,(L)}_{n,k}(Z)$ be the set of points $(Y, \mathcal{L})$ in $Gr_{k,k+m;2^\ell}$ such that*

$$(1) \qquad \langle Y, i, i+1, j, j+1 \rangle > 0 \tag{7.3.5}$$

$$\textit{and } \operatorname{var}(\langle Y1234\rangle, \langle Y1235\rangle, \ldots \langle Y123n\rangle) = k. \tag{7.3.6}$$

$$(2) \quad \langle Y\mathcal{L}^{(\ell)}, i, i+1\rangle > 0, \ \langle Y\mathcal{L}^{(\ell)}\mathcal{L}^{(s)}\rangle > 0, \tag{7.3.7}$$

$$\textit{and } \operatorname{var}(\langle Y\mathcal{L}^{(\ell)}12\rangle, \langle Y\mathcal{L}^{(\ell)}13\rangle, \ldots \langle Y\mathcal{L}^{(\ell)}1\,n\rangle) = k+2. \tag{7.3.8}$$

Then $\mathcal{A}_{n,k}^{(L)}(Z) = \overline{\mathcal{J}_{n,k}^{\circ(L)}(Z)}$.

Remark 7.3.9 Understanding the $m = 2$ amplituhedron is also relevant for planar $\mathcal{N} = 4$ SYM amplitudes for the following reasons. We observe that if $(Y, \mathcal{L}) \in \mathcal{A}_{n,k}^{(L)}(Z)$, then

$$Y \in \mathcal{A}_{n,k,4}(Z), \ (Y, \mathcal{L}^{(\ell)}) \in \mathcal{A}_{n,k+2,2}(Z), \quad \ell \in [L]. \tag{7.3.10}$$

In particular, the one-loop MHV integrand (i.e. $L = 1$ and $k = 0$) of $\mathcal{N} = 4$ SYM is encoded in the amplituhedron for $m = 2$ and $k = 2$:

$$\mathcal{A}_{n,0}^{(1)}(Z) = \mathcal{A}_{n,2,2}(Z). \tag{7.3.11}$$

As explained in Sect. 2.2.1, one can go from twistor coordinates to physical momentum twistors in $\mathrm{Gr}_{4,n}$. In the following, with abuse of notation, we will sometime denote both twistor coordinates $\langle Y a\, b\, c\, d\rangle$ and brackets $\langle a\, b\, c\, d\rangle$ in momentum twistors as $\langle a\, b\, c\, d\rangle$ and it will be clear from the context.

7.3.2 Leading Singularities from $\mathcal{A}_{n,k}^{(L)}$

We review here the concept of *Leading Singularities*. In particular, we show how Leading Singularities for $\mathcal{N} = 4$ SYM can be computed more geometrically via a Grassmannian approach and via the loop amplituhedron.

Leading Singularities. The concept of leading singularities was originally introduced within the *Analytic Bootstrap Programme* in the 1960s [43]. Since the beginning of this century, with the advent of novel on-shell techniques such as *generalised unitarity*, the concept of Leading Singularities has been broadly employed and exploited in computation of scattering amplitudes, in particular in Yang-Mills [68].

Loop amplitudes in planar $\mathcal{N} = 4$ SYM are computed from *integrands*, which are rational functions of external kinematics and loop momenta, by integration over particular real-contours in the $4L$-dimensional loop momentum space. However, in general this contour is known not to preserve the symmetries of the theory and leads, for example, to IR-divergences. In this regards, it might seem natural to choose complex contours corresponding to computing residues of the integrand. Leading singularities are then the residues of the integrand computed around tori encircling

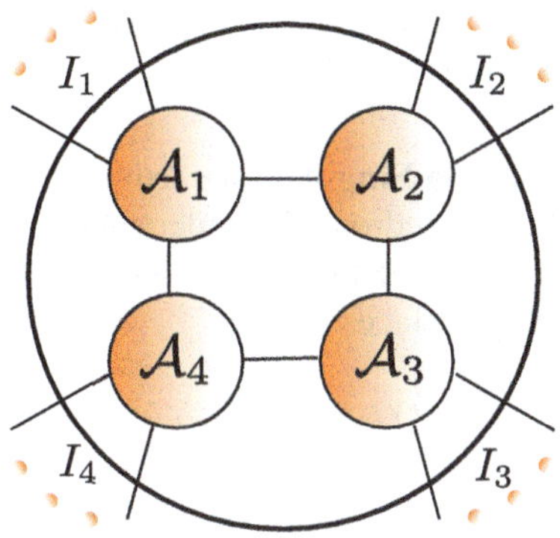

Fig. 7.4 Depiction of a general Leading Singularity $\mathcal{A}_1(I_1) \otimes \mathcal{A}_2(I_2) \otimes \mathcal{A}_3(I_3) \otimes \mathcal{A}_4(I_4)$ at one loop

the loci where a maximal set of internal propagators (e.g. four for one-loop) go on-shell.

Given a 1-loop n-points scattering amplitude $\mathcal{A}(1, \ldots, n)$ and a partition of $\{1, \ldots, n\}$ into 4 disjoint subsets $I_1, \ldots, I_4$, then the *Leading Singularity* of the amplitude is defined as

$$\int \prod_{a=1}^{4} d^4\eta_a \, d^4\ell_a \, \delta(\ell_a^2) \prod_{a=1}^{4} \mathcal{A}_a \left(\{\ell_a, \eta_a\}, I_a, \{-\ell_{a+1}, \eta_{a+1}\}\right) \qquad (7.3.12)$$

where the index a is mod 4, the integral over ℓ is localised over the solutions of the delta function and the integral over the Grassmann coordinates η_a (cf. See Sect. 1.3.3) amounts to sum over all possible internal states flowing between the different sub-amplitudes $\{\mathcal{A}_1, \ldots, \mathcal{A}_4\}$, see Fig. 7.4. Since the four internal propagators are forced to vanish by the delta function, the internal particles can be taken on-shell. Therefore, leading singularities are in general simply the products of tree-amplitudes, summed over all the internal particles which can be exchanged, and integrated over the on-shell phase space of each.

Leading Singularities from the Grassmannian. In [58], leading singularities were proposed as the complete set of IR-finite quantities that contains all the information needed to compute the S-Matrix of planar $\mathcal{N} = 4$ SYM. Beautifully, both in momentum space and in momentum twistor space [59], all the n-points leading singularities of the theory, at any loop order, can be computed by a contour integral over the Grassmannian $\mathrm{Gr}_{k,n}$. Here $k \leq n$ is the helicity sector of the amplitude in the tree-level case. Remarkably, in [32], it was shown that only the positive Grassmannian $\mathrm{Gr}_{k,n}^{\geq 0}$ is relevant for scattering amplitudes. In particular, the integration contour providing Leading Singularities is performed on some of its positroid cells (cf. Sect. 2.1.1). Positroid cells are in bijection with many nice combinatorial objects, including equivalence classes of reduced plabic graphs (cf. Appendix A), also known as *on-shell diagrams* in the context of scattering amplitudes. A comprehensive summary about on-shell diagrams in the physical context can be found in [32]. Formulae for 1-loop

Leading Singularities for planar $\mathcal{N} = 4$ SYM in momentum twistor variables are reported in [69] using on-shell diagrams, and will be used in an example in Sect. 7.3.2.

Leading Singularities from $\mathcal{A}_{n,k}^{(L)}$. Let us now consider the boundaries of the loop amplituhedron and understand how they are related to Leading Singularities. It is conjectured that the boundaries[17] of the tree amplituhedron lie on the following hypersurfaces:

$$\langle Y i_1\, i_1+1\, j_1\, j_1+1\rangle = 0, \ldots, \langle Y i_d\, i_d+1\, j_d\, j_d+1\rangle = 0 \tag{7.3.14}$$

for some $d > 0$ and all indices (considered cyclically) in $\{1, \ldots, n\}$. In order to make connection with Leading Singularities, we will not focus on this tree-level type of boundaries. Instead, we will consider boundaries whose points $(Y, \mathcal{L}^{(1)}, \ldots, \mathcal{L}^{(L)})$ satisfy *on-shell conditions*, which we define to be a subset of the following conditions:

$$\langle Y \mathcal{L}^{(l_a)}\, i_a\, i_a+1\rangle = 0,\ \langle Y \mathcal{L}^{(s_1)} \mathcal{L}^{(s_2)}\rangle = 0, \tag{7.3.15}$$

for $l_a, s_1, s_2 \in \{1, \ldots, L\}$ and $i_a \in \{1, \ldots, n\}$, and Y does not lie on any of the tree-level type boundaries given by the equation in (7.3.14). Each set $\mathcal{C}$ of on-shell conditions has a certain number of boundaries lying on the respective hypersurfaces. Following the terminology of [63], boundaries of the type (7.3.15) are called *$\mathcal{L}$-boundaries*. Boundaries associated to the same on-shell conditions $\mathcal{C}$ are called *branches*. Loop momenta in a specific branch will be denoted as $\{\mathcal{L}_a^*\}_{\mathcal{C}}$, with $a \in [L]$. If it exists, we denote as $\mathcal{B}[\mathcal{C}, \mathcal{L}^*]$ the boundaries of the loop amplituhedron, which are $\mathcal{L}$-boundaries determined by the set of on-shell conditions $\mathcal{C}$ and are in the branch corresponding to the solution $\mathcal{L}^*$. In [63], it has been showed that, once we fix $\mathcal{C}$ and $\mathcal{L}^*$, there exists a minimum[18] k_{min} such that the loop amplituhedron $\mathcal{A}_{n,k}^{(L)}$ has the boundaries $\mathcal{B}[\mathcal{C}, \mathcal{L}^*]$ for all $k \geq k_{min}$.

Finally, we will focus on the $\mathcal{L}$-boundaries that are relevant for Leading Singularities, which correspond to *maximal cuts*. If $\mathcal{C}$ is a set of on-shell conditions, then $\mathcal{C}$ is a *maximal-cut* if it is maximal by inclusion, i.e. we can not add more on-shell conditions to $\mathcal{C}$ with Y not being on tree-level type boundaries of Eq. (7.3.14). In particular, an $\mathcal{L}$-boundary associated to a maximal-cut has codimension $4L$ and the solutions in each branch have loop momenta localised in points $\{\mathcal{L}^*\}$.

For a maximal cut $\mathcal{C}$, boundaries $\mathcal{B}[\mathcal{C}, \mathcal{L}^*]$ of the loop amplituhedron correspond to Leading Singularities of the L-loop amplitude $\mathcal{A}_{\mathrm{N}^k\mathrm{MHV}_n}^{(L)}$. In particular, as one can extract tree-level amplitudes $\mathcal{A}_{\mathrm{N}^k\mathrm{MHV}_n}$ from the canonical form of the tree amplituhedron, one can extract the Leading Singularities $\mathrm{LeS}[\mathcal{C}, \mathcal{L}^*]$ from the canonical form of the codimension-$4L$ boundaries $\mathcal{B}[\mathcal{C}, \mathcal{L}^*]$ of the loop amplituhedron .

---

[17] Abusing notation, in this section we will use 'boundaries' to refer to *faces* defined in Definition 2.1.8.

[18] Using *parity*, which is a symmetry of scattering amplitudes and of the amplituhedron, one can also establish an upper bound as: $k \leq n - \bar{k}_{min} - 4$, where $\bar{k}_{min}$ is the minimal value of k for which the parity-conjugated branch appears.

It is known that all Leading Singularities of an amplitude $\mathcal{A}^{(L)}_{\mathrm{N}^k\mathrm{MHV}_n}$ can be expressed as a sum of n-particles N^kMHV Yangian invariants, i.e., for a certain Leading Singularity, LeS there is a collection of $4k$-dimensional cells $\{S_a\}$ in $\mathrm{Gr}^{\geq 0}_{k,n}$, such that

$$\mathrm{LeS} = \sum_a \mathcal{Y}_{S_a}. \tag{7.3.16}$$

This is just a rephrasing of the conjecture that the Grassmannian integral representation of scattering amplitudes provides Leading Singularities if integrated over proper contours, such as the one[19] provided by the above collection of cells $\{S_a\}$. The sum in (7.3.16) is the geometrical equivalent of 'tiling' the boundary of the loop amplituhedron, corresponding to the Leading Singularity. As different representations of a scattering amplitude (tree-level or loop integrand) are just different ways to tile the amplituhedron (tree or loop), different representations of a Leading Singularity LeS as sum of Yangian invariants correspond to different tilings of the corresponding boundary of the loop amplituhedron.

We will now exploit the geometric definition of the loop amplituhedron to compute *all* Yangian invariants which can be part of a tiling of a given boundary of the loop amplituhedron, i.e. all Yangian invariants which can be used to express a given Leading Singularity. Moreover, we will see how this connects to the Landau analysis in the next section.

Yangian Invariants from the loop amplituhedron. For an L-loop, n-point $\mathrm{N}^k\mathrm{MHV}$ amplitude, let $\mathcal{C}$ be a maximal-cut and $\mathcal{L}^*$ one branch of its solutions. Then we can directly use the definition of the loop amplituhedron to determine whether a given Yangian invariant $\mathcal{Y}$ can be used to express the corresponding Leading Singularity $\mathrm{LeS}[\mathcal{C}, \mathcal{L}^*]$. In geometric terms this means, that the image of corresponding positroid cell S is inside the codimension-$4L$ boundary $\mathcal{B}[\mathcal{C}, \mathcal{L}^*]$ of the loop amplituhedron $\mathcal{A}^{(L)}_{n,k}$.

As mentioned in Eq. (7.3.39), on the maximal-cut $\mathcal{C}$ and on the branch of solutions $\mathcal{L}^*$, the loop momentum twistors are localised in terms of twistors of external kinematic:

$$\mathcal{L}^{(l)^*} = D^{(l)^*}(\langle z_{i_1} z_{i_2} z_{i_3} z_{i_4}\rangle) \cdot z, \quad l = 1, \ldots, L, \tag{7.3.17}$$

where $D^{(l)^*}$ are $2 \times n$ matrices depending on Pluckers of the twistors of external kinematics and z is the $n \times 4$ matrix whose rows are z_i.

On the amplituhedron side, on the boundary $\mathcal{B}[\mathcal{C}, \mathcal{L}^*]$, the loop momentum twistors are localised as:

$$\mathcal{L}^{(l)^*} = D^{(l)^*}(\langle Y Z_{i_1} Z_{i_2} Z_{i_3} Z_{i_4}\rangle) \cdot Z, \quad l = 1, \ldots, L, \tag{7.3.18}$$

---

[19] With suited orientation of each cell.

where $D^{(l)^*}$ are the same as in Eq. (7.3.17), however their dependence on $\langle z_{i_1} z_{i_2} z_{i_3} z_{i_4}\rangle$ has been uplifted in the ampliutedron into a dependence on twistor coordinates $\langle Y Z_{i_1} Z_{i_2} Z_{i_3} Z_{i_4}\rangle$.

Let S be a $4k$-dimensional cell in $\mathrm{Gr}_{k,n}^{\geq 0}$ with full-dimensional image Z_S via the $\tilde{Z}$ map. Then Z_S belongs to the boundary $\mathcal{B}[\mathcal{C}, \mathcal{L}^*]$ if the positivity conditions in Definition 7.3.2 are satisfied, i.e.

$$\begin{pmatrix} C \\ D^{(i_1)^*}|_{Y=C\cdot Z} \\ \vdots \\ D^{(i_s)}|_{Y=C\cdot Z}{}^* \end{pmatrix} \tag{7.3.19}$$

are totally positive matrices, with $0 \leq s \leq L$, for all representative matrices C in the cell S. In (7.3.19), we denoted $D^{(i_a)^*}|_{Y=C\cdot Z}$ as the matrix which depends on $\langle Y Z_{i_1} Z_{i_2} Z_{i_3} Z_{i_4}\rangle$, with Y in the image of the cell S, i.e. $Y = C \cdot Z$. As in previous sections, in order to handle positroid cells in the positive Grassmannian, we use the Mathematica package `positroids.m`.

Using this procedure, by selecting the $4k$-dimensional cells in $\mathrm{Gr}_{k,n}^{\geq 0}$ that have full-dimensional image, we get a list $\{S_i\}_i$ such that their images $\{Z_{S_i}\}_i$ under the loop amplituhedron map are all in the boundary $\mathcal{B}[\mathcal{C}, \mathcal{L}^*]$. Finally, this means that we obtain the list of Yangian invariants $\{\mathcal{Y}_{S_i}\}_i$ which can appear as summands to represent the Leading Singularity $\mathrm{LeS}[\mathcal{C}, \mathcal{L}^*]$.

Leading Singularities at one loop. In this section, we will provide an illustrative example on how to compute Leading Singularities from the Grassmannian for one-loop NMHV, following [69] (in particular, see Table 3). We will consider only some maximal-cuts which will be relevant for our analysis. We will briefly comment on the N^{2}MHV case.

Given a cut $\mathcal{C} = \{I_1, \ldots, I_4\}$ for the loop amplitude $\mathcal{A}^{(L=1)}_{\mathrm{N}^k\mathrm{MHV}_n}$, on-shell diagrams with tree sub-amplitudes $\mathcal{A}_{n_1,k_1}(I_1) \otimes \ldots \otimes \mathcal{A}_{n_4,k_4}(I_4)$ (see Fig. 7.5), such that:

$$\sum_{a=1}^{4} k_a = k - 2, \quad \sum_{a=1}^{4} n_a = n + 8, \tag{7.3.20}$$

correspond to Leading Singularities of $\mathcal{A}^{(1)}_{\mathrm{N}^k\mathrm{MHV}_n}$. Here we denoted sub-amplitudes as $\mathcal{A}_{n',k'}(I')$, where k', with $0 \leq k' \leq n-4$, is its $\mathrm{N}^{k'}$MHV helicity sector,[20] n' the number of legs, and I' denotes the indices the external particles contained.

Example 7.3.21 (*Leading Singularities: NMHV @ 1 loop*). Let us now list the types of Leading Singularities which can appear at NMHV at one-loop. By Eq. (7.3.20) we must have:

[20] For $n = 3$, we also admit $k' = -1$, which corresponds to $\overline{\mathrm{MHV}}$, i.e. a *white* vertex. Moreover, note that $\mathcal{A}_0(\ldots) = 1$, since we are in the momentum twistor space.

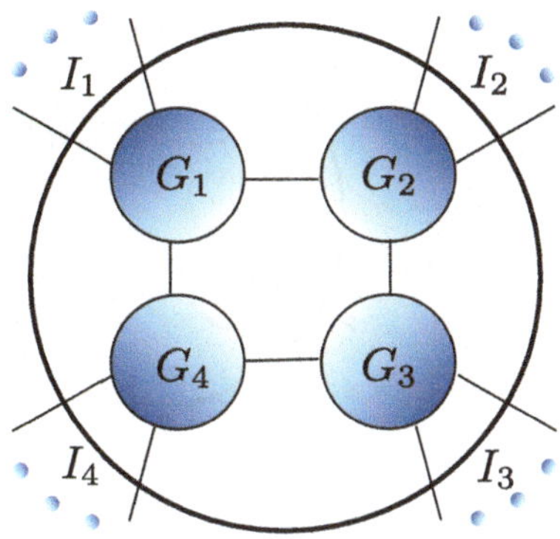

Fig. 7.5 General form of a plabic graph $G_1(I_1) \otimes G_2(I_2) \otimes G_3(I_3) \otimes G_4(I_4)$ corresponding to a Leading Singularity at one loop

$$k_1 + k_2 + k_3 + k_4 = 1 - 2 = -1 \tag{7.3.22}$$

Since we can have $k_a = -1$ only when one of the sub-amplitude is a 3 point amplitude, otherwise k_a are positive, then we must have at least a 3-point subamplitude to satisfy Eq. (7.3.20). Given a subamplitude $\mathcal{A}_{n',k'}(I')$, in the following we will omit the dependence of the sub-amplitudes on n' and we will use ... for some or all indices in I'. They can be easily inferred from the context. All indices will be cyclically ordered $i < i+1 < j < j+1 < k < k+1$. We refer to Fig. 7.6.

1. The *Two-mass easy box* C^E_{ij} is a maximal cut with the following on-shell conditions:

$$\langle \mathcal{L}\, i-1, i\rangle = \langle \mathcal{L}\, i, i+1\rangle = \langle \mathcal{L}\, j-1, j\rangle = \langle \mathcal{L}\, j, j+1\rangle = 0. \tag{7.3.23}$$

There are two possible on-shell diagrams contributing to this cut, whose Leading Singularities are:

$$\text{LeS}\left[\mathcal{A}_{-1}(i) \otimes \mathcal{A}_0(\ldots) \otimes \mathcal{A}_{-1}(j) \otimes \mathcal{A}_1(\ldots)\right] = \mathcal{A}_{NMHV}(j, \ldots, i), \tag{7.3.24}$$
$$\text{LeS}\left[\mathcal{A}_{-1}(i) \otimes \mathcal{A}_1(\ldots) \otimes \mathcal{A}_{-1}(j) \otimes \mathcal{A}_0(\ldots)\right] = \mathcal{A}_{NMHV}(i, \ldots, j). \tag{7.3.25}$$

2. The *two-mass hard box* C^H_{ij} is a maximal cut with the following on-shell conditions:

$$\langle \mathcal{L}\, i-1, i\rangle = \langle \mathcal{L}\, i, i+1\rangle = \langle \mathcal{L}\, i+1, i+2\rangle = \langle \mathcal{L}\, j, j+1\rangle = 0. \tag{7.3.26}$$

There are two possible on-shell diagrams contributing to this cut, whose Leading Singularities are:

$$\text{LeS}\left[\mathcal{A}_{-1}(i) \otimes \mathcal{A}_0(i+1) \otimes \mathcal{A}_0(\ldots, j) \otimes \mathcal{A}_0(\ldots)\right] = R_{i,i+1,i+2,j,j+1}, \tag{7.3.27}$$
$$\text{LeS}\left[\mathcal{A}_0(i) \otimes \mathcal{A}_{-1}(i+1) \otimes \mathcal{A}_0(\ldots, j) \otimes \mathcal{A}_0(\ldots)\right] = R_{i-1,i,i+1,j,j+1}. \tag{7.3.28}$$

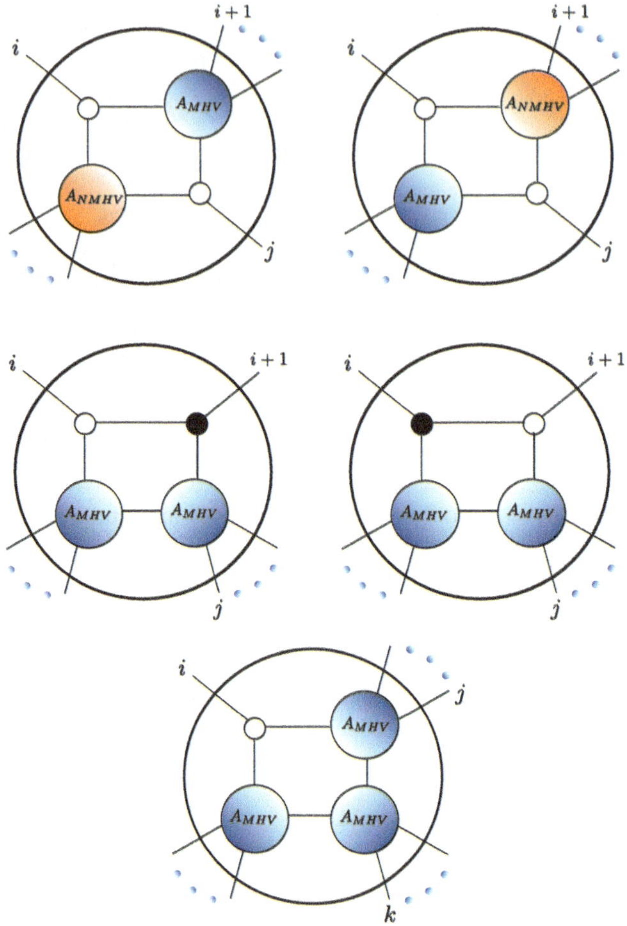

Fig. 7.6 Plabic graph corresponding to Leading Singularities of the following maximal cuts: 2-mass easy box $\mathcal{C}_{ij}^{E}$ (first row); 2-mass hard box $\mathcal{C}_{ij}^{H}$ (second row); 3-mass box $\mathcal{C}_{ijk}$ (third row)

3. The *three-mass box* $\mathcal{C}_{ijk}$ is a maximal cut with the following on-shell conditions:

$$\langle \mathcal{L}\, i-1, i\rangle = \langle \mathcal{L}\, i, i+1\rangle = \langle \mathcal{L}\, j, j+1\rangle = \langle \mathcal{L}\, k, k+1\rangle = 0. \qquad (7.3.29)$$

There is only one on-shell diagrams contributing to this cut, whose Leading Singularity is:

$$\mathrm{LeS}\left[\mathcal{A}_{-1}(i) \otimes \mathcal{A}_0(\ldots, j) \otimes \mathcal{A}_0(\ldots, k) \otimes \mathcal{A}_0(\ldots)\right] = R_{i,j,j+1,k,k+1}. \qquad (7.3.30)$$

◇

Example 7.3.31 (*Leading Singularities: N^{2}MHV @ 1 loop*) For N^{2}MHV at one-loop, we have all cuts of the type appearing at NMHV, and in addition the *four-mass box cut* appears from 8 points. This is associated to Leading Singularities which contains non-rational Yangian invariants, and, by Landau analysis, to algebraic singularities of the loop amplitude. We leave these cases for explorations in future works.

N^{2}MHV Leading Singularities are in general expressed as:

$$R_I \cdot R_J, \;\; R_I \cdot \mathcal{A}_{NMHV}(J), \;\; \varphi R_I \cdot R_J \tag{7.3.32}$$

where I, J are lists of twistors (in general, expressed as intersection of lines or planes defined from z_i), R_I are R-invariants with twistors in the list I, and φ is an extra function, not relevant for our purposes. Nevertheless, as discussed in (7.3.16) all of them are just combinations of N^{2}MHV Yangian invariants. For our purposes, we are not interested in representations of Leading Singularities like (7.3.32). We would like the full list $\{\mathcal{Y}_a\}$ of Yangian invariants which can be used to express a given Leading singularity. We saw that such a list can be obtained directly from the geometry of the loop amplituhedron. ◇

7.3.3 Landau Singularities from $\mathcal{A}_{n,k}^{(L)}$

We will briefly review how the Landau analysis can be used to infer singularities of the *integral*, from the poles of the *integrand*. First, we will review the original definition in terms of Feynman diagrams and then following [63] we will review the role amplituhedron plays in this analysis.

Landau Singularities. The concept of Landau singularities was originally introduced in 1959, when Landau stated a set of equations, called *Landau equations*, whose solutions parametrises the locus in the space of kinematic data, where a given Feynman integral can have branch points [70].

Given a Feynman integral I contributing to an L loop scattering amplitude in D spacetime dimensions, we can always bring it to the following form by using Feynman parametrisation:

$$I = c \int \prod_{a=1}^{L} d^D \ell_a \int_{\Delta_{\nu-1}} \frac{\mathcal{N}(\{\ell_a\}, \{p_r\})}{\mathcal{D}^\nu(\alpha; \{q_i\})}, \quad \mathcal{D} = \sum_{i=1}^{\nu} \alpha_i (q_i^2 - m_i^2), \tag{7.3.33}$$

where c is just constant which does not enter our analysis, the integration is performed over the simplex $\Delta_{\nu-1}$, i.e. $\alpha_1 + \ldots + \alpha_\nu = 1$ and $\alpha_i \geq 0$, q_i is the momentum flowing along the corresponding propagator i, $\{p_r\}$ are the momenta of external particles, and $\mathcal{N}$ is a function of the kinematic data. It is known that the physical amplitude from (7.3.33) is obtained by performing the integral over a particular contour defined by the $i\epsilon$ prescription in the propagators. However, in order to understand the analytic continuation outside the physical sheet in the space of kinematic, one has to study arbitrary contours.

The Landau analysis establishes that the integral I in (7.3.33) *can* develop singularities if the following conditions are satisfied:

$$\sum_{i \in \text{loop}} \alpha_i q_i = 0, \quad \text{for all loops} \quad \text{and} \quad \alpha_i (q_i^2 - m_i^2) = 0, \ \forall i. \tag{7.3.34}$$

In order to capture the analytic structure of I away from the physical sheet, one allows solutions of the following equations with α_i and ℓ_a away from the physical contour as well. When some of the α_i are different than zero, the second second case in Eq. (7.3.34) corresponds to putting some internal propagators on-shell, and these will be related to 'cuts'. In the following we will be interested only when solutions exist on codimension-one subspaces of the external kinematic space, i.e. when they are parametrised by the vanishing locus of a certain function of external kinematic.

We notice that the power of this method seems to be affected by two major inconveniences. Firstly, this analysis does not know about the numerator $\mathcal{N}$ in Eq. (7.3.33), which can cancel some singularities of the denominator. Secondly, even when the numerator does not affect the singularities, singularities of individual Feynman integrals can cancel in the summation to get the full amplitude. In summary, the Landau analysis, even if predicts all potential singularities of the amplitude, in general it predicts many 'spurious' singularities as well, which are not actual singularities of the amplitude. In [61], it was suggested that one can circumvent these issues by directly appealing to the geometry of the amplituhedron.

There are branches in mathematics—such as the ones studying *parametric integrals* and *singularity theory*—with more homological approaches to understand singularities of (Feynman) integrals. Relevant references[21] are [71–74].

Landau Singularities from the loop amplituhedron. Given a Landau singularity corresponding to setting to zero a certain number of internal propagators, i.e. a *cut*, this is an actual singularity of the amplitude if the cut corresponds to a boundary of the loop amplituhedron.

In order to make the connection with the amplituhedron more explicitly, as shown in [63], one can re-write the Landau equation in momentum twistors. If a cut C is a collection of constraints of the type:

$$f_j(\mathcal{L}, z) = 0, \tag{7.3.35}$$

where $\mathcal{L}$ collectively denotes the momentum twistors associated to loops $\mathcal{L}^{(1)}, \ldots, \mathcal{L}^{(L)}$, and $\{z\}$ are momentum twistors encoding the kinematic data of external particles. Then the Landau equations for this set of on-shell constraints include the above equations together with a set of equations of the type:

[21] We thank Erik Panzer and Marko Berghoff for pointing us at some of the literature.

$$\sum_{j=1}^{d} \alpha_j \frac{\partial f_j(\mathcal{L}(\beta), z)}{\partial \beta_s} = 0, \quad s = 1, \ldots, 4L, \tag{7.3.36}$$

where the β's are $4L$ coordinates used to parametrise $\mathcal{L}^{(1)}, \ldots, \mathcal{L}^{(L)}$. This latter equations are often referred to as *Kirchhoff conditions*. We observe that the Landau equations are $d + 4L$ equations in $d + 4L - 1$ variables (since we can always rescale all the α's in the Kirchhoff equations). Therefore, one might expect that they do not admit solutions for general kinematics. For the purpose of this analysis, one is then interested in knowing what the codimension-one loci in kinematic space of z's are, for which Landau equations admit solutions (with α's not all zero). Let such loci be the vanishing set of the function

$$\text{LaS}[\mathcal{C}, \mathcal{L}^*](z) = \prod_{t=1}^{N} a_t(z) = 0, \tag{7.3.37}$$

where $\mathcal{C}$ is the cut associated to the Landau Equations (7.3.35), $\mathcal{L}^*$ is one branch of solutions of the on-shell conditions we are considering, and $a_t(z)$ are certain irreducible polynomials of Plücker coordinates of z. In the following, we will refer to $\text{LaS}[\mathcal{C}, \mathcal{L}^*]$ as the *Landau singularity* associated to the cut $\mathcal{C}$ in the branch $\mathcal{L}^*$. With a slight abuse of terminology, we will also refer to $a_1, \ldots, a_N$ as corresponding Landau singularities.

Finally, a given Landau singularity $\text{LaS}[\mathcal{C}, \mathcal{L}^*]$ is a true singularity of the amplitude $\mathcal{A}_{n,k}^{(L)}$ if the loop amplituhedron has a boundary of the type $\mathcal{B}[\mathcal{C}, \mathcal{L}^*]$ [61].

In summary, on one hand, the Landau analysis can connect the geometry of boundaries of the amplituhedron to the location of singularities of integrated amplitudes. On the other, the amplituhedron can tell which are the true singularities of the integrand, and therefore select the true Landau singularities, among the spurious ones coming from summing over Feynman diagrams.

Example 7.3.38 (*Landau Singularities at 1 loop*) We report below the Landau singularities corresponding to some maximal cuts that will be relevant for our analysis. These can be found in [63], Table 1. We also report the points where the loop momenta localise on different cuts. In particular, for a maximal-cut $\mathcal{C}$, there are 2 solutions $\mathcal{L}_1^*, \mathcal{L}_2^*$ (parity conjugate to each other), each of which can be expressed in term of momentum twistors of external kinematic as:

$$\mathcal{L}_a^* = D[\mathcal{C}, \mathcal{L}_a^*] \cdot z, \tag{7.3.39}$$

where $D[\mathcal{C}, \mathcal{L}_a^*]$ is a $2 \times n$ matrix depending on Plückers of momentum twistors of the external kinematics $\{z_i\}_{i \in [n]}$ and z is the $n \times 4$ matrix whose rows are z_i. In the following, only the non-zero columns of D will be displayed explicitly. Moreover, we consider cyclically ordered indices $i < i+1 < j < j+1 < k < k+1$.

1. The *two-mass hard box cut* $\mathcal{C}_{ij}^{E}$ in Eq. (7.3.23) has in general 2 solutions:

$$\mathcal{L}_1^* = (ij), \quad \mathcal{L}_2^* = \bar{i} \cap \bar{j}, \tag{7.3.40}$$

the first is valid for $0 \leq k \leq n-6$ and the second for $2 \leq k \leq n-4$. The corresponding matrices are[22]

$$D[C_{ij}^E, \mathcal{L}_1^*] = \begin{pmatrix} \overset{i}{1} & \overset{j}{0} \\ 0 & 1 \end{pmatrix}, \; D[C_{ij}^E, \mathcal{L}_2^*] = \begin{pmatrix} \overset{i-1}{\langle i\bar{j}\rangle} & \overset{i}{-\langle i-1, \bar{j}\rangle} & \overset{i+1}{0} \\ 0 & -\langle i+1, \bar{j}\rangle & \langle i\bar{j}\rangle \end{pmatrix}. \tag{7.3.41}$$

For this cut and both of the branches,[23] we have the following Landau singularities:

$$\mathrm{LaS}[C_{ij}^E, \mathcal{L}_1^*](z) = \mathrm{LaS}[C_{ij}^E, \mathcal{L}_2^*](z) = \langle i\bar{j}\rangle\langle \bar{i} j\rangle. \tag{7.3.42}$$

2. The *two-mass easy box cut* C_{ij}^H in Eq. (7.3.26) has in general 2 solutions:

$$\mathcal{L}_1^* = \overline{i+1} \cap (ijj+1), \quad \mathcal{L}_2^* = \bar{i} \cap (i+1, jj+1) \tag{7.3.43}$$

They are both valid for $1 \leq k \leq n-5$. The corresponding matrices are:

$$D[C_{ij}^H, \mathcal{L}_1^*] = \begin{pmatrix} \overset{i}{1} & \overset{i+1}{0} & \overset{i+2}{0} \\ 0 & -\langle i, i+2, j, j+1\rangle & \langle i, i+1, j, j+1\rangle \end{pmatrix} \tag{7.3.44}$$

$$D[C_{ij}^H, \mathcal{L}_2^*] = \begin{pmatrix} \overset{i-1}{0} & \overset{i}{0} & \overset{i+1}{1} \\ -\langle i, i+1, j, j+1\rangle & \langle i-1, i+1, j, j+1\rangle & 0 \end{pmatrix} \tag{7.3.45}$$

For this cut we have the following Landau singularities:

$$\mathrm{LaS}[C_{ij}^H, \mathcal{L}_1^*](z) = \mathrm{LaS}[C_{ij}^H, \mathcal{L}_2^*](z) = \langle i, i+1, j, j+1\rangle. \tag{7.3.46}$$

3. The *three-mass easy box* C_{ijk} in Eq. (7.3.29) has in general 2 solutions:

$$\mathcal{L}_1^* = (ijj+1) \cap (ikk+1), \quad \mathcal{L}_2^* = (\bar{i} \cap (jj+1), \bar{i} \cap (kk+1)). \tag{7.3.47}$$

The first is valid for $1 \leq k \leq n-6$ and the second for $2 \leq k \leq n-5$. The corresponding matrices are:

---

[22] They are determined up to $GL(2)$ (and up to adding rows of C, see Definition 7.3.3).

[23] In general, we can have different Landau singularities for different branches of the same cut. However, this does not happen at one loop [63].

$$D[\mathcal{C}_{ijk}, \mathcal{L}_1^*] = \begin{pmatrix} \overset{i}{1} & \overset{i+1}{0} & \overset{j}{0} \\ 0 & \langle i, j, k, k+1\rangle & -\langle i, j+1, k, k+1\rangle \end{pmatrix} \tag{7.3.48}$$

$$D[\mathcal{C}_{ijk}, \mathcal{L}_2^*] = \begin{pmatrix} \overset{j}{-\langle \bar{i}, j+1\rangle} & \overset{j+1}{\langle \bar{i}, j\rangle} & \overset{k}{0} & \overset{k+1}{0} \\ 0 & 0 & -\langle \bar{i}, k+1\rangle & \langle \bar{i}, k\rangle \end{pmatrix}. \tag{7.3.49}$$

For this cut, we have the following Landau singularities:

$$\mathrm{LaS}[\mathcal{C}_{ijk}, \mathcal{L}_1^*](z) = \mathrm{LaS}[\mathcal{C}_{ijk}, \mathcal{L}_2^*](z) = \langle i(i-1, i+1)(j, j+1)(k, k+1)\rangle\,, \tag{7.3.50}$$

We recall our notation for twistor coordinates and momentum twistor brackets from Sect. 7.1.3. and Eq. 7.1.22. ◇

7.4 Clusters, Leading and Landau Singularities

In this section we present a conjecture about a new cluster structure which connects Leading Singularities (see Sect. 7.3.2) and Landau Singularities (see Sect. 7.3.3): *cluster adjacency between Leading and Landau singularities*, which we abbreviate as 'LL-cluster adjacency'. In Sect. 7.4.1 we formulate this conjecture and then in Sect. 7.4.2, for one loop, we present the proof for NMHV amplitudes any points and checks for N^2MHV amplitudes up to 9 points. In Sect. 7.4.3 we show the one-loop NMHV 7-point amplitude in a representation which is uniquely fixed by LL-cluster adjacency. Finally, in Sect. 7.4.4 we end with summary and directions for future works.

7.4.1 LL-Cluster Adjacency Conjecture

In this section, we enhance the tree-level cluster adjacency of Yangian invariants explored in Sect. 7.1.3, to include information of loop-level singularities, i.e. Landau singularities. In particular, we provide evidence that cluster variables corresponding to poles of a Yangian invariant in a given cut, and cluster variables of the corresponding Landau singularity can be found together in a cluster.

Cluster adjacency seems to know about compatibility between different singularities, or equivalently, between boundaries. We have seen at tree-level how the collection of poles of a Yangian invariant (or, equivalently, of their boundaries) corresponds to cluster variables in a common cluster. One can naturally extend this compatibility, thinking of a Yangian invariant as being located 'inside' a given Leading Singularity. Algebraically, this means that it can be used as an addend to express the Leading Singularity. Geometrically, this means that the image of the cell corresponding to the Yangian invariant is contained in the codimension-$4L$ boundary of the loop amplituhedron associated to the maximal cut of the Leading Singularity, as explained in Sect. 7.3.2. Given a Landau singularity which corresponds to branch points of the amplitude, we can list the maximal cuts of the integrand which via the Landau analysis are responsible for these singularities. The Leading Singularities of these maximal cuts will be then expressed in terms of Yangian invariants with certain poles. Cluster algebras seem to tell us that we can find the Landau singularity *and* all the poles of a given Yangian as above in a common cluster. Let us state our conjecture more explicitly.

Conjecture 7.4.1 (LL-Cluster Adjacency) *Let $\mathcal{C}$ be a maximal cut of an L-loop n-point N^{k}MHV amplitude and $\mathcal{L}^*$ a branch of its solutions. Moreover, let* $\mathrm{LaS}[\mathcal{C}, \mathcal{L}^*]$ *and* $\mathrm{LeS}[\mathcal{C}, \mathcal{L}^*]$ *be the corresponding Landau singularity and Leading Singularity, as defined in Sects. 7.3.3 and 7.3.2, respectively, expressed as:*

$$\mathrm{LaS}[\mathcal{C}, \mathcal{L}^*](z) = \prod_{t=1}^{N} a_t(z), \quad \mathrm{LeS}[\mathcal{C}, \mathcal{L}^*] = \sum_i \mathcal{Y}_i, \tag{7.4.2}$$

where $\{a_1(z), \ldots, a_N(z)\}$ and the poles of the n-points N^{k}MHV rational Yangian invariants $\mathcal{Y}_i$ are cluster variables of the $\mathrm{Gr}_{4,n}$ *cluster algebra. Then the set of poles of $\mathcal{Y}_i$ together with $a_1, \ldots, a_N$ is a collection of compatible cluster variables for* $\mathrm{Gr}_{4,n}$.

Remark 7.4.3 This refers to any *rational* Yangian invariant $\mathcal{Y}$ (see Sect. 7.1.3) that can be used to represent the given Leading Singularity LeS. Geometrically, this is a statement about the facets of the positroid tiles which can be used to tile the $4L$-codimenisional face of $\mathcal{A}_{n,k}^{(L)}$ corresponding to LeS.

We will refer to the cluster adjacency predicted by these conjectures as the *LL-Cluster Adjacency* (i.e. Leading and Landau singularities Cluster Adjacency).

7.4.2 LL-Cluster Adjacency for One-Loop Amplitudes

In this section we prove LL-cluster adjacency Conjecture 7.4.1 for one-loop amplitudes—at NMHV for any number n of scattering particles, and for N^{2}MHV up to $n = 9$. We leave the proof for higher loops or n to future work.

LL-Cluster Adjacency for one-loop NMHV Amplitudes. Let us consider the case of one-loop n-points NMHV amplitudes and state the expected LL-cluster adjacencies by matching Yangian invariants in representations of a Leading Singularity with the corresponding Landau singularities associated to the same maximal-cut. We note that our studies focus on the non-trivial cases when Landau singularities are not only product of frozen variables in $\mathrm{Gr}_{4,n}$, which are the ones presented in Example 7.3.38. We will refer to Example 7.3.21 for the corresponding Leading Singularities.

The Landau singularity for the two easy-mass box cut $\mathcal{C}^E_{ij}$ is given by the product of the cluster variables $\langle i\,\bar{j}\rangle$ and $\langle \bar{i}\,j\rangle$. Whereas the Leading Singularities for the cut $\mathcal{C}^E_{ij}$ are $\mathcal{A}_{\text{NMHV}}(i,\ldots,j)$ and $\mathcal{A}_{\text{NMHV}}(j,\ldots,i)$. It is straightforward to see that the R-invariants which can appear in a representation of these Leading Singularities are just the ones of the type R_I, with I a 5-element subset of the set $\{j,\ldots,i\}$ or of $\{i,\ldots,j\}$. The two hard-mass box cut $\mathcal{C}^H_{ij}$ has a Landau singularity which is just the cluster variable $\langle i, i+1, j, j+1\rangle$, whereas its Leading Singularities are $R_{i-1,i,i+1,j,j+1}$ and $R_{i,i+1,i+2,j,j+1}$. Since $\langle i, i+1, j, j+1\rangle$ is already a pole of both the latter two R-invariants, LL-cluster adjacency is trivially satisfied in this case. Finally, the three-mass box $\mathcal{C}_{ijk}$ has an associated Landau singularity which is $\langle i(i-1\,i+1)(jj+1)(kk+1)\rangle$ and the Leading Singularity is $R_{i,j,j+1,k,k+1}$.

In summary, LL-cluster adjacency for all points one-loop NMHV reads as follows:

Proposition 7.4.4 (LL-Adjacency for one-loop NMHV_n Amplitudes) *The following are collections of compatible cluster variables for* $\mathrm{Gr}_{4,n}$*:*

$$(1)\quad \{\partial R_I, \langle i\,\bar{j}\rangle, \langle \bar{i}\,j\rangle\},\quad I\in\binom{[j,i]}{5},\binom{[i,j]}{5} \tag{7.4.5}$$

$$(2)\quad \{\partial R_{i,j,j+1,k,k+1}, \langle i(i-1\,i+1)(jj+1)(kk+1)\rangle\}, \tag{7.4.6}$$

where by ∂R_J we denoted the list of all poles of the R-invariant R_J, i.e. $\langle M\rangle$, $M\in\binom{J}{4}$, and $i<i+1<j<j+1<k<k+1$ are cyclically ordered indices in $\{1,\ldots,n\}$.

Proof **Proof of (1)**. When I is a five-element subset of $[i,j]$ or $[j,i]$, it is straightforward to check that the poles of R_I and $\langle i\,\bar{j}\rangle$, $\langle \bar{i}\,j\rangle$ form a collection of weakly separated Plücker coordinates of $\mathrm{Gr}_{4,n}$, which are therefore compatible cluster variables for $\mathrm{Gr}_{4,n}$.

Proof of (2). Without loss of generality, we can fix $j=1$ and assume $k+1<i-1$. All other cases are related to this by cyclic symmetry.

We will show this by explicitly constructing such a cluster, closely following [9] where the cluster-adjacency of any R-invariant was proved based on partial rotations.

We first find a cluster that contains the poles of R_{12467} and $\langle 4(23)(12)(67)\rangle$ in the $\mathrm{Gr}_{4,7}$ cluster algebra. This cluster can be obtained after a sequence of mutations, which we shall denote by Σ_0. The resulting cluster has the quiver diagram displayed in Fig. 7.8.

The cluster we aim to find is just a relabelling of the cluster above, and this can be achieved through partial cyclic rotations. In particular we need to find a sequence of rotations that maps the labels of $(1,2,3,4,5,6)$ to $(k,k+1,i-1,i,i+1,1,2)$. These rotations are

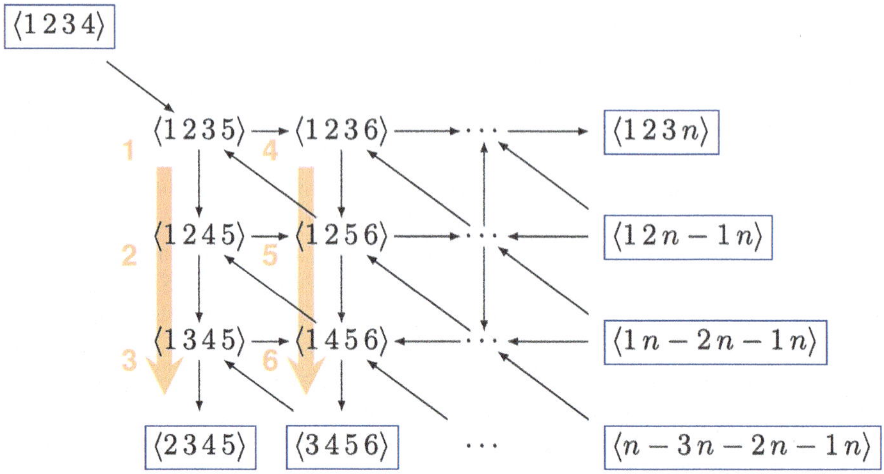

Fig. 7.7 Initial cluster for $\text{Gr}_{4,n}$ and a numbering of its nodes to encode mutation sequences

$$\begin{aligned}(1, 2, 3, 4, 5, 6, 7) &\xrightarrow{-2\big|_{k+4}} (k+3, k+4, 1, 2, 3, 4, 5) \\ &\xrightarrow{-3\big|_{i+1}} (k, k+1, i-1, i, i+1, 1, 2)\,, \end{aligned} \tag{7.4.7}$$

where $r\big|_m$ denotes r rotations in the $\text{Gr}_{4,m}$ algebra. In [9] it was explained how to find a mutation sequence that realises such a transformation, and we denote this sequence by Σ_m^r. For negative r, it is understood that the mutation sequence is applied in reverse.

If we then apply the mutation sequence Σ_0 to the appropriately relabelled cluster. In other words, if we mutate the $\text{Gr}_{4,n}$ initial cluster into the sequence

$$\Sigma_{i+1}^{-3}\,\Sigma_{k+4}^{-2}\,\Sigma_0, \tag{7.4.8}$$

we obtain a cluster which contains all the poles of the R-invariant $R_{1\,2\,k\,k+1\,i}$ as well as the letter $\langle i(i-1\,i+1)(12)(k\,k+1)\rangle$. □

In the following we report examples of LL-cluster adjacencies up to 9-points. In order to shorten the notation, let us denote as $(i_1, \ldots, i_{n-5})$ the R-invariant $R_{[n]/\{i_1,\ldots,i_{n-5}\}}$. In order to check the cluster adjacency properties, we look for clusters containing these poles by performing a sequence of mutations from the initial cluster in $\text{Gr}_{4,n}$. We enumerate the active nodes of the initial cluster, starting from ⟨1235⟩ going downwards and continuing in the second column starting with ⟨1236⟩, numbered 4 (see Fig. 7.7). We note that the mutation sequence that relates two clusters is not unique.

Example 7.4.9 (*LL-Cluster Adjacency: NMHV*$_7$ @ 1 *loop*). Up to cyclic shift, for $n = 7$ there only 3 types of R-invariants: (12), (13), (14). The adjacencies between Landau and Leading Singularities reads:

$$\begin{aligned} &(12) \text{ is compatible with } \langle 3\bar{7}\rangle, \langle \bar{3}7\rangle \\ &(13) \text{ is compatible with } \langle 2(13)(45)(67)\rangle . \end{aligned}$$

Both adjacencies are manifested in the cluster polytope by the presence of a subpolytope made out of clusters that contain all poles corresponding to mutable cluster variables of the Yangian invariants and the Landau singularities. The R-invariant (12) has three poles corresponding to mutable cluster variables, and together with the Landau singularity, these give to four mutable variables. The remaining two degrees of freedom correspond to a pentagonal face of the cluster polytope, i.e. an A_2 subalgebra.

The pentagons that correspond to $\{(12), \langle 1367\rangle\}$ and $\{(12), \langle 2347\rangle\}$ share an edge, i.e. the subpolytope of two cluster that contain the poles of (12) as well as both parity conjugate Landau singularities. This will be a recurring feature for higher n: when the Landau analysis predicts the product of two parity conjugate Plücker coordinates, there is a cluster that contains both of them as well as the Yangian invariant. We will omit the parity conjugate singularity to keep notation short.

The R-invariant (13) has 4 poles corresponding to mutable cluster variables, and together with the Landau singularity $\langle 2(13)(45)(67)\rangle$, this adjacency corresponds to a line segment. ◇

Example 7.4.10 (*LL-Cluster Adjacency: NMHV*$_8$ @ 1 *loop*). There are 7 cyclically inequivalent R-invariants in this case, and we find that all associated Landau singularities are cluster adjacent. In Table 7.1 we provide the checks for the cases that are not implied by the $n = 7$ case. We also omit the cases where the Landau singularity is a pole of the NMHV Yangian invariant.

For example, the cluster-adjacency statements that

$$(123) \text{ is compatible with } \langle 4\bar{8}\rangle, \langle \bar{4}8\rangle \tag{7.4.11}$$

$$(124) \text{ is compatible with } \langle 3(24)(56)(78)\rangle \tag{7.4.12}$$

follow from the adjacency of (12) to $\langle 3\bar{7}\rangle$ and the ajacency of (13) to $\langle 2(13)(45)(67)\rangle$.

◇

Example 7.4.13 (*LL-Cluster Adjacency: NMHV*$_9$ @ 1 *loop*). There are 14 R-invariants up to cyclic symmetry. The adjacencies we need to check along with the verifications are listed in Table 7.2. We note that cluster adjacencies between Landau singularities and Leading Singularities at $n = 7$ and $n = 8$ are embedded in $n = 9$, as we should expect. ◇

LL-Cluster Adjacency for one-loop N^{2}MHV Amplitudes. We now use results from Sect. 7.1.3, in particular the lists of poles of (rational) N^{2}MHV Yangian invariants expressed as cluster coordinates, to test LL-Cluster Adjacency at one-loop for

Table 7.1 Checks of LL-cluster adjacency for NMHV$_8$. The mutation sequences describe the mutations needed to get the given cluster starting from the initial cluster (see Fig. 7.7 for labelling conventions). The sub-algebras are obtained from $\mathrm{Gr}_{4,n}$ by freezing the set of poles in the R-invariant and respective Landau singularity

R-Inv.	Landau sing.	Mutation sequence	Subalgebra
(123)	$\langle 3\bar{8}\rangle\langle\bar{3}8\rangle$	(1,5,9,1,2,8,1,2,5,7,1,2,4,2,3,1,5,6)	D_5
	$\langle 4\bar{1}\rangle\langle 1\bar{4}\rangle$	(1,4,7,5,9,1,5,8,1,4,5)	D_4
(124)	$\langle 3\bar{8}\rangle\langle 8\bar{3}\rangle$	(2,6,5,8,9,4,5,4,7,3,1,2,4)	A_3
(125)	$\langle 3\bar{8}\rangle\langle 8\bar{3}\rangle$	(2,5,8,6,9,1,2,4,5,4,7,3,1,4)	$A_1 \times A_1$
(126)	$\langle 3\bar{8}\rangle\langle\bar{3}8\rangle$	(2,5,8,2,4,6,2,4,5,9,1,2,1,7,1,3)	$A_1 \times A_1$
(127)	$\langle 3\bar{8}\rangle\langle\bar{3}8\rangle$	(2,5,8,5,7,9,6,1,2,5,1,2,4,1,2,3)	A_3
	$\langle 8(71)(34)(56)\rangle$	(7,8,9,2,6,1,2,5,4,7,1,2)	D_4
(147)	$\langle 8(71)(23)(56)\rangle$	(2,3,6,7,8,9,2,5,4,7)	$A_1 \times A_1 \times A_1$

Table 7.2 Checks of LL-cluster adjacency for NMHV$_9$

R-Inv.	Landau sing.	Mutation sequence	Subalgebra
(1234)	$\langle 1\bar{4}\rangle\langle 4\bar{1}\rangle$	(4,7,10)	E_7
(1235)	$\langle\bar{1}4\rangle\langle 1\bar{4}\rangle$	(2,1,4,4,10)	E_6
	$\langle 3\bar{9}\rangle\langle 9\bar{3}\rangle$	(2,1,4,4,10,11,12)	E_7
(1236)	$\langle 4\bar{1}\rangle\langle 1\bar{4}\rangle$	(3,6,9,10,11,12,1,2,1,3)	E_7
	$\langle 3\bar{9}\rangle\langle 9\bar{3}\rangle$	(3,6,9,10,11,12,1,6,1)	E_6
(1237)	$\langle 4\bar{1}\rangle\langle 1\bar{4}\rangle$	(7,1,2,1,3,6)	E_6
	$\langle 3\bar{9}\rangle\langle 9\bar{3}\rangle$	(7,1,3,1)	E_6
(1238)	$\langle 4\bar{1}\rangle\langle 1\bar{4}\rangle$	(1,2,3,5,10)	$\tilde{D}_5$
	$\langle 3\bar{9}\rangle\langle 9\bar{3}\rangle$	(1,2,3,5,11,7,10,12)	E_6
	$\langle 4\bar{9}\rangle\langle 9\bar{4}\rangle$	(1,2,3,5,7,10)	D_6
	$\langle 9(81)(45)(67)\rangle$	(1,2,3,5,6,5)	$\tilde{E}_6$
(1278)	$\langle 9\bar{6}\rangle\langle 6\bar{9}\rangle$	(5,1,2,4,7,8,5,10,2,5,2,7,8,10)	E_6
	$\langle 3\bar{9}\rangle\langle 9\bar{3}\rangle$	(5,1,2,4,7,8,5,10,2,8)	E_6
	$\langle 9(81)(34)(56)\rangle$	(5,1,2,4,7,8,5,10,2,5)	E_7
(1246)	$\langle 3\bar{9}\rangle\langle 9\bar{3}\rangle$	(2,5,7,8,1,3,6)	$A_2 \times A_2$
(1247)	$\langle 3\bar{9}\rangle\langle 9\bar{3}\rangle$	(3,4,6,8,9,1,5,1)	$A_3 \times A_2 \times A_1$
	$\langle 3(24)(56)(89)\rangle$	(3,4,6,8,9,1,6)	$D_6 \times A_1$
(1248)	$\langle 3\bar{9}\rangle\langle 9\bar{3}\rangle$	(2,6,7,4,5,8,9,12,2,5,2,6,7,9)	$D_6 \times A_1$
(1267)	$\langle 8\bar{5}\rangle\langle 5\bar{8}\rangle$	(2,5,6,1,3,1,2)	$A_4 \times A_1$
(1257)	$\langle 3\bar{9}\rangle\langle 9\bar{3}\rangle$	(3,6,7,8,9,1,6,1)	$A_3 \times A_2 \times A_2$
	$\langle 6(57)(89)(34)\rangle$	(1,4,5,3,6,3)	$A_5 \times A_1$
(1258)	$\langle 3\bar{9}\rangle\langle 9\bar{3}\rangle$	(3,4,5,6,1)	$A_3 \times A_1 \times A_1$
(1268)	$\langle 3\bar{9}\rangle\langle 9\bar{3}\rangle$	(2,5,8,7,10,1,3)	A_5

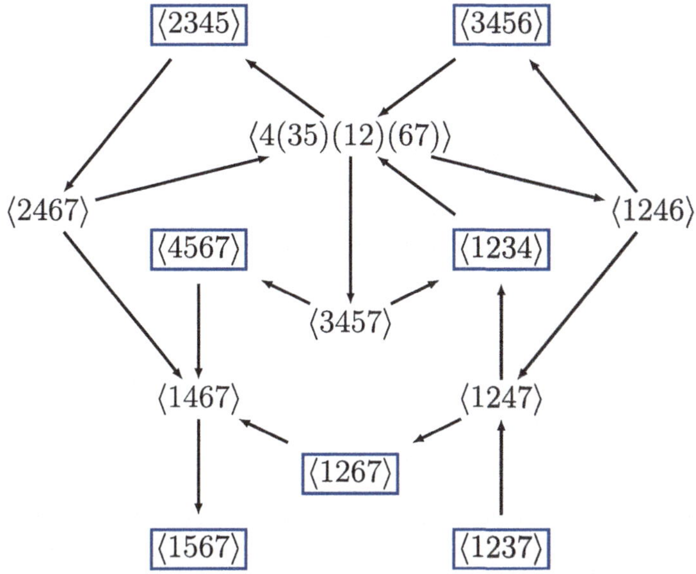

Fig. 7.8 The quiver diagram of a $\mathrm{Gr}_{4,7}$ cluster containing the poles of the R-invariant R_{12467} and the cluster variable $\langle 4(23)(12)(67)\rangle$. This cluster is enough to prove the LL-cluster adjacency for all one-loop NMHV amplitudes

$\mathrm{N}^2\mathrm{MHV}$ amplitudes. Moreover, for a given cut, we use positivity as in Eq. (7.3.19) to get the full list of $\mathrm{N}^2\mathrm{MHV}$ Yangian invariants which appear in a representation of the corresponding Leading Singularity. For example, if we consider the two-mass hard box cut $\mathcal{C}^H_{15}$ and the branch as in Eq. (7.3.45), then for 8 points we find a list of six $\mathrm{N}^2\mathrm{MHV}$ Yangian invariants, whereas for $n = 9$ and the same cut and branch we find there are 21. Each of these $\mathrm{N}^2\mathrm{MHV}$ Yangian invariants has poles which can be found in the same cluster together with the Landau singularity of the corresponding cut in Eq. (7.3.46), i.e. $\langle 1256\rangle$.

Example 7.4.14 (*LL-Cluster Adjacency: N^2MHV_7 @ 1 loop*) This case is related to $n = 7$ NMHV case discussed above through parity conjugation. Up to cyclic shift, for $n = 7$ there only 3 types of Yangian invariants, which are parity conjugates of (12), (13), (14) respectively. Their poles have been explicitly presented in Sect. 7.1.3.

For these Yangian invariants, we only have the following associated Landau singularity that is not implied by a case worked out earlier:

Y-Inv.	Landau sing.	Mutation sequence	Subalgebra
$\mathcal{Y}_a$	$\langle 3\bar{6}\rangle, \langle 6\bar{3}\rangle$	(3)	A_1
$\mathcal{Y}_b$	$\langle 4(12)(35)(67)\rangle$	(6,5,4,1,2)	A_1

where the poles of $\mathcal{Y}_a$ are:

$$\text{Facet}(\mathcal{Y}_a) = \{\langle 2\bar{6}\rangle, \langle 2367\rangle, \langle \bar{3}7\rangle, \langle 2\rangle, \langle 3\rangle, \langle 4\rangle\}$$

and $\mathcal{Y}_b = \mathcal{Y}_2$ of Sect. 7.1.3. ◇

Example 7.4.15 (*LL-Cluster Adjacency: N²MHV₈ @ 1 loop*). There are 24 Yangian invariants up to cyclic symmetry. We see that some LL-cluster adjacencies are implied either by checks for $n = 6$ and $n = 7$, or the by the cluster adjacency of the Yangian invariant itself, when the Landau singularity is a pole of the former. The remaining cases are listed in the table below together with the clusters that contain of all said singularities:

Y-Inv.	Landau sing.	Mutation sequence	Subalgebra
$\mathcal{Y}_c$	$\langle 1378\rangle$	(2,5,8,6,9,1,2,4,5,1,4,7)	$A_3 \times A_2$
$\mathcal{Y}_d$	$\langle 8(71)(23)(56)\rangle$	(7,8,9,3,6,2,5,7,1,4)	$A_3 \times A_1$
$\mathcal{Y}_e$	$\langle 7(68)(23)(45)\rangle$	(1,5,9,3,5,6,1,2,3,4,5,4,8,7)	$A_1 \times A_1$
$\mathcal{Y}_f$	$\langle 8(71)(23)(45)\rangle$	(7,8,9,4,5,6,3,1,2,7,8,1,4,7)	$A_1 \times A_1$

where we have:

$$\begin{aligned}
\text{Facet}(\mathcal{Y}_c) &= \{\langle 3\bar{7}\rangle, \langle 3478\rangle, \langle \bar{7}8\rangle, \langle 3\rangle, \langle 4\rangle, \langle 5\rangle\}\\
\text{Facet}(\mathcal{Y}_d) &= \{\langle 2\bar{6}\rangle, \langle 2368\rangle, \langle 2348\rangle, \langle 2\rangle, \langle 3\rangle, \langle 4\rangle\}\\
\text{Facet}(\mathcal{Y}_e) &= \{\langle 4\bar{7}\rangle, \langle 8(23)(45)(67)\rangle, \langle 6(78)(23)(45)\rangle, \langle 4\rangle, \langle 3\bar{7}\rangle, \langle 5\rangle, \langle\langle 2\bar{7}\rangle, \langle 2378\}\\
\text{Facet}(\mathcal{Y}_f) &= \{\langle 2458\rangle, \langle 8(23)(45)(67)\rangle, \langle \bar{3}8\rangle, \langle 4578\rangle, \langle 2\rangle, \langle 4\rangle, \langle \bar{5}8\rangle, \langle \bar{4}8\rangle, \langle 5\rangle\}.
\end{aligned}$$

◇

We have also checked the cluster-adjacency properties for Landau singularities and Yangian invariants relevant for N²MHV $n = 9$ one-loop amplitudes. These are too lengthy to present here and we refer to the ancillary file `LLCAk2n9.m` attached to [25].

7.4.3 Bootstrap via LL-Cluster Adjacency

In [9], it was argued that to make the adjacency of the final entries of symbols with the R-invariants of NMHV loop amplitudes, one has to expand the symbol of the amplitude over the full set of R-invariants, which satisfy linear, six-term relations among them. This way of writing the amplitude is certainly not unique due to these identities, and cluster-adjacency between the R-invariants and the final entries can be made manifest in a number of ways. We suggest that LL-cluster adjacency between Landau singularities and R-invariants can put a further constraint on the final entry condition. More precisely, we can rule out any pair of final entry and R-invariant that are not related to each other through LL-cluster adjacency. We note that standard cluster adjacency relies on compatibility of cluster variables in $\mathrm{Gr}_{4,n}$, and therefore does not contain information on the helicity sector k. On the other hand, LL-cluster adjacency gives sets of compatible cluster variables depending on Landau and Leading singularities, which depends by the helicities of the scattering particles.

We checked that the 1-loop NMHV amplitude can be uniquely fixed in a form in which it obeys the adjacency discussed above:

$$\begin{aligned}\mathcal{A}_{\mathrm{NMHV}_7}^{\mathrm{BDS-like},(1)} = &-(13)\left[a_{11}\otimes a_{62}+a_{13}\otimes a_{62}\right]\\ &+(14)\left[-a_{11}\otimes a_{11}+a_{11}\otimes a_{14}+a_{14}\otimes a_{11}-a_{14}\otimes a_{14}\right]\\ &+(12)\left[a_{11}\otimes a_{15}-a_{11}\otimes a_{22}-a_{11}\otimes a_{31}+a_{12}\otimes a_{15}-a_{12}\otimes a_{22}\right.\\ &\qquad\left.-a_{12}\otimes a_{31}-2a_{15}\otimes a_{15}+a_{15}\otimes a_{22}+a_{15}\otimes a_{31}\right]\\ &+\text{ cyclic},\end{aligned} \tag{7.4.16}$$

where the notation a_{ij} for the symbol letters follows the literature on seven-point amplitudes [75]. It would be interesting to show that any NMHV 1-loop amplitude can be written in such a way. Then one may speculate whether this leads to a final entry condition for a given MHV degree and loop order. This exciting directions are left for future work.

7.4.4 Summary and Outlook

Cluster phenomena have become increasingly relevant in understanding singularities of scattering amplitudes. The remarkable observation that building blocks of scattering amplitudes in planar $\mathcal{N}=4$ SYM satisfy a property called *cluster adjacency*, provided both paths to understand their deep mathematical structures and tools to perform computations which otherwise would be beyond reach. At tree level, scattering amplitudes are rational functions whose singularity structure is encoded in the location of its poles. They can be expressed as sums of Yangian invariants, which have their own singularities, some of which are 'spurious' as they do not appear in the final amplitude. Nevertheless, collections of all poles of each Yangian invariant which can appear in a representation of tree-level scattering amplitudes seem themselves to be part of a beautiful mathematical story.

In this section, we argued for an enhancement of the phenomenon of cluster adjacency of Yangian invariants to include singularities of loop amplitudes in $\mathcal{N} = 4$ SYM. In particular, via an amplituhedron-based approach we observed a new manifestation of cluster adjacency for Leading and Landau singularities, which we called 'LL-cluster adjacency' for brevity. Given a maximal-cut of a loop amplitude, the corresponding Landau singularities are found in the same cluster as the poles of each Yangian invariant which can appear in a representation of the Leading Singularity related to the cut. Moreover, we proved LL-cluster adjacencies for all one-loop NMHV amplitudes, and N^2MHV up to 9 points. Interestingly, one-loop NMHV 7-points amplitude are uniquely fixed by LL-cluster adjacencies, once these are interpreted as a final-entry conditions. On the way, we proved that all N^2MHV Yangian invariants corresponding to positroid tiles are cluster adjacent, confirming the conjectures of [9, 10]. We also show that, for the non rational Yangian invariants of the four-mass box type, the poles of the rational sum of their algebraic terms do not satisfy cluster adjacency.

Studies of cluster adjacency of Yangian invariants [10, 24] motivate the question of whether this phenomenon should be regarded as a built-in mathematical feature of Yangian invariants, or whether it is a physical constraint. While cluster-adjacency may be a mathematical feature for rational Yangian invariants rooted in the geometry of the amplituhedron (see Sect. 7.1.1), the inclusion of Landau singularities is certainly a new, 'physical' information, just like extended-Steinmann conditions on symbol letters. Our conjecture calls for a geometric understanding that unifies all the related incarnations of cluster-adjacency and makes manifest how the more physical ones are implied by the mathematical ones and vice versa.

At each loop L and helicity sector k, LL-cluster adjacencies provide a natural set of pairings between Yangian invariants and symbol letters of the corresponding Landau Singularities. This set is in general much smaller than the full set of adjacencies obtained from $\mathrm{Gr}_{4,n}$ cluster algebras, which is loop- and helicity- agnostic. It would be interesting to see if LL-cluster adjacencies could be interpreted as *refined* set one can use to contain amplitudes at fixed loop order and MHV degree, instead of relying on the full cluster algebra. We provided an example of how this principle can be interpreted as a final-entry condition and fix the $n = 7$ one-loop amplitude. Such a principle would be even more restrictive than the recently proposed truncations of infinite cluster algebras [11, 13, 35], but without further examples with higher multiplicity and loop order, it is merely a wishful speculation.

On this regard, extending our analysis to non-rational Yangian invariants and Landau singularities corresponding to algebraic letters could be natural direction to pursue. There have been many promising results on how to understand algebraic letters in a cluster algebra fashion with the help of tropical positive Grassmannians [11, 13, 35]. However, a full understanding of $\mathrm{Gr}_{4,n}$ infinite cluster algebras is still missing, e.g. notions of 'adjacencies' involving algebraic letters have still to be defined.

Our work is in the direction of making steps towards answering the long-standing question of how the cluster structure of integrands in $\mathcal{N} = 4$ SYM theory is related to the cluster structure of the integrated amplitudes. The manifestation of the physical information carried by LL-cluster adjacencies which relates Leading Singularities with Landau Singularities shows further evidence that cluster phenomena know about the mathematical structure encoding (the singularities of) loop amplitudes.

References

1. S. Fomin, A. Zelevinsky, Cluster algebras I: foundations (2001)
2. S. Fomin, L. Williams, A. Zelevinsky, *Introduction to Cluster Algebras*. Chapters 1–3 (2016)
3. S. Fomin, L. Williams, A. Zelevinsky, *Introduction to Cluster Algebras*. Chapter 6 (2020)
4. M. Gross, P. Hacking, S. Keel, Birational geometry of cluster algebras. Algebr. Geom. **2**(2), 137–175 (2015)
5. J.S. Scott, Grassmannians and Cluster Algebras, in *Proceedings of the London Mathematical Society*, vol. 92, pp. 345–380, 03 2006
6. J. Golden, A.B. Goncharov, M. Spradlin, C. Vergu, A. Volovich, Motivic amplitudes and cluster coordinates. JHEP **01**, 091 (2014)
7. S. Caron-Huot, L.J. Dixon, J.M. Drummond, F. Dulat, J. Foster, O. Gürdoğan, M. von Hippel, A.J. McLeod, G. Papathanasiou, The Steinmann cluster bootstrap for $N = 4$ super Yang-Mills amplitudes. PoS **CORFU2019**, 003 (2020)
8. J. Drummond, J. Foster, Ö. Gürdoğan, Cluster adjacency properties of scattering amplitudes in $N = 4$ supersymmetric Yang-Mills theory. Phys. Rev. Lett. **120**(16), 161601 (2018)
9. J. Drummond, J. Foster, Ö. Gürdoğan, Cluster adjacency beyond MHV. JHEP **03**, 086 (2019)
10. J. Mago, A. Schreiber, M. Spradlin, A. Volovich, Yangian invariants and cluster adjacency in $\mathcal{N} = 4$ Yang-Mills. JHEP **10**, 099 (2019)
11. J. Drummond, J. Foster, Ö. Gürdoğan, C. Kalousios, Algebraic singularities of scattering amplitudes from tropical geometry, 12 2019
12. N. Arkani-Hamed, S. He, T. Lam, H. Thomas, binary geometries, generalized particles and strings, and cluster algebras (2019)
13. N. Henke, G. Papathanasiou, How tropical are seven- and eight-particle amplitudes? 12 2019
14. N. Henke, G. Papathanasiou, Singularities of eight- and nine-particle amplitudes from cluster algebras and tropical geometry. JHEP **21**, 007 (2020)
15. S. He, Z. Li, A note on letters of Yangian invariants. JHEP **02**, 155 (2021)
16. J. Mago, A. Schreiber, M. Spradlin, A. Volovich, Symbol alphabets from plabic graphs. JHEP **10**, 128 (2020)
17. J. Mago, A. Schreiber, M. Spradlin, A. Yelleshpur Srikant, A. Volovich, Symbol alphabets from plabic graphs II: rational letters. JHEP **04**, 056 (2021)
18. J. Mago, A. Schreiber, M. Spradlin, A. Yelleshpur Srikant, A. Volovich, Symbol Alphabets from Plabic Graphs III: $n = 9$, 6 2021
19. A. Herderschee, Algebraic branch points at all loop orders from positive kinematics and wall crossing. JHEP **07**, 049 (2021)
20. L. Ren, M. Spradlin, A. Volovich, Symbol Alphabets from Tensor Diagrams, 6 2021
21. S. He, Z. Li, Q. Yang, Truncated cluster algebras and Feynman integrals with algebraic letters, 6 2021
22. D. Chicherin, J.M. Henn, G. Papathanasiou, Cluster algebras for Feynman integrals. Phys. Rev. Lett. **126**(9), 091603 (2021)
23. M. Parisi, M. Sherman-Bennett, L. Williams, The $m = 2$ amplituhedron and the hypersimplex: signs, clusters, triangulations, Eulerian numbers (2023). Accepted and to be published in *Communications of the American Mathematical Society*

24. T. Łukowski, M. Parisi, M. Spradlin, A. Volovich, Cluster adjacency for $m = 2$ Yangian invariants. JHEP **10**, 158 (2019)
25. O. Gürdoğan, M. Parisi, Cluster patterns in Landau and leading singularities via the amplituhedron. Ann. Inst. Henri Poincaré Comb. Phys. Interact., 03 2023
26. G. Papathanasiou, The Steinmann Cluster Bootstrap for N = 4 SYM Amplitudes, *Talk at Amplitudes 2017* (2017)
27. S. Caron-Huot, L.J. Dixon, F. Dulat, M. Von Hippel, A.J. McLeod, G. Papathanasiou, The cosmic Galois group and extended Steinmann relations for planar $\mathcal{N} = 4$ SYM amplitudes. JHEP **09**, 061 (2019)
28. N. Arkani-Hamed, Y. Bai, T. Lam, Positive geometries and canonical forms. JHEP **11**, 039 (2017)
29. S. Fomin, A. Zelevinsky, Cluster algebras ii: finite type classification. Inventiones mathematicae **154**, 63–121 (2003)
30. C. Even-Zohar, M. Parisi, M. Sherman-Bennett, R.J. Tessler, T. Lakrec, L. Williams, Cluster algebras and Tilings for the $m = 4$ amplituhedron, In preparation
31. V. Del Duca, S. Druc, J. Drummond, C. Duhr, F. Dulat, R. Marzucca, G. Papathanasiou, B. Verbeek, Multi-Regge kinematics and the moduli space of Riemann spheres with marked points. J. High Energy Phys. **2016**, 152 (2016)
32. N. Arkani-Hamed, J.L. Bourjaily, F. Cachazo, A.B. Goncharov, A. Postnikov, J. Trnka, *Scattering Amplitudes and the Positive Grassmannian* (Cambridge University Press, 2012)
33. J.L. Bourjaily, Positroids, Plabic Graphs, and Scattering Amplitudes in Mathematica, 12 2012
34. N. Arkani-Hamed, J. Bourjaily, F. Cachazo, A. Goncharov, A. Postnikov, J. Trnka, *Grassmannian Geometry of Scattering Amplitudes* (Cambridge University Press, 2016)
35. N. Arkani-Hamed, T. Lam, M. Spradlin, Non-perturbative geometries for planar $\mathcal{N} = 4$ SYM amplitudes, 12 2019
36. P. Cao, F. Li, The enough g-pairs property and denominator vectors of cluster algebras. Mathematische Annalen **377**, 1547–1572 (2020)
37. R.J. Marsh, K. Rietsch, Parametrizations of flag varieties, in *Represent. Theory*, vol. 8, pp. 212–242 (2004)
38. K. Talaska, L. Williams, Network parametrizations for the Grassmannian. Algebra Number Theory **7**(9), 2275–2311 (2013)
39. T. Lam, Amplituhedron cells and Stanley symmetric functions. Comm. Math. Phys. **343**(3), 1025–1037 (2016)
40. S. Fomin, A. Zelevinsky, Cluster algebras. I. Foundations. J. Am. Math. Soc. **15**(2), 497–529 (electronic) (2002)
41. S. Fomin, L. Williams, A. Zelevinsky, *Introduction to Cluster Algebras*. Chapters 4–5 (2017)
42. A. Berenstein, S. Fomin, A. Zelevinsky, Cluster algebras. III. Upper bounds and double Bruhat cells. Duke Math. J. **126**(1), 1–52 (2005)
43. R.J. Eden, P.V. Landshoff, D.I. Olive, J.C. Polkinghorne, *The Analytic S-Matrix* (Cambridge University Press, Cambridge, 1966)
44. J. Mago, A. Schreiber, M. Spradlin, A. Volovich, A Note on One-loop Cluster Adjacency in N = 4 SYM, 5 2020
45. L.J. Dixon, J.M. Drummond, J.M. Henn, Bootstrapping the three-loop hexagon. JHEP **11**, 023 (2011)
46. L.J. Dixon, J.M. Drummond, J.M. Henn, Analytic result for the two-loop six-point NMHV amplitude in N = 4 super Yang-Mills theory. JHEP **01**, 024 (2012)
47. L.J. Dixon, J.M. Drummond, M. von Hippel, J. Pennington, Hexagon functions and the three-loop remainder function. JHEP **12**, 049 (2013)
48. L.J. Dixon, J.M. Drummond, C. Duhr, J. Pennington, The four-loop remainder function and multi-Regge behavior at NNLLA in planar N = 4 super-Yang-Mills theory. JHEP **06**, 116 (2014)
49. L.J. Dixon, M. von Hippel, Bootstrapping an NMHV amplitude through three loops. JHEP **10**, 065 (2014)

50. J.M. Drummond, G. Papathanasiou, M. Spradlin, A symbol of uniqueness: the cluster bootstrap for the 3-loop MHV heptagon. JHEP **03**, 072 (2015)
51. L.J. Dixon, M. von Hippel, A.J. McLeod, The four-loop six-gluon NMHV ratio function. JHEP **01**, 053 (2016)
52. S. Caron-Huot, L.J. Dixon, A. McLeod, M. von Hippel, Bootstrapping a five-loop amplitude using steinmann relations. Phys. Rev. Lett. **117**(24), 241601 (2016)
53. S. Caron-Huot, L.J. Dixon, F. Dulat, M. von Hippel, A.J. McLeod, G. Papathanasiou, Six-Gluon amplitudes in planar $\mathcal{N} = 4$ super-Yang-Mills theory at six and seven loops. JHEP **08**, 016 (2019)
54. J. Drummond, J. Foster, Ö. Gürdoğan, G. Papathanasiou, Cluster adjacency and the four-loop NMHV heptagon. JHEP **03**, 087 (2019)
55. N. Arkani-Hamed, A. Hodges, J. Trnka, Positive amplitudes in the amplituhedron. JHEP **08**, 030 (2015)
56. R. Kojima, C. Langer, Sign flip triangulations of the amplituhedron. J. High Energ. Phys. **2020**, 121 (2020). https://doi.org/10.1007/JHEP05(2020)121
57. D.E. Speyer, L.K. Williams, The tropical totally positive Grassmannian. J. Algebraic Comb. **22**, 189–210 (2003)
58. N. Arkani-Hamed, F. Cachazo, C. Cheung, J. Kaplan, A duality for the S matrix. JHEP **03**, 020 (2010)
59. L. Mason, D. Skinner, Dual superconformal invariance, momentum Twistors and Grassmannians. JHEP **0911**, 045 (2009)
60. N. Arkani-Hamed, J. Trnka, JHEP **10**, 030 (2014)
61. T. Dennen, I. Prlina, M. Spradlin, S. Stanojevic, A. Volovich, Landau Singularities from the amplituhedron. JHEP **06**, 152 (2017)
62. I. Prlina, M. Spradlin, J. Stankowicz, S. Stanojevic, Boundaries of amplituhedra and NMHV symbol alphabets at two loops. JHEP **04**, 049 (2018)
63. I. Prlina, M. Spradlin, J. Stankowicz, S. Stanojevic, A. Volovich, All-helicity symbol alphabets from unwound amplituhedra. JHEP **05**, 159 (2018)
64. N. Arkani-Hamed, J. Trnka, Into the amplituhedron. JHEP **1412**, 182 (2014)
65. N. Arkani-Hamed, J.L. Bourjaily, F. Cachazo, S. Caron-Huot, J. Trnka, The all-loop integrand for scattering amplitudes in planar N = 4 SYM. JHEP **1101**, 041 (2011)
66. G. Lusztig, Singularities, character formulas, and a q-analog of weight multiplicities, in *Analyse et topologie sur les espaces singuliers (II-III) - 6 - 10 juillet 1981*, no. 101–102 in Astérisque, Société mathématique de France (1983)
67. N. Arkani-Hamed, H. Thomas, J. Trnka, Unwinding in binary. Preprint, http://arxiv.org/abs/1704.05069
68. R. Britto, F. Cachazo, B. Feng, Generalized unitarity and one-loop amplitudes in N = 4 super-Yang-Mills. Nucl. Phys. B **725**, 275–305 (2005)
69. J.L. Bourjaily, S. Caron-Huot, J. Trnka, Dual-conformal regularization of infrared loop divergences and the chiral box expansion. JHEP **01**, 001 (2015)
70. L. Landau, On analytic properties of vertex parts in quantum field theory. Nucl. Phys. **13**(1), 181–192 (1960)
71. F. Pham, *Singularities of Integrals - Homology, Hyperfunctions and Microlocal Analysis* (Springer, London, 2011)
72. R. Hwa, V. Teplitz, Homology and Feynman integrals. Nuclear Phys. A **98**(3), 627 (1967)
73. J.B. Boyling, A homological approach to parametric Feynman integrals. Il Nuovo Cimento A **1965–1970**(53), 351–375 (1968)
74. J.B. Boyling, Construction of vanishing cycles for integrals over hyperspheres. J. Math. Phys. **7**(10), 1749–1763 (1966)
75. J.M. Drummond, G. Papathanasiou, M. Spradlin, A symbol of uniqueness: the cluster bootstrap for the 3-loop MHV heptagon. J. High Energy Phys., **2015**(3), 1–28 (2015)

Chapter 8
Conclusions

Around a decade ago, physicists discovered that the *positive Grassmannian* was needed to describe the interaction of particles in the 'planar limit' of $\mathcal{N} = 4$ super-Yang Mills theory. In 2009, the works of Arkani-Hamed–Cachazo–Cheung–Kaplan [1] and Bullimore–Mason–Skinner [2] introduced Grassmannian formulations for scattering amplitudes in this theory. Remarkably, this led to the discovery that the positive Grassmannian encodes most of the physical properties of amplitudes [3]. In turn, physicists generated new cutting-edge ideas for combinatorialists themselves. Building on these developments and on Hodges' idea that scattering amplitudes might be 'volumes' of some geometric object [4], Arkani-Hamed and Trnka arrived at the definition of the *amplituhedron* [5] in 2013. The above geometrisation programme hinges on the idea that quantum mechanical observables in particle physics and cosmology come from underlying novel mathematical objects. Physical properties, such as *locality* and *unitarity*, purely emerge from geometry and combinatorics. Understanding this process advances our grasp of the basic principles of Quantum Field Theories and allows us to perform calculations which were previously beyond reach. Crucially, it also cross-fertilises ideas in pure mathematics, such as in Combinatorics.

The major thread of this process of geometrization is captured by the framework of *positive geometries*. Associated to a physical observable we have a space, or better pair (X, X_+), where X is a complex projective variety and X_+ is a semi-algebraic set included in the real part of X (i.e. determined by a finite set of real polynomial equations and inequalities). Moreover, we have a meromorphic top form called *canonical form* $\mathbf{\Omega}(X_+)$. The combinatorial and geometric properties of the positive geometry are encoded in the analytic structure of such canonical form. The latter reflects physical properties of the physical observable coming from basic physics principles. For example, locality is encoded in where the poles of $\mathbf{\Omega}(X_+)$ are, which reflects where the faces of X_+ are; unitarity is encoded in what the *residues* of $\mathbf{\Omega}(X_+)$ are, which reflects what the geometries of the faces of X_+ are. A way to compute scattering amplitudes or other physical observables is by summing over terms obtained via certain expansions, e.g., BCFW recursions, Feynman diagrams, etc. This corresponds

M. Parisi, *Combinatorial Aspects of Scattering Amplitudes*, Springer Theses,
https://doi.org/10.1007/978-3-031-41069-7_8

to subdividing the associated positive geometry into elementary pieces—i.e., finding *tilings*. By studying tilings, one can: (i) get different representations of the physical observable, some of which not derivable from any standard physical expansions; (ii) manifest new properties and symmetries of the physical observable; (iii) shed light on dualities—e.g., if we find geometries which are different, but have the same combinatorics of tilings. This would correspond to physical observables related by some duality, such as *T-duality*—one of the main topic of this work.

The $m = 4$ amplituhedron $\mathcal{A}_{n,k,4}$ is the positive geometry of tree-level scattering amplitudes in planar $\mathcal{N} = 4$ SYM. However, the amplituhedron is a well-defined and interesting mathematical object for any m. For example, the $m = 1$ amplituhedron $\mathcal{A}_{n,k,1}$ can be identified with the complex of bounded faces of a cyclic hyperplane arrangement [6]. The $m = 2$ amplituhedron $\mathcal{A}_{n,k,2}$, which is one of the main subjects of this work, also has a beautiful combinatorial structure. From the point of view of physics, $\mathcal{A}_{n,k,2}$ is not only a good toy-model for the $m = 4$ case. It enters the geometry of *loop amplituhedra* (Sect. 7.3.1), which encodes integrands of loop amplitudes in planar $\mathcal{N} = 4$ SYM. Moreover, it is related to correlators of determinant operators and to tree-level MHV *form factors* of half-BPS operators in $\mathcal{N} = 4$ SYM [8, 9].

Physicists have observed a duality between the formulations of scattering amplitudes $\mathcal{N} = 4$ SYM in momentum space[1] and in momentum twistor space. This is possible because of the so-called 'Amplitude/Wilson loop duality' [10–16], which was shown to arise from a more fundamental duality in String Theory called 'T-duality' [17]. There is beautiful way this duality emerges from the framework of positive geometries, in particular, in connection with tilings. In this work we conjecture that collections of $4k$-dimensional cells of $\mathrm{Gr}_{k,n}^{\geq 0}$ give a tiling of the the amplituhedron $\mathcal{A}_{n,k,4}$ if and only if the corresponding T-dual collections of $(2n-4)$-dimensional cells of $\mathrm{Gr}_{k+2,n}^{\geq 0}$ give a tiling of the *momentum amplituhedron* $\mathcal{M}_{n,k+2,4}$. The latter object is newly introduced in this work, and it is the sought-after positive geometry of tree-level scattering amplitudes of $\mathcal{N} = 4$ SYM in momentum (spinor-helicity) space.

We show that this duality—we have evocatively called *T-duality*—extends beyond $m = 4$. In particular, for $m = 2$, T-duality connects two seemingly unrelated objects which have been at the center of attention of both mathematicians and physicists: the *hypersimplex* $\Delta_{k+1,n}$—a $(n-1)$-dimensional polytope and the $m = 2$ amplituhedron $\mathcal{A}_{n,k,2}$—a $2k$-dimensional subset of the Grassmannian. While the latter has been recently defined by physicists and then studied by mathematicians, the study of the hypersimplex dates back to the foundational work of Gelfand-Goresky-MacPherson-Serganov [18], which related the geometry of torus orbits in the Grassmannian to matroid polytopes. Employing T-duality for $m = 2$, we draw striking parallels between the two objects, some of which are illustrated in Table 4.1.

We use *twistor coordinates* and the geometry of the hypersimplex and positroid polytopes to obtain a deeper understanding of the amplituhedron. We classify *positroid tiles*, full-dimensional images of positroid cells which map injectively into

[1] In 'spinor helicity' space, or—related by half-Fourier transform—in twistor space. See [3, Sect. 8].

the amplituhedron $\mathcal{A}_{n,k,2}$. We then give a new characterization of them in terms of the signs of their twistor coordinates. We use this result to prove a conjecture of Arkani-Hamed–Thomas–Trnka that $\mathcal{A}_{n,k,2}$ can be characterized using sign flips of twistor coordinates.

Moreover, we find that the inequalities describing positroid polytopes are T-dual to sign conditions on twistor coordinates characterizing the corresponding Grasstopes. And we show that the sign patterns on twistor coordinates naturally subdivide the amplituhedron into *chambers*. We prove that the ones which are *realizable* are exactly enumerated by the Eulerian numbers $E_{k,n-1}$, just as the hypersimplex can be subdivided into simplices enumerated by $E_{k,n-1}$. We use these properties to prove one of the main results: that a collection of positroid polytopes is a tiling of $\Delta_{k+1,n}$ if and only if the collection of T-dual Grasstopes is a tiling of $\mathcal{A}_{n,k,2}$ for all Z. Along the way, we also prove new properties of the hypersimplex $\Delta_{k+1,n}$, such that its regular positroid subdivisions are governed by the *positive tropical Grassmannian* $\text{Trop}^+ \text{Gr}_{k+1,n}$. By T-duality, this implies $\text{Trop}^+ \text{Gr}_{k+1,n}$ induces positroid tilings for $\mathcal{A}_{n,k,2}$! In general, we conjecture that $\text{Trop}^+ \text{Gr}_{k+1,n}$ serves as the *secondary fan* for $\mathcal{A}_{n,k,2}$ with respect to these 'nice' positroid subdivisions. This means that the face poset of $\text{Trop}^+ \text{Gr}_{k+1,n}$ should match the poset of the 'nice' positroid subdivisions of $\mathcal{A}_{n,k,2}$. For example, maximal cones of $\text{Trop}^+ \text{Gr}_{k+1,n}$ would give positroid tilings and faces correspond to coarser subdivisions of $\mathcal{A}_{n,k,2}$. The connection between $\text{Trop}^+ \text{Gr}_{k+1,n}$ and $\mathcal{A}_{n,k,2}$ was rather surprising: it was exactly where our explorations started. Remarkably, we saw that T-duality goes beyond this correspondence: it relates *any* positroid tiling and, conjecturally, any dissection of $\Delta_{k+1,n}$ and $\mathcal{A}_{n,k,2}$; it relates the geometry of positroid tiles, and also of the 'finest' pieces—the w-simplices and w-chambers.

In our work, we also introduce a generalization $\mathcal{M}_{n,k',m}$ of the momentum amplituhedron $\mathcal{M}_{n,k',4}$, which we expect to play a role in generalizing T-duality story for higher m. It is fascinating that T-duality relates objects which seem very different, but have the same *secondary geometry*, i.e., the same combinatorics of tilings. By reverting the perspective, it seems that thinking about tilings can shed light on deep and surprising dualities—both in mathematics and physics.

Our discovery that Eulerian numbers count sign chambers of the $m = 2$ amplituhedron is intriguing because Eulerian numbers have also come up in the context of *scattering equations* [19]. Scattering equations connect the singularity structure of scattering amplitudes of n-particles to that of the boundaries of the moduli space of *Riemann spheres with n punctures*. For $\mathcal{N} = 4$ SYM, the number of solutions of the 'N^kMHV' sector of the theory is exactly the Eulerian number $E_{k,n-3}$ [19, 20]. Moreover, [21] provided an explicit bijection between such solutions and permutations on $[n-3]$ with k descents. Finally, in the case of certain scalar quantum field theories, the authors of [22] formulated a generalization of scattering equations. By studying 'arrays of Feynman diagrams', they made connections to the positive tropical Grassmannian, and, by results of [23], to the hypersimplex. It would be fascinating to explore possible relations between (generalized) scattering equations, simplices of the hypersimplex, and chambers of the amplituhedron.

Meanwhile, in recent years physicists have been increasingly interested in understanding how *cluster algebras* encode the analytic properties of scattering amplitudes, both at tree- and loop- level. In this work we provide many results relating the amplituhedra to cluster algebras.

First, we prove *cluster adjacency* for $\mathcal{A}_{n,k,2}$, which says that the Plücker coordinates labeling facets of a given positroid tile consist of compatible cluster variables. We also state and prove a generalization of this by showing that twistor coordinates of a positroid tile associated to Plücker coordinates compatible with the ones labelling its facets have constant sign. This hints at further cluster structures inside the amplituhedron. Moreover, we give a conjectural formula for all *Yangian invariants* of the $m = 2$ toy theory whose positive geometry is $\mathcal{A}_{n,k,2}$—i.e., the canonical forms of positroid tiles—which manifest their pole structures and their cluster adjacency properties.

Second, we associate a cluster variety to each positroid tile in $\mathcal{A}_{n,k,2} \subset \mathrm{Gr}_{k,k+2}$, and show that the positroid tile is the totally positive part of that cluster variety. We then have the strange phenomenon that the amplituhedron $\mathcal{A}_{n,k,2}$ can be tiled into positroid tiles, each of which is the totally positive part of a cluster variety. Moreover, there are of course many such tilings. It would be fascinating to explore this phenomenon for general m, in particular for $m = 4$ which is physically relevant for tree-level amplitudes in $\mathcal{N} = 4$ SYM. See [24] for recent related work. The cluster adjacency conjecture and its generalization (Conjecture 7.1.4) strongly hint at hidden cluster structures inside amplituhedra $\mathcal{A}_{n,k,m}$—as we found for $m = 2$. The cluster algebras involved are of Grassmannian type $\mathrm{Gr}_{m,n}$, which are in general not well understood (for $m \geq 4$ and $n \geq 8$). Leveraging the connections provided by T-duality, one could employ the momentum amplituhedron to understand such cluster structures. These could reduce to ones inside products of simpler Grassmannians, e.g., $\mathrm{Gr}_{2,n} \times \mathrm{Gr}_{2,n}$ for the physical $m = 4$.

Finally, we discover *LL-cluster adjacency* of scattering amplitudes in planar $\mathcal{N} = 4$ SYM, which relates *Leading Singularities*—the 'deepest' singularities of integrands—and *Landau Singularities*—singularities of the integrated amplitudes. Our work is therefore in the direction of making steps towards answering the long-standing question of how the cluster structure of integrands in $\mathcal{N} = 4$ SYM theory is related to the cluster structure of the integrated amplitudes. LL-cluster adjacency enhances the cluster adjacency conjectured at tree-level by including cluster variables arising from Landau singularities—which are singularities of loop amplitudes. This opens the question whether also cluster phenomena at loop level could be understood in terms of the underlying geometry. Loop amplitudes have a very complicated analytic structure—such as branch cuts—and the Landau analysis, performed either on individual Feynman integrals or on the *whole* integrand encoded in the loop amplituhedron $\mathcal{A}_{n,k}^{(L)}$, connects to the correspondingly rich underlying geometric structures of *Landau varieties* (see [25] for a review). The interplay between the framework of positive geometries, Landau analysis and cluster algebras could lead to new ways of understanding singularities of scattering amplitudes. It would be fascinating to see all cluster phenomena at loop-level emerge from a unified, more fundamental geometric canvas, and the impact of this on bootstrap programs, and beyond.

References

1. N. Arkani-Hamed, F. Cachazo, C. Cheung, J. Kaplan, A duality for the S matrix. JHEP **03**, 020 (2010)
2. M. Bullimore, L.J. Mason, D. Skinner, Twistor-strings. Grassmannians and leading singularities. JHEP **03**, 070 (2010)
3. N. Arkani-Hamed, J. Bourjaily, F. Cachazo, A. Goncharov, A. Postnikov, J. Trnka, *Grassmannian Geometry of Scattering Amplitudes*. (Cambridge University Press, 2016)
4. A. Hodges, Eliminating spurious poles from gauge-theoretic amplitudes. JHEP **05**, 135 (2013)
5. N. Arkani-Hamed, J. Trnka, The amplituhedron. J. High Energy Phys. **10**, 33 (2014)
6. S.N. Karp, L.K. Williams, *The $m = 1$ amplituhedron and cyclic hyperplane arrangements Int* (Math. Res. Not, IMRN (to appear), 2017)
7. R. Kojima, C. Langer, Sign flip triangulations of. **1** (2020)
8. B. Basso, A.G. Tumanov, Wilson Loop Duality and OPE for Super Form Factors of Half-BPS Operators. arXiv preprint arXiv:2308.08432 (2023)
9. S. Caron-Huot, F. Coronado, B. Muhlmann, Determinants in self-dual N= 4 SYM and twistor space. arXiv preprint arXiv:2304.12341 (2023)
10. L.F. Alday, R. Roiban, Scattering amplitudes, Wilson loops and the string/gauge theory correspondence. Phys. Rept. **468**, 153–211 (2008)
11. L.F. Alday, J.M. Maldacena, Gluon scattering amplitudes at strong coupling. JHEP **06**, 064 (2007)
12. J.M. Drummond, G.P. Korchemsky, E. Sokatchev, Conformal properties of four-gluon planar amplitudes and Wilson loops. Nucl. Phys. B **795**, 385–408 (2008)
13. A. Brandhuber, P. Heslop, G. Travaglini, MHV amplitudes in $\mathcal{N} = 4$ super Yang-Mills and Wilson loops. Nucl. Phys. B **794**, 231–243 (2008)
14. J.M. Drummond, J. Henn, G.P. Korchemsky, E. Sokatchev, On planar gluon amplitudes/Wilson loops duality. Nucl. Phys. B **795**, 52–68 (2008)
15. Z. Bern, L.J. Dixon, D.A. Kosower, R. Roiban, M. Spradlin, C. Vergu, A. Volovich, The two-loop six-gluon MHV amplitude in maximally supersymmetric Yang-Mills theory. Phys. Rev. D **78**, 045007 (2008)
16. J.M. Drummond, J. Henn, G.P. Korchemsky, E. Sokatchev, Hexagon Wilson loop = six-gluon MHV amplitude. Nucl. Phys. B **815**, 142–173 (2009)
17. N. Berkovits, J. Maldacena, Fermionic T-Duality, Dual superconformal symmetry, and the amplitude/Wilson loop connection. JHEP **09**, 062 (2008)
18. I.M. Gelfand, R.M. Goresky, R.D. MacPherson, V.V. Serganova, Combinatorial geometries, convex polyhedra, and Schubert cells. Adv. Math. **63**(3), 301–316 (1987)
19. F. Cachazo, S. He, E.Y. Yuan, Scattering in three dimensions from rational maps. JHEP **10**, 141 (2013)
20. M. Spradlin, A. Volovich, From twistor string theory to recursion relations. Phys. Rev. D **80**, 085022 (2009)
21. F. Cachazo, S. Mizera, G. Zhang, Scattering equations: Real solutions and particles on a Line. JHEP **03**, 151 (2017)
22. F. Cachazo, N. Early, A. Guevara, S. Mizera, Scattering equations: from projective spaces to tropical grassmannians. JHEP **06**, 039 (2019)
23. T. Łukowski, M. Parisi, L.K. Williams, The positive tropical Grassmannian, the hypersimplex, and the $m = 2$ amplituhedron. Int. Math. Res. Not. **03**, rnad010 (2023)

24. C. Even-Zohar, T. Lakrec, P. Matteo, M. Sherman-Bennett, R.J. Tessler, L. Williams, Cluster algebras and Tilings for the m = 4 amplituhedron. In preparation
25. F. Pham, *Singularities of Integrals-Homology, Hyperfunctions and Microlocal Analysis* (Springer-Verlag, London, 2011)

Appendix A
Combinatorics of the Positive Grassmannian

In [1], Postnikov defined several families of combinatorial objects which are in bijection with cells of the positive Grassmannian, including *decorated permutations*, and equivalence classes of *reduced plabic graphs*. He also used these objects to give concrete descriptions of the cells. Here we review some of this technology.

Definition A.0.1 A *decorated permutation* on $[n]$ is a bijection $\pi : [n] \to [n]$ whose fixed points are each coloured either black (loop) or white (coloop). We denote a black fixed point i by $\pi(i) = \underline{i}$, and a white fixed point i by $\pi(i) = \overline{i}$. An *anti-excedance* of the decorated permutation π is an element $i \in [n]$ such that either $\pi^{-1}(i) > i$ or $\pi(i) = \overline{i}$. We say that a decorated permutation on $[n]$ is of *type* (k, n) if it has k anti-excedances.

For example, $\pi = (3, \underline{2}, 5, 1, 6, 8, \overline{7}, 4)$ has a loop in position 2, and a coloop in position 7. It has three anti-excedances, in positions 4, 7, 8.

Decorated permutations can be equivalently thought of as affine permutations [2]. An *affine permutation* on $[n]$ is a bijection $\pi : \mathbb{Z} \to \mathbb{Z}$ such that $\pi(i + n) = \pi(i) + n$ and $i \leq \pi(i) \leq i + n$, for all $i \in \mathbb{Z}$. It is additionally (k, n)*-bounded* if $\sum_{i=1}^{n}(\pi(i) - i) = kn$.

There is a bijection between decorated permutations of type (k, n) and (k, n)-bounded affine permutations. Given a decorated permutation π_d we can define an affine permutation π_a by the following procedure: if $\pi_d(i) > i$, then define $\pi_a(i) := \pi_d(i)$; if $\pi_d(i) < i$, then define $\pi_a(i) := \pi_d(i) + n$; if $\pi_d(i)$ is a loop then define $\pi_a(i) := i$; if $\pi_d(i)$ is a coloop then define $\pi_a(i) := i + n$. For example, under this map, the decorated permutation $\pi_d = (3, \underline{2}, 5, 1, 6, 8, \overline{7}, 4)$ in the previous example gives rise to $\pi_a = (3, 2, 5, 9, 6, 8, 15, 12)$.

Given a $k \times n$ matrix $C = (c_1, \ldots, c_n)$ written as a list of its columns, we associate a decorated (equivalently, affine) permutation π as follows. Given $i, j \in [n]$, let $r[i, j]$ denote the rank of $\langle c_i, c_{i+1}, \ldots, c_j \rangle$, where we list the columns in cyclic order, going from c_n to c_1 if $i > j$. We set $\pi(i) := j$ to be the label of the first column j such that $c_i \in \text{span}\{c_{i+1}, c_{i+2}, \ldots, c_j\}$. If c_i is the all-zero vector, we call

M. Parisi, *Combinatorial Aspects of Scattering Amplitudes*, Springer Theses,
https://doi.org/10.1007/978-3-031-41069-7

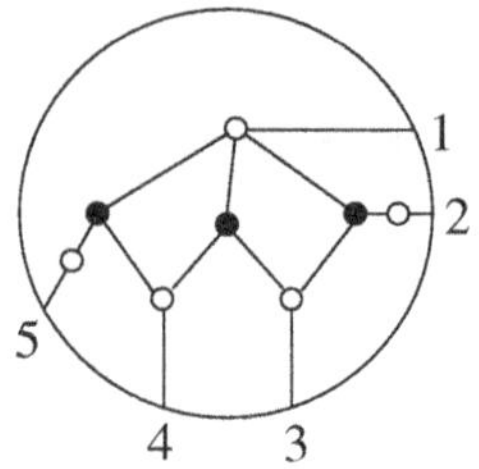

Fig. A.1 A plabic graph

i a loop or black fixed point (and set $\pi(i) = i$ in the affine permutation), and if c_i is not in the span of the other column vectors, we call i a coloop or white fixed point (and set $\pi(i) = i + n$ in the affine permutation).

The map $C \mapsto \pi$ extends to a map on positroid cells. Moreover, Postnikov showed that the positroids for $Gr_{k,n}^{\geq 0}$ are in bijection with decorated permutations of $[n]$ with exactly k anti-excedances (equivalently, by (k, n)-bounded affine permutations) [1, Sect. 16]. One may read off the dimension of the cell S_π from the affine permutation π as follows. Let $\text{inv}(\pi)$ be the number of pairs (i, j) such that $i \in [n]$, $j \in \mathbb{Z}$, $i < j$, and $\pi(i) > \pi(j)$. Then the dimension of S_π equals $k(n-k) - \text{inv}(\pi)$.

Definition A.0.2 A *planar bicolored graph* (or "plabic graph") is a planar graph G properly embedded into a closed disk, such that each internal vertex is colored black or white; each internal vertex is connected by a path to some boundary vertex; there are (uncolored) vertices lying on the boundary of the disk labeled $1, \dots, n$ for some positive n; and each of the boundary vertices is incident to a single edge. See Fig. A.1 for an example.

If the connected component of G attached to a boundary vertex i is a path ending at a black (resp., white) leaf, we call this component a black (resp., white) *lollipop. We will require that our plabic graphs have no internal leaves except for lollipops.*

There is a natural set of local transformations (moves) of plabic graphs:

(M1) *Square move* (or *urban renewal*). If a plabic graph has a square formed by four trivalent vertices whose colors alternate, then we can switch the colors of these four vertices.

(M2) *Contracting/expanding a vertex.* Two adjacent internal vertices of the same color can be merged. This operation can also be reversed.

(M3) *Middle vertex insertion/removal.* We can remove/add degree 2 vertices.

See Fig. A.2 for depictions of these three moves.

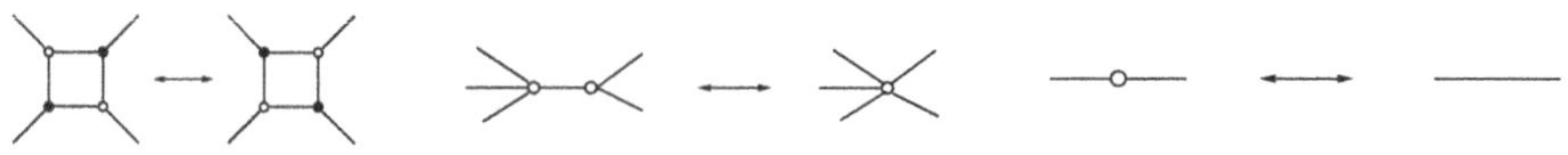

Fig. A.2 Local moves (M1), (M2), (M3) on plabic graphs

Definition A.0.3 Two plabic graphs are called *move-equivalent* if they can be obtained from each other by moves (M1)-(M3). The *move-equivalence class* of a given plabic graph G is the set of all plabic graphs which are move-equivalent to G. A plabic graph is called *reduced* if there is no graph in its move-equivalence in which two adjacent vertices u and v are connected by more than one edge.

Note that given a plabic graph G, we can always apply moves to G to obtain a new graph G' which is bipartite.

Definition A.0.4 Let G be a reduced plabic graph as above with boundary vertices $1, \ldots, n$. For each boundary vertex $i \in [n]$, we follow a path along the edges of G starting at i, turning (maximally) right at every internal black vertex, and (maximally) left at every internal white vertex. This path ends at some boundary vertex $\pi(i)$. By [1, Sect. 13], the fact that G is reduced implies that each fixed point of π is attached to a lollipop; we color each fixed point by the color of its lollipop. In this way we obtain the *decorated permutation* $\pi_G = \pi$ of G. We say that G is of *type* (k, n), where k is the number of anti-excedances of π_G.

The decorated permutation of the plabic graph G of Fig. A.1 is $\pi_G = (3, 4, 5, 1, 2)$, which has $k = 2$ anti-excedances.

Definition A.0.5 Let G be a bipartite plabic graph. Use move (M3) to ensure that each boundary vertex is incident to a white vertex. An *almost perfect matching* M of a plabic graph G is a subset M of edges such that each internal vertex is incident to exactly one edge in M (and each boundary vertex i is incident to either one or no edges in M). We let $\partial M = \{i \mid i \text{ is incident to an edge of } M\}$.

We associate to each graph G as above a collection of subsets $\mathcal{M}(G) \subset [n]$ as follows.

Proposition A.0.6 *[1, Propostion 11.7, Lemma 11.10] Let G be a plabic graph as in Definition A.0.5, and let $\mathcal{M}(G) = \{\partial M \mid M \text{ an almost perfect matching of } G\}$. Then $\mathcal{M}(G)$ is the set of bases of a positroid on $[n]$. Its rank is $\#\{\text{white vertices of } G\} - \#\{\text{black vertices of } G\}$, which is the size of ∂M for any almost perfect matching M of G.*

Postnikov used plabic graphs to give parameterizations of cells of $\mathrm{Gr}_{k,n}^{\geq 0}$. These parameterizations of cells can be recast as a variant of a theorem of Kasteleyn, as was made explicit in [3]. We follow the exposition there.

Theorem A.0.7 *[3] Let G be a bipartite graph with boundary embedded in a disk, such that all of the boundary vertices are black. Suppose there are $N + k$ white vertices $W_1, \ldots, W_{N+k}$, N internal black vertices $B_1, \ldots, B_N$, and n boundary vertices $B_{N+1}, \ldots, B_{N+n}$, labeled in clockwise order. Let $w : \mathrm{Edges}(G) \to \mathbb{R}_{>0}$ be any weighting function; if there is an edge between vertices i and j, we denote the weight on this edge by w_{ij}. For a perfect matching M, define $w(M) = \prod_{e \in M} w(e)$ and define ∂M to be the indices of the boundary vertices covered by an edge in M. For a subset I of $\{B_{N+1}, \ldots, B_{N+n}\}$, define $\mathbb{D}(G, I, w) = \sum_{\partial M = I} w(M)$.*

Then there is a real $k \times n$ Kasteleyn matrix *L such that for each k-element subset I of ∂G, the determinant* $\det L_I$ *of the $k \times k$ submatrix of L using the columns indexed by I is* $\det L_I = \mathbb{D}(G, I, w)$*. In particular, all Plücker coordinates of L are non-negative.*

The *positroid cell* $S_G \subset \mathrm{Gr}_{k,n}$ associated to the plabic graph G is the set of all k-planes in $\mathbb{R}^n$ spanned by matrices L as in Theorem A.0.7. If G is a tree, we call S_G a *tree positroid.*

Remark A.0.8 The Kasteleyn matrix L is constructed as follows. First construct an $(N + k) \times (N + n)$ matrix K, with rows indexed by white vertices and columns indexed by black vertices, with $K_{ij} = \pm w_e$ if there is an edge e between vertices i and j (otherwise $K_{ij} = 0$). Then, assuming G has at least one perfect matching, we can apply row operations to transform K into a matrix of block form $\begin{pmatrix} \mathrm{Id}_N & \star \\ 0 & L \end{pmatrix}$.

Appendix B
Positive Geometries and Tilings

In recent years, positive geometries are frequently arising as the underlying mathematical structures for quantum mechanical observables of many theories in particle physics and cosmology. Even though positive geometries have a priori no reference to any physics notions, the physical principles and properties of physical observables can be seen as emergent from the mathematical properties of these objects. Therefore, they play a fundamental role in understanding the physics they encode.

We refer to [4–6] as references for this appendix and for a more detailed definition and examples of positive geometries. A positive geometry is a pair $(X_{\mathbb{C}}, X_+)$ of a complex algebraic variety $X_{\mathbb{C}}$ defined over $\mathbb{R}$ and a semialgebraic subset $X_+ \subset X_{\mathbb{R}}$ such that there exists a unique[1] meromorphic form $\boldsymbol{\Omega}(X)$ on $X_{\mathbb{C}}$, called *canonical form*, which has simple poles or *logarithmic singularities* along the (complexified) boundaries of X_+ and it is regular everywhere else. A canonical form of a positive geometry explicitly encodes its associated physical observables. Given a positive geometry $(X_{\mathbb{C}}, X_+)$ endowed with a canonical form $\boldsymbol{\Omega}(X)$, a natural way to compute $\boldsymbol{\Omega}(X)$ is by *tiling* the geometry and summing over the canonical forms $\boldsymbol{\Omega}(X_a)$ of the cells X_a in a tiling $\{X_a\}$ of X_+. Since the forms $\boldsymbol{\Omega}(X_a)$ are usually straightforward to compute, the problem is reduced to finding the tiling of positive geometries. Moreover, different tiling correspond to different representations of $\boldsymbol{\Omega}(X)$ and lead to different expressions of the associated physical observable. Hence, it is desirable to develop a theory of *secondary positive geometry* to encode the interrelations of different representations of physical observables.

The combinatorics of *subdivisions* of polytopes is very rich and has been extensively studied in polyhedral geometry (see, e.g., [7] and references therein). More importantly, by the seminal works of Gelfand, Kapranov and Zelevinsky [8, 9], given any polytope P there exists a so-called secondary polytope whose face lattice is isomorphic to the poset of regular subdivisions of P. In a subsequent work [10], Billera and Sturmfels introduced the notion of *fiber polytopes* which contain secondary

[1] It is unique up to multiplication by a scalar. This can be fixed e.g., by prescribing a residue at one of the highest-codimension boundaries.

M. Parisi, *Combinatorial Aspects of Scattering Amplitudes*, Springer Theses,
https://doi.org/10.1007/978-3-031-41069-7

polytopes as examples. This approach provides a representation of the secondary polytope of P as *Minkowski sum* over the fibers of a certain projection on P at some points in the polytope. An interesting family of polytopes whose subdivisions are well-understood is *cyclic polytopes* (see, e.g., [11–14]).

The study of subdivisions of objects beyond polytopes is quite underdeveloped. One particular family of interest is *amplituhedra*, which is one of the main topics of this work. We recall they are geometric objects defined by Arkani-Hamed and Trnka [15] whose subdivisions have profound uses in physics, especially in computing scattering amplitudes of particles. Cyclic polytopes are examples of amplituhedra. However, a general amplituhedron is *not* necessarily a polytope.

B.0.1 Canonical Forms

The geometry of the positive Grassmannian and amplituhedron can be supplemented with certain differential forms called *canonical* forms. These forms are uniquely[2] defined by their property of having simple (logarithmic) poles along the boundary of the space they are associated to. They were first introduced for the positive Grassmannians, and then were expanded for amplituhedra (see, e.g., [17] and references therein). In particular, Arkani-Hamed and Trnka in [15] showed how such forms algebraically encode scattering amplitudes in $\mathcal{N} = 4$ super Yang-Mills theory.

We now briefly recall the definition of *residue*. We refer to [18] as a standard literature as well as [19, 20, §2.2] for more introductory texts.

Definition B.0.1 (*Residue*). Let $Y \subset X$ be an irreducible subvariety of codimension k, and Ω a rational form on X. Let $\alpha_1, \alpha_2, \ldots, \alpha_d$ be local coordinates of X and let Y be locally determined by the vanishing locus $\alpha_1 = \cdots = \alpha_k = 0$. Further, suppose that $\mathbf{\Omega} = \frac{\mathrm{d}\alpha_1}{\alpha_1} \wedge \cdots \wedge \frac{\mathrm{d}\alpha_k}{\alpha_k} \wedge \mathbf{\Omega}'$ with $\mathbf{\Omega}'$ of the form $f(\alpha_1, \alpha_2, \ldots, \alpha_d)\, \mathrm{d}\alpha_{k+1} \wedge \mathrm{d}\alpha_{k+2} \wedge \cdots \wedge \mathrm{d}\alpha_d$ for some rational function f which is locally analytic on Y. Then the *residue* of $\mathbf{\Omega}$ along Y is defined as $\mathrm{Res}_Y \Omega := \Omega'|_Y$. Note that the $\mathrm{Res}_Y \mathbf{\Omega}$ is only defined up to sign, which depends on the order of the functions $\alpha_1, \ldots, \alpha_k$.

We now introduce the standard volume form on the Grassmannian as follows.

Definition B.0.2 Let $C = (C_{\alpha i})$ be a generic matrix of indeterminates in $\mathrm{Gr}_{k,n}$. The *standard volume form* of $\mathrm{Gr}_{k,n}$ is defined as[3]

$$\mu_{\mathrm{Gr}_{k,n}}(C) := \bigwedge_{\alpha=1}^{k} \langle C \mathrm{d}^{n-k} C_\alpha \rangle, \quad \text{where} \quad \langle C\, \mathrm{d}^{n-k}\, C_\alpha \rangle := \sum_{I \in \binom{[n]}{k}} \epsilon_{I,[n]\setminus I}\, p_I(C) \bigwedge_{\bar{i} \in [n]\setminus I} \mathrm{d}\, C_{\alpha \bar{i}}.$$

[2] The existence of such form on a variety with boundary divisors with these properties is non-trivial. See [16, §4].

[3] See e.g., [21, Appendix C].

Proposition B.0.3 *([5, 17]). Let C be a generic matrix of indeterminates in $\mathrm{Gr}_{k,n}^{\geq 0}$. Then the differential form*

$$\boldsymbol{\Omega}\left(\mathrm{Gr}_{k,n}^{\geq 0}\right) = \frac{\mu_{\mathrm{Gr}_{k,n}}(C)}{p_{[1,k]}(C)p_{[2,k+1]}(C)\dots p_{[n,k-1]}(C)}$$

is the canonical form of $\mathrm{Gr}_{k,n}^{\geq 0}$.

Note that the formula above is invariant under GL_k-action, hence the form is well-defined on $\mathrm{Gr}_{k,n}$.

The amplituhedron $\mathcal{A}_{n,k,m}$ is conjectured to be a positive geometry [5, 15]. In particular, one can obtain its canonical form $\boldsymbol{\Omega}\left(\mathcal{A}_{n,k,m}\right)$ by knowing its positroid tiling.

Conjecture B.0.4 ([5, 15]). Let $\{Z_\pi\}$ be a positroid tiling of $\mathcal{A}_{n,k,m}$. Then

$$\boldsymbol{\Omega}\left(\mathcal{A}_{n,k,m}\right) = \sum_{\{Z_\pi\}} \boldsymbol{\Omega}\left(Z_\pi\right). \tag{B.0.5}$$

In particular, the sum does not depend on the positroid tiling.

Remark B.0.6 As explained in [22, §8], the signs of $\boldsymbol{\Omega}\left(Z_\pi\right)$ in the right hand side of (B.0.5) are chosen according to a fixed orientation of $\mathrm{Gr}_{k,k+m}$. More concretely, let us fix a top-degree form on $\mathrm{Gr}_{k,k+m}$ which is non-vanishing on $\mathcal{A}_{n,k,m}$. As an example of such form we can take $\mu_{\mathrm{Gr}_{k,k+m}}$ from Definition B.0.2. Then the sign of $\boldsymbol{\Omega}\left(Z_\pi\right)$ is chosen such that

$$\frac{\boldsymbol{\Omega}\left(Z_\pi\right)(Y)}{\mu_{\mathrm{Gr}_{k,k+m}}(Y)} > 0, \text{ for any } Y \in Z_\pi.$$

Remark B.0.7 We can compute $\boldsymbol{\Omega}\left(\mathcal{A}_{n,k,m}\right)$ from $\boldsymbol{\Omega}\left(\mathrm{Gr}_{k,n}^{\geq 0}\right)$ by knowing a positroid tiling of $\mathcal{A}_{n,k,m}$. More precisely, the form $\boldsymbol{\Omega}\left(Z_\pi\right)$ in (B.0.5) can be obtained from $\boldsymbol{\Omega}\left(S_\pi\right)$ using the diffeomorphism $\tilde{Z}$ between the positroid cell S_π and its image in $\mathcal{A}_{n,k,m}$ (see [22, Conjecture 8.4]). In turn, as described before, the canonical form of a positroid cell S_π can be obtained as: $\boldsymbol{\Omega}\left(S_\pi\right) = \mathrm{Res}_{S_\pi}\left(\boldsymbol{\Omega}\left(\mathrm{Gr}_{k,n}^{\geq 0}\right)\right)$.

Definition B.0.8 *([15]).* Let $\boldsymbol{\Omega}(\mathcal{A}_{n,k,m})$ be the canonical form of the amplituhedron $\mathcal{A}_{n,k,m}$ and let $\Omega(\mathcal{A}_{n,k,m})$ be a function of weight $-(m+k)$ on $\mathrm{Gr}_{k,k+m}$ such that

$$\boldsymbol{\Omega}(\mathcal{A}_{n,k,m}) = \Omega(\mathcal{A}_{n,k,m})\,\mu_{\mathrm{Gr}_{k,k+m}}. \tag{B.0.9}$$

Then we call $\Omega(\mathcal{A}_{n,k,m})$ the *canonical function* of the amplituhedron $\mathcal{A}_{n,k,m}$.

B.0.2 Positive Geometries, Fibers and Tilings

Finally, we report a framework which puts together the philosophy of positive geometries and secondary geometries. This was explored in [4, Sect. 7].

Let (X, X_+) and (Y, Y_+) be two positive geometries such that $\dim(X) \geq \dim(Y)$ and let $\pi : X \to Y$ and $\pi_+ : X_+ \to Y_+$ be projections such that the fibers $(\pi^{-1}(y), \pi_+^{-1}(y))$ are also positive geometries for all $y \in Y_+$. Analogous to Definition 5.1.1 and Definition 3.2.1, we define *tilings* of Y_+ induced by the map π_+ as follows:

Definition B.0.10 (*Tilings/Dissections*). Given a finite collection $\{S\}$ of strata of X_+, let us denote $\pi_S = \overline{\pi_+(S)}$ and $\pi_S^\circ = \pi_+(S)$. The collection $\{\pi_S\}$ is a π_+*-dissection* of Y_+ if we have that:

1. π_S is full-dimensional, i.e., $\dim \pi_S = \dim(Y)$, for each S;
2. the interiors π_S° and $\pi_{S'}^\circ$ are pair-wise disjoint, for each $S \neq S'$;
3. $\bigcup \pi_S = Y_+$, i.e., the union equals the whole space Y_+.

A π_+-dissection $\{\pi_S\}$ of Y_+ is a π_+*-tiling* of Y_+ if for each π_S, π_+ is injective on S. In this case, π_S is called a π_+*-tile* of Y_+.

Remark B.0.11 In [4], 'π_+-tilings' were referred to as 'π_+-triangulations'. See also Remark 5.0.2.

Then, as in Conjecture B.0.4, we would like to have a compatibility between tilings and canonical forms of positive geometries, as follows:

Property B.0.12 (Canonical Forms and Tilings). *Let $\{\pi_S\}$ be a π_+-tiling of Y_+. Then its canonical form can be computed as:*

$$\mathbf{\Omega}\left(Y_+\right) = \sum_{\{\pi_S\}} \mathbf{\Omega}\left(\pi_S\right).$$

In particular, the sum does not depend on the π_+-tiling.

Let μ_Y be a standard (covariant) volume form on Y. The canonical form of Y_+ can be written as:

$$\mathbf{\Omega}\left(Y_+\right) = \Omega(Y_+)\, \mu_Y,$$

where we call $\Omega(Y_+)$ the *canonical function* of Y_+. Moreover, we define:

Definition B.0.13 (*Fiber Volume Form*). Let us fix $y \in Y$. Let ω and α be differential forms on X such that:

$$\mathbf{\Omega}(X_+) = \omega \wedge \pi^*(\mu_Y) + \alpha,$$

and the restriction of α on $\pi^{-1}(y)$ vanishes. Then we define $\omega_\pi(y)$ to be differential form of degree $\dim(X) - \dim(Y)$ obtained by restricting ω to the fiber $\pi^{-1}(y)$ of Y:

$$\omega_\pi(y) = \omega|_{\pi^{-1}(y)}.$$

We call $\omega_\pi(y)$ the *volume form on the fiber* $\pi^{-1}(y)$ of Y.

As in [4, Theorem 4.8], we would like to relate ω to the tilings of the positive geometry, as follows:

Property B.0.14 *Let $y \in Y_+$ and let $\omega_\pi(y)$ be the volume form on the fiber $\pi^{-1}(y)$. Then:*

(i) *Every π_+- tile π_S of Y_+ corresponds to a pole q_S of $\omega_\pi(y)$, where $q_S = \pi^{-1}(y) \cap \overline{S}$.*
(ii) *The residue*[4] *of $\omega_\pi(y)$ at the pole q_S computes the canonical function of the generalized π_+-tile triangle:*

$$\Omega(\pi_S)(y) = \mathrm{Res}_{q_S}\omega_\pi(y).$$

(iii) *In particular, given a π_+-tiling $\{\pi_S\}$ of Y_+, its canonical form is given by:*

$$\Omega(Y_+)(y) = \mu_Y \cdot \sum_{\{S\}} \mathrm{Res}_{q_S}\omega_\pi(y).$$

Remark B.0.15 In the physics literature, this procedure of computing the canonical function via fiber volume forms, would be summarised in the following formula:

$$\Omega(Y_+)(y) = \int_\gamma \boldsymbol{\Omega}(X_+)(x)\, \delta_X^{(\dim(Y))}(\pi(x); y), \tag{B.0.16}$$

where $\delta_X^{(\dim(Y))}(\pi(x); y))$ is a delta function restricting x on the subspace of X where $\pi(x) = y$, i.e., on the fiber $\pi^{-1}(y)$. Finally γ is a suitable integration contour in X—e.g., a *BCFW* contour in case of the amplituhedron. The integrand would give $\omega_\pi(y)$ and the integration over γ would be the residue procedure in (iii) of Property B.0.14.

The above properties could provide useful tools to study the notion of *fiber positive geometry*—a positive geometry encoding the combinatorics of π_+-tilings of Y_+. Moreover, the algebraic-analytical counterpart of this problem would be finding a suited residue procedure on the fiber volume form ω_π which computes the canonical form $\boldsymbol{\Omega}(Y_+)$ and reflects the combinatorics of π_+-tilings (dissections) of Y_+. Note that in the case of polytopes, the fiber positive geometry is the construction of *fiber polytopes* [10] and the residue procedure is the *Jeffrey-Kirwan residue* [4, 23].

References

1. A. Postnikov, Total positivity, Grassmannians, and networks. Preprint http://math.mit.edu/~apost/papers/tpgrass.pdf

[4] The residues have to be performed with suitable signs, as explained in Remark B.0.6.

2. A. Knutson, T. Lam, D.E. Speyer, Positroid varieties: juggling and geometry. Compositio Math. **149**(10), 1710–1752 (2013)
3. D.E. Speyer, Variations on a theme of Kasteleyn, with application to the totally nonnegative Grassmannian. Electron. J. Combin. **23**(2), 2.24–7 (2016)
4. F. Mohammadi, L. Monin, M. Parisi, Triangulations and canonical forms of amplituhedra: A fiber-based approach beyond polytopes. Commun. Math. Phys. **8** (2021)
5. T. Lam, *Totally Nonnegative Grassmannian and Grassmann Polytopes* (2015)
6. N. Arkani-Hamed, Y. Bai, T. Lam, Positive geometries and canonical forms. JHEP **11**, 039 (2017)
7. J. De Loera, J. Rambau, F. Santos, *Triangulations: Structures for Algorithms and Applications. Algorithms and Computation in Mathematics* (Springer, Berlin Heidelberg, 2010)
8. I.M. Gelfand, A.V. Zelevinskii, M.M. Kapranov, Newton polyhedra of principal A-determinants. Sov. Math. Dokl. **40**(2), 278–281 (1989)
9. I. Gelfand, M. Kapranov, A. Zelevinsky, *Discriminants, Resultants, and Multidimensional Determinants* (Modern Birkhäuser Classics, Birkhäuser Boston, 2008)
10. L.J. Billera, B. Sturmfels, Fiber polytopes. Ann. Math. 527–549 (1992)
11. P.H. Edelman, V. Reiner, The higher Stasheff-Tamari posets. Mathematika **43**(1), 127–154 (1996)
12. J. Rambau, Triangulations of cyclic polytopes and higher Bruhat orders. Mathematika **44**(1), 162–194 (1997)
13. C.A. Athanasiadis, J. De Loera, V. Reiner, F. Santos, Fiber polytopes for the projections between cyclic polytopes. Eur. J. Combinat. **21**(1), 19–47 (2000)
14. G.M. Ziegler, Higher bruhat orders and cyclic hyperplane arrangements. Topology **32**(2), 259–279 (1993)
15. N. Arkani-Hamed, J. Trnka, The amplituhedron. J. High Energy Phys. **10**, 33 (2014)
16. F. Brown, C. Dupont, Single-valued integration and double copy. J. für die reine und angewandte Mathematik **1** (2020)
17. N. Arkani-Hamed, J. Bourjaily, F. Cachazo, A. Goncharov, A. Postnikov, J. Trnka, *Grassmannian Geometry of Scattering Amplitudes*. (Cambridge University Press, 2016)
18. R. Hartshorne, *Residues and Duality*, vol. 20. (Springer, 1966)
19. P.A. Griffiths, Poincaré and algebraic geometry. Bull. Am. Math. Soc. **6**(2), 147–159 (1982)
20. B. Khesin, R. Wendt, *The Geometry of Infinite-dimensional Groups*, vol. 51. (Springer Science & Business Media, 2008)
21. N. Arkani-Hamed, Y. Bai, T. Lam, Positive geometries and canonical forms. J. High Energy Phys. **2017**(11), 39 (2017)
22. P. Galashin, T. Lam, Parity duality for the amplituhedron. Compositio Math. **156**(11), 2207–2262 (2020)
23. L. Ferro, T. Lukowski, M. Parisi, Amplituhedron meets Jeffrey-Kirwan Residue. J. Phys. A **52**(4), 045201 (2018)

About the Author

Dr. Matteo Parisi is a member at the Institute for Advanced Study in Princeton and a joint Postdoctoral Fellow at the Harvard Center of Mathematical Sciences and Applications. He is also a Lecturer in Physics at Princeton University. He obtained his DPhil (Ph.D) in Mathematics at the University of Oxford. He has been recipient of several honors and awards, among which the Daniel M. Sachs and the Della Riccia scholarships during his DPhil at Oxford, the DAAD scholarship during his MSc in Munich, and two Certificates of Merit by the Rector of his Alma Mater University of Bologna.

His research lies at the intersection of High Energy Physics and Algebraic Combinatorics. He has been working on a new program in which quantum mechanical observables in particle physics and cosmology arise from underlying novel mathematical objects. Physical properties purely emerge from their combinatorics and geometry. In particular, his research focuses on scattering amplitudes in quantum field theories, in relation to the (positive) Grassmannian, Amplituhedra, tropical geometry and cluster algebras. Being one of the leading contributors in this program, he produced over ten publications in major peer-reviewed journals during his Ph.D. These ranged from theoretical physics to mathematical physics and pure mathematics, with more than 250 total citations.

Teaching, outreach and advocating for inclusivity and diversity, within and beyond academia, are a key part of his identity. During his PhD, his teaching experiences ranged from holding Lectureships at two Oxford colleges (St. Peter's and Keble), to organizing and teaching teenagers at the first ever Math Camp in Cameroon. He has worked with equality and diversity units of the university; he is a trained LGBT+ Role Model and organized conferences and interacted with charities on LGBTIQ+ issues. In his free time, he loves playing the piano, playing volleyball, experiencing the beauty of art (e.g., concerts, ballets, museums), travelling, and forging and fostering lifelong friendships.

Credit to the photographer *Dan Komoda, Institute for Advanced Study.*

M. Parisi, *Combinatorial Aspects of Scattering Amplitudes*, Springer Theses,
https://doi.org/10.1007/978-3-031-41069-7

The manufacturer's authorised representative in the EU is Springer Nature Customer Service Centre GmbH, Europaplatz 3, 69115 Heidelberg, Germany. If you have any concerns regarding our products, please contact ProductSafety@springernature.com

Printed and bound by CPI Group (UK) Ltd, Croydon, CR0 4YY
30/09/2026
02225322-0002